LIFE

THE PLAY OF LIFE ON THE STAGE OF THE WORLD
IN FINE ARTS, STAGE-PLAY, AND LITERATURE

ANALECTA HUSSERLIANA

THE YEARBOOK OF PHENOMENOLOGICAL RESEARCH

VOLUME LXXIII

For sequel volumes see the end of this volume.

LIFE

THE PLAY OF LIFE ON THE STAGE OF THE WORLD IN FINE ARTS, STAGE-PLAY, AND LITERATURE

Edited by

ANNA-TERESA TYMIENIECKA

The World Phenomenology Institute

Published under the auspices of
The World Institute for Advanced Phenomenological Research and Learning
A.-T. Tymieniecka, President

KLUWER ACADEMIC PUBLISHERS
DORDRECHT / BOSTON / LONDON

Library of Congress Cataloging-in-Publication Data is available.

ISBN 0-7923-7032-5

Published by Kluwer Academic Publishers,
P.O. Box 17, 3300 AA Dordrecht, The Netherlands.

Sold and distributed in North, Central and South America
by Kluwer Academic Publishers,
101 Philip Drive, Norwell, MA 02061, U.S.A.

In all other countries, sold and distributed
by Kluwer Academic Publishers,
P.O. Box 322, 3300 AH Dordrecht, The Netherlands.

Printed on acid-free paper

Printed in The Netherlands.

TABLE OF CONTENTS

SECTION IV

SECTION V

ACKNOWLEDGEMENTS

The present collection, continuing our research into the philosophy/phenomenology of the fine arts, literature and aesthetics, gathers papers presented at our 6th annual convention held at the Harvard Divinity School in April 1999. First of all we want to express our appreciation to the authors who have provided the material for this philosophical feast, and to Professors Marlies Kronegger, President of the International Society for Phenomenology, Aesthetics, and the Fine Arts, and its Secretary General Patricia Trutty-Coohill, for their inspirational organization of this meeting.

Our thanks go to Isabelle Houthakker for expert editing of the papers, to Robert Wise Jr. for preparing the Index, and to Jeffrey Hurlburt for help in organising this event and the present volume.

A-T. T.

ALL LIFE UPON THE STAGE

Art has often been considered to mirror human life. The metaphor "theatrum mundi," signifying that all of life takes place on the stage of the world, goes back as far as Democritus (460 B.C., see the paper by Scolnicov, *infra*, p. 3). It gained universal currency in the early modern era when sounded by Juan Luis Vives, William Shakespeare, and Sir Walter Raleigh. It remains valid today. We may surmise that the concept that the puzzling existence of the human being in the world is a play has been intuitively held since humanity's first artistic grappling with reality.

"Bringing life to the stage," as if setting a mirror in front of the public so that it may see itself as being represented, applies – as we will see in the present collection of studies – not only to the stage play, but to all art, to art at large. As we look at it more closely, art is, neither for the artist, nor for the spectator, reader, listener, a depiction of "real" life, its representation. The mirror of art is "the magic mirror of the witch," in which the kitchen maid may see herself as a princess, a pretentious benefactor as a calculating miser, an unknown soldier as a heroic figure, etc. The artist's intention is not to depict the obvious, the surface of givenness, to merely reproduce that which is conventionally taken as given to the eye and mind. Percipients on their side might be pleased and content just to see well-known landscapes or their own faces as they are familiar to them, but this surface semblance ultimately does not satisfy. Even when looking in a mirror put to the face, a human being will seek something of his "true self" that is not ordinarily obvious in his appearance. And it is this "reality" that one hopes will be discovered and conjured by the artist.

To create this magic mirror, artists, playwrights, dancers, etc. immerse themselves in the sought-after dimension of reality and rescue findings relevant to their own tastes, moods, preoccupations, or quests. They fashion their own lenses and choose their own vantage points. Only by throwing his own net so prepared onto this depth may the artist harvest the material out of which to conjure an image that answers to the human interrogation of reality. Thus the recipients in order to decipher the magic image have to plunge into the intricacies of the pluridimensional construct confronting them, be it a stage play or a painting, and distill from it the magic image. Thus "mirroring" entails a most complex scrutiny of the "true reality" to be conjured in the mirror over against the reality of the pedestrian facts of life.

The sphere of the play between factual statements and imagination is already an "enchanted" realm. In that resides the attraction of portraits and plays. The innermost of personalities as depicted by great artists in portraits, dramas, and comedies enchant us even if the resemblance to the model lies beyond our recognition.

Is this "true reality" a reality in itself? We do not come to witness it in everyday life or presume that it is there except in extremely rare glimpses. Thus the artistic presentation's colorful array of aspects of and perspectives on life's protagonists, events, and interactions enhances regular life. In its power to enchant us, it gives us a novel vision of life and ourselves. The significance of heroism, nobility, generosity, courage, villainy, etc. that it brings to light impinges on our heart and mind sustaining this glorious vision of the otherwise pedestrian course of existence. Art as well sustains the poetic inwardness of nature – as in that vision of the rose in which the poet Rilke conjures a mystical depth, thus bringing beauty and the sublime into nature's sphere.

Yet in discovering this hidden "reality," this "true reality," is not art creating an illusion? How can it stand the test of the "real" facts? Delacroix's pictorial dramas make us see the entanglements of historical situations magnifying the aesthetic and moral values of the protagonists standing there in front of us as "real," in their "true" character. But scrutinized against the "real facts" of historical research, these may not stand up to the test. Is the depiction not illusory then? What is real and what is an illusion, albeit an illusion that can play so great a role in our real existence?

Strangely enough, despite all these considerations, it is in art, especially in the gripping art of the dramatic stage play, that we seek the very key to understanding the factual reality of life. We seek in art the clues, the key by which to open the entrance into the enigmatic sources where lie life's hidden reasons. From this issues our fascination with bringing the drama of life to the stage.

Life, treading forcefully the furrows that it digs for its course, fully captivates the attention of living beings, and it moves so rapidly step after step – too rapidly to allow us to grasp its intricacies, to disentangle its spontaneous concatenations and so bring to light its obscure connections, its astounding development. That would require a pause in its course, the achievement of some distance from the pulp of the life we remain drawn into. It would demand a thorough and in principle impossible investigation of all the ins and outs of our actions, feelings, desires and of those too of all the others with whom we deal. Such a thing is impossible. Yet the project and its enigma still fascinate us, grasping at least some fraction of the presumed causal chain in our life somewhat into an equally elusive future. Desires, projects, plans, expectations, predictions, hopes require some sense of the plot of life. When that future arrives we may fail in our plans, be disappointed in our expectations, recognize the vanity of our hopes, yet we undertake to start all over again.

Or at a loss to grasp life's course, we seek to retain its most significant instants through art. Hence we have depictions of great national moments in historical paintings and sculptures, their commemoration in festive musical performances, compositions celebrating victories or charters of liberty, great epic literary works that bring forth the life and habits of a nation at an important period of its history, and rites, folkloric dance, song, architecture. All these reveal to us the profound reasons for preserving in our memory, whether personal, familial, communal, or national, certain deeds that would otherwise fall into oblivion, deeds that inspire pride in us and that give our intentions and dreams direction.

All art brings the drama of life to the stage for all to behold and for each to find his or her role in, whether it be a tragedy or a farce performed on a real stage, or a depiction in splashes of color or in forms, or an epic narration. All art aims at clarifying or celebrating human life, now by exulting in it and now by deploring it. It aims to put us face to face with ourselves, not with the selves we want to see, but with what we really believe, appreciate, love, and hate under the pretences we create to conceal these. Art brings out our hidden motives, our hidden pride, our follies, and our wisdom. It uses all means to despoil us of vain pretense. It invents innumerable means of disguise in order to lead us to discovery – masks, costumes, plays within a play, chiaroscuro, changing rhythms – all to make us see in the magic mirror what we really are.

What is real, and what is illusory? What is true, and what is fictitious? What is obvious, and what surges out from hiding? Art is witness to all dreams and deceits. On the grand stage of the world all these intermingle and complement each other in the grand drama of the human being living out life in the world.

A-T. T.

ANNA-TERESA TYMIENIECKA

INAUGURAL STUDY: THE SPECTACLE OF LIFE UPON THE STAGE OF THE WORLD

Our life is a constant succession of events, feelings, and desires amid changing situations, aims, partners, struggles.... Despite our efforts to arrange events to some degree, contrary things always happen. Incalculable elements figure in the unfolding of events, so that we are always taken somewhat by surprise. We then try reflectively, searchingly to disentangle the chains of events that came together to thwart our plans, but being "in the heat of the battle" we are in no position to isolate them all and pursue them to the end. Nor can we dwell on this scrutiny too long for the stream of life's flow engages us in yet other pursuits. Thus, we never know the reasons for events, facts, for the failures, or for successes too, that we are privy to. We do not even know the very nature of our own feelings and attitudes, nor the motivations of our desires, not in their origins or real nature.

And yet caught in this stream of life, we nurture a profound desire to stop for a while in order to ponder the seemingly haphazard continuity of our existence and the ground upon which our meanders and connections might be grasped and elucidated. While we will not submit to the fate of being the product of the play of circumstances, we are in no position to pursue this search while in action. Captivated by the intense current of life, we are too absorbed by our pursuits to seek answers to the numerous questions that will always tantalize us, questions about the sense of what we are doing, about the direction to take, about the criteria to adopt for our judgement and conduct, and finally about the meaning of our life and the destiny toward which we ignorantly move.

It takes distance from these life entanglements as well as creative power to penetrate into the hidden springs of life and then grasp the whole of it in a synthetic, representational fashion. As simply existing persons, even if we ponder the secrets of our destiny, we never break out of the narrow corridor in which we live our round. That existence is a closed one; we never get even a glimpse of the entire spectacle of the world, life as such.

To offer us this spectacle is the privilege of art. The artist with his or her inquisitive, penetrating, and representational powers brings us face to face with the mysteries of our existence. The plastic arts, literature, the theatre offer us this spectacle of the world and life revealing the intrinsic realities of the existence with which we deal. In representing the world and life the artist seeks to ascertain (estimate) our human situation, inquiring into our real place and status in the conundrum of life within the world, seeks to disentangle the knots of reality in order to check whether there is a definitive status for human life. While absorbed by our everyday concerns, we may only see some glimpses of the whole, but the artist with his detachment and vision encompasses the world, human actions, and existence at "one glance."

It is especially the theatre that, next to painting and the novel, undertakes best the role of offering us the spectacle of human life. The theatre stage is the stage of the world. This is a spectacle that does not stop at "external" presentation, but brings to light the inner workings of human existence and destiny.

Already in the Greek theatre this spectacle dissected human life as played out on the stage of the world, depicting with the greatest depth the human predicament, human aspirations, and human submission to higher forces. With such a revelatory intent the Greek theatre had more the character of ritual and less the character of entertainment than ours.

Theatre assumes a cultural role for therein human beings see in front of themselves their own situations interpreted within the entire spread of their questioning of their lot, with answers and explanations then being proposed that take into account not only the wisdom of submission but individuals' higher aspirations, nostalgia, dreams, foreboding.

The Greek tragedy always presents protagonists of heightened stature, which gives the ordinary human being a measure of his or her own response to situations. For these heroes, whether of divine or human parentage, always remain earthbound. And, as we see exemplified in Aeschylus' play *Prometheus Bound*,[1] it is *the human world* that is the proper stage for the peripeties of men and gods. The action takes place in prehistoric times. Hephaestus is chaining Prometheus, a fallen Titan, to a desolate rock high above the sea as a punishment ordained by Zeus for having contravened an order not to share fire with human beings. It is at this undetermined place, where there is no action or any trace of life, that the ruling power of the gods and the subservient condition of the human race on the stage of the world are investigated.

Where our own theatre displays, the Greek theatre depicts in words. Thus we learn from Prometheus himself about his refusal to follow the will of the gods. He, whose name means "forethought" or "providence," had created mankind and then, in response to Zeus' neglect of this race, had undertaken to instruct the race, first by giving men fire, which the gods had wanted to keep for themselves alone. He had continued to educate human beings and help them develop. In Prometheus' words,

> After all my benisons to men,
> here I am caught beneath this yoke – compelled:
> I the one who snared within a fennel stalk
> The source of fire –
> Man's great teacher of the arts, his universal boon.
> This is the sin for which I pay the price,
> Clamped beneath the naked sky and shackled here.[2]

From his own words we learn of the happenings in Olympus: of how Zeus killed his father Cronus, taking the throne for himself; of the Titans, "children of heaven and earth." He brings us into the hidden dealings of Zeus, who on becoming ruler had apportioned to the gods "proper perquisites and powers" but had not given anything to humans, planning even to wipe them away and put in their place a race of new beings. This doom only Prometheus had had the courage to oppose. We thus see human destiny held by reckless hands. These are the hands of the highest ruler, but Prometheus decries this recklessness and prophesies its end.

From out of the prehistory of mankind, never breaking out of the bonds that fix him to that solitary rock and receiving but a few visitors – among them Hermes, the messenger of the gods, who had transmitted to him their sentence – Prometheus, by his recital and by his prophecies, throws rays of light into the nature of the human condition, a condition he participates in, that being the price he has to pay for his noble deeds on behalf of mankind. After having denounced Zeus' reckless deeds, Prometheus prophesies with almost eschatological breadth, that after ages and ages the fall of Zeus' unwarranted tyranny over gods and men will come, a liberation that will include his own liberation from chains.

In the meantime Prometheus imparts to humans a piece of wisdom. He avers what his mother Themis – Earth – had advised, that someday, not by force, "but only by sheer brain the master race would win."[3]

He prophesies sufferings that are to be inflicted by Zeus on various countries – Asia, Arabia, Scythia, etc. He underscores the common lot of humans by an appeal to the chorus:

> Let yourselves – oh, let yourselves – share pain
> With one who mourns today;
> For suffering walks the world – alas the same –
> And sits beside us all in turn.[4]

At which the chorus of nymphs descends to the crag to dance around him.

For Prometheus the most significant "good deed" he has performed for humankind is his having made humans ignorant of the pains to come, in order to prevent their succumbing to terror and going extinct in the face of terrible sufferings and torments. As he puts it, "Blind hopes I lodged within their breasts."

We see here how human life is taken out of its narrow corridors and situated within a vast panorama. Human limitation, dependency, ignorance of our lot and destiny are set within the panorama of the struggle of higher forces, liberating or despotic. The panorama extends from Hades to Olympus, from the prehistory of humankind to immeasurable future ages. Such a presentation of a great panorama probing and enlightening our individual lives is not characteristic of the Greek theatre only.

This is a way to show humanity, as in a mirror, all of its plight, its triumphs, struggles, its situation and its prospects. The grand spectacle of the manifestation of the world and the human being within it pervades the cultural development of the Occident, with variations according to changes in the cultural climate.

Over time the emphasis shifted to human conflicts. There emerged already in Greece the metaphor of "the theatre of the world," which has the human being at its center. This juxtaposition of the "world play" and human life is already present in Pythagoras, who according to Diogenes Laertius compared human life to a festival "as some come in order to fight, others to buy or to sell goods, and others, who are the best, just to look; in the same vein in life some are already born as slaves of glory, others hunters of goods, and others philosophers, lovers of truth."[5]

In Epictetus there is a direct appeal to humans in which we have the indication of an even vaster horizon extending before and after us:

> Remember that you are an actor in a drama and such a one as it pleases the Author to make; one having a short part, if he desires it short, and a long one if he desires it long. If he wishes you to assume the role of a beggar, a lame invalid, a sovereign, or a simple subject, use your capacities to represent well your role. It is your job to interpret well the personage that has been entrusted to you. To choose it belongs to another. (*Enchiridion* 17)

We find this theme in Seneca,[6] Plotinus,[7] and numerous other authors.

The human being and his life are the focus of the spectacle of the world. The medieval drama of the mystery plays and the religious processions on Corpus Christi and other feast days found culmination in the great "Theatre of the World" of Baroque Spain. With its secular content much expanded, this theatre spread throughout Europe. It is enough here to mention the name of Shakespeare's theatre, the Globe.

Pedro Calderón de la Barca's plays *El gran teatro del mundo* and *El gran mercado del mundo* are to be situated in the midst of the Spanish Golden Age and the Baroque era. Life had already been called a theatre by Quevedo in his 1635 translation *Epicteto y Phocílides en español con consonantes*.[8] He declared, "Do not forget that your life is a comedy and a farce [*teatro de farsa*] of the world, all of which changes apparel instantaneously; realize that God is the author of this comedy with such a grand and spread out argument; it is He who made it and composed it."

With this we are introduced into the heart of the seventeenth-century *theatrum mundi*, of which the most representative work is the aforementioned *auto sacramental* of Calderón and its sequel in his *El gran mercado del mundo*.

In the opening scene of *El gran teatro del mundo* there first appears the Author, who calls upon and dialogues with the World (who also comes *in persona*).[9] With this direct focus on the human being, the orientation is changed. One could expect that in this situation the human being would come to understand his own life and his dealings in the world and that enlightened by religious teaching he would also change, that the Divine precepts for human conduct supported by the constant intervention of the "voice" that prompts appropriate acts and discourages wayward ones will yield – nay, guarantee – the continuity and sense of our human concrete life dealings. But none of this is so, as we will see!

The theatre of life is, like the Greek theatre, suspended between the furthest horizons of the Divine at the one extreme and the destiny of the human being at the other. At the one limit are the rules, devices, laws of the transcendent Creator who will judge the outcome of each human peregrination in life; at the other is the human being living in the world.

Accordingly, the scene represents two realms. Two spheres are set on the stage, a celestial sphere and the terrestrial realm, between which doors open and close. The terrestrial realm has two doors, birth and death, the cradle and the coffin. The celestial sphere has a ladder to be climbed by those invited to dine with the Creator. The play opens with the Author personally calling forth the World and asking in return for a feast to be offered to Him. Then the Author of All calls forth the mortals to be.

> Mortales que aún no vivís
> y ya os llamo yo mortales,
> pues en mi concepto iguales
> antes de ser asistís;
> aunque mis voces no oís,
> venid a aquestos vergeles,
> que ceñido de laurales,

cedros y palma os espero,
porque yo entre todos quiero
repartir estos papeles.[10]

(I.e., "to hand out parts" to the players.)

In His appeal, the Author addresses human beings generically: the king, the farmer, the rich man, the poor man, the unborn child. He also addresses discretion (piety), the law of Grace (whom He makes the prompter in the play), and Beauty as being self-aware of their functions. Beauty responds, "Sólo en tu concepto estamos, ni animamos, ni vivimos." To them the Author entrusts the World, properly outfitting them for the roles that each has to play. It is the World that as a stage manager, as it were, then outfits them according to their entrusted stations and dismisses them, recalling them when they have finished performing their parts. They are then ready to leave the scene of the world.

Humans were not left by the Author entirely to their own devices. Calling himself "Justicia distributiva," the Author states that He knows best which role to entrust to each person. Yet even though He could determine how each plays his or her role, He gives to him or her, on the contrary, the "freedom to decide and to choose" – "albedrío" – leaving them only the Divine Law to guide conduct. Then too, He implants a "voice" to be heard by all through their span of time in the orbit of the world. This voice responds to the singular situations of each person, prompting the following of that law. Still, within the confines of life, each has to decide on his or her own whether or not to follow.

The Author, who from the arc of Heaven surveys the play that is life on earth, will ultimately judge the deeds of all according to their merits and in accord with His law, "el apunto a mi Ley."[11]

And so the Author admonishes men once more to remember that they will have to render account and that they know not when they will finish their roles.

Thus we have a play in a play. The name given this play within life by its Author is "Act well because God is God" ("Obra bien pues Dios es Dios"). The theatre of the world is, therefore, a play within the great play involving the Creator and the World.

All of the characters then leave to play the roles on which their destiny hangs. They do not depend on external forces or on the Divine Will, but on their own conduct, which has for its orientation the Divine Law, the ever prompting voice, and the great device, "Act well because God is God."

We see in the end that human beings properly outfitted are immersed in their roles in the game of life and identified entirely with their predicaments.

The poor man laments that he is the poorest of all; the farmer avers that nobody works as hard as he; the king thinks that his is absolute power in the world; the rich man flaunts his riches as the only worthy goal in life. Each is so identified with his concrete life stream that the general orientation for conduct given to them by the Law and the persistent voice, the sense of purpose, the remembrance of destiny, escape them.

None wants to leave the scene when the time of death comes. But in the final account, the poor may redeem themselves by renunciation; farming folk, by humility; the king, by defending the Church; and the miscarried baby goes to Limbo. When the Author reenters the scene, only the rich man cannot understand anything and has done nothing to make amends and so is condemned to eternal damnation in Hell. Then all the rest of the company are invited to climb the celestial stairs to partake of the Eucharistic meal that never ends, with the Author presiding.

All in all, the Author's benevolence might have arranged things for the best at the end of this play within a play. But while acting upon the stage of the world – in the outer play – none of the protagonists understand what he is really after and what he is doing.

According to interpreters there are two ways of seeing the *teatro del mundo*. I favor that which sees the contrast between the two plays, that which sees the opening and closing scenes in which the Author appears as presenting reality and the intervening scenes depicting "the play of life" as passing diversion. The human being wanders in this life uncertain of his destiny, lacking understanding, comically chasing illusions, finding no meaning, except in the redeeming grace intuitively followed by some. The play on the plane of reality stands over the play of the phantoms of "real life" and points to the contrast between truth and falsity.

The *teatro del mundo* sounds the great theme of Erasmus' *Praise of Folly*, which expostulated on the many ways in which the human being fools himself all through life, not really knowing what he is after, unless, of course, he refers to the truth *transcending* the world of phantoms and illusion.

The contrast between truth and falsity, reality and illusion, is one of the major issues informing sixteenth-century drama and literature, at least as seen from our perspective. The revelation of the truth was sought through the use of the technical device of the play within a play. *Hamlet* provides the best example of such a search after the truth of facts. The play within the play that depicts the murder of Duke Gonzago is meant to place a mirror before the king. In Shakespeare's *Pericles, Prince of Tyre*, death is symbolized by a mirror brought to life.[12]

In brief, the human being is depicted as forging his destiny amid confusion over what he is doing, unable to always rely on his senses and apt to be misled by his imagination, ensnared by deceit, caught up in hypocrisy, carried away by folly, or subject to outright delusion.

Strange to say, we may find all of the main themes and devices of the theatre of the world in contemporary literature. This way of seeing life, as a passing dream, as a game that is senseless unless there is another stage on which it is played, some absolute frame within which it has meaning, is a major theme of modern literature. It is enough to mention Kafka and Sartre.

But this vision is embodied in the particularly elaborate form of the *teatro del mundo* in Umberto Eco's novel *The Name of the Rose*. Here, while the common man follows his life course willy-nilly among the entanglements of larger conflicts and trends, religious, political, etc., the scholar, as the philosopher driven by an unquenchable passion for truth, occupies center stage. The library, with its treasure of knowledge to be mined by those seeking to unravel the mysteries of life, is the stage of the play within a play. It is framed by the larger stage of the abbey with its liturgically ordained surface life and its subsurface brewing passions. This enclave, autonomous, living a life of its own, is not unlike the crag to which Prometheus was bound in that it is at the center of the world's influences and conflicts – the rivalry between religious orders, the political rivalry between emperor and pope, and whatever stirs the local populace with which it maintains vitally significant contact.

Eco, a philosopher-semiotician, shows us first the middle sphere of reality and the world by going beyond its surface to decipher the hidden meanings deposited by nature and societal life in every item of the world, there to be marveled at and quaffed to satisfy our thirst for the beauty and truth of things. But the rapid development of the action of life shows that this is not enough for the human mind, for the scholar, the philosopher.

The library as the stage on which the play within a play is acted out is precisely a metaphor for the depository of these ciphers as recovered by the human mind, the ciphers deposited by nature, society, the workings of the human spirit so that the human mind may progress in grasping the specifically human significance of life.

But the library is also a metaphor for the labyrinth of the human mind as it searches out the passion to seek and find the truth of things and life. We enter this realm as if enchanted, so much does it differ from surface, everyday life, from survival-dominated existence, and we do not easily or at all find a way back out of this realm. There is no Minotaur lying in wait for the adventurous, but, as we all know, the links between the ciphered messages, the pointers for further elucidation and elaboration of the signs deciphered, the associations, the interpretations, lead us on in all directions, with the exit of a final grasp of the meaning of things and of life eluding us.

The logos turns upon itself to recover its workings in an inventory of the spectacle, driven by renewed passion for pursuit of the truth of the things it has already established. The logos of vital and societal unfolding knows no end, no halt. Its drive, its impetus engenders our human struggles to discover the truth. And so we are always abandoning one wild goose chase and taking up another.

The exit from the labyrinth forever eludes us. Our inquisitive, retrieving, re-presenting logos taunts us with promises but never yields the ultimate answer to our questions. We sail in eternal pursuit of fortune upon tempestuous seas, without a compass or definite bearings.

Despite the library's lofty and beautiful significance and the role that the abbot sees it as playing within Christendom, the play within a play sees it destroyed in five days' time, after which the protagonists in this play are either dead or dispersed. This convulses the larger, external play, the life of the abbey and its village, which then becomes distorted.

This is the tragic story of a human being torn between the sublime passion for the truth and the crude libidinal drives of human beings. His struggle to enter, at any price, the labyrinth of knowledge in order to extract from it the philosopher's stone leads to the destruction of all.

This story involves the whole set of actors, those of the inner play staged in the library and those of the outer play staged in society. All of them vanish from the scene, and the inner arena is itself destroyed. Even the memory of that arena would have vanished had not a witness to the events recorded its existence, a young monk who sets the story in a further horizon by providing an interpretive schema drawing on the history and religious thinking of the times.

Umberto Eco's *The Name of the Rose* apprises us that we can never encompass the entire spectacle, the entire truth "at one glance." Each name, each concept, each theory is enmeshed within a multidirectional weave of ciphers and meanings and in its significance draws on them all. True, from the horizon of the human mind the panorama lures us. When we philosophize we can follow these intricate interconnections to the end of the human mind's unfolding in a given historical period since everything thought out draws on all the strings of the work of the logos. We may get a glimpse of the entire spectacle by sitting in the front row, as it were.

Along with our uncertain footsteps in life there is the thread of its rationale, which we attempt to find. On the basis of those limited segments of that rationale that we think to have discovered, we plot future plans, we try to keep from falling into the traps we can imagine, we project ourselves in expectations and with hope undertake projects. These may fail and then fall into oblivion with the rush of oncoming events. But the puzzle of the status of reality, of its mysterious rationale will still engage us all the same. The search for its solution constitutes life's loftiest, noblest pursuit.

NOTES

1 Aeschylus, *Prometheus Bound*, trans. Paul Roche (New York: New American Library, 1964).

2 Ibid., p. 30.

3 Ibid., p. 35.

4 Ibid., p. 38.

5 Quoted after Eugenio Frutos Cortés in his introduction to Pedro Calderón de la Barca, *El gran teatro del mundo; El gran mercado del mundo*, Letras Hispánicas 15 (Madrid: Cátedra, 1989), p. 25.

6 Seneca, *Epistolae morales ad Lucilium* 75–6; "quomodo fabula sic vita."

7 Plotinus, *Enneads* II, III, XVI, XVII.

8 Francisco Quevedo Villegas, *Obras en prosa* (Madrid: 1653).

9 John J. Allen and Domingo Ynduráin find the idea of God as an author already in Francisco Sánchez de las Brozas' influential work *Doctrina de estoico filósofo Epicteto que se llama comúnmente Enquiridión* (Madrid: 1612). See Pedro Calderón de la Barca, *El gran teatro del mundo*, ed., prologue, and annot. John J. Allen and Domingo Ynduráin, with a preliminary study by Domingo Ynduráin, Biblioteca clásica 72 (Barcelona: Crítica, 1997), p. xxvi.

10 Ibid., p. 12.

11 Ibid., p. 18.

12 The reflection of reality in a mirror had its place in Greek myth and history; it is enough just to mention Perseus' use of Athena's shield to reflect back the literally petrifying visage of the Gorgon Medusa, and the belief that collaborators used a burnished shield to flash a signal of reflected sunlight to the Persian fleet after the Battle of Marathon, which perception caused Miltiades to march his troops back to Athens in time to confront the fleet when it arrived there.

SECTION I

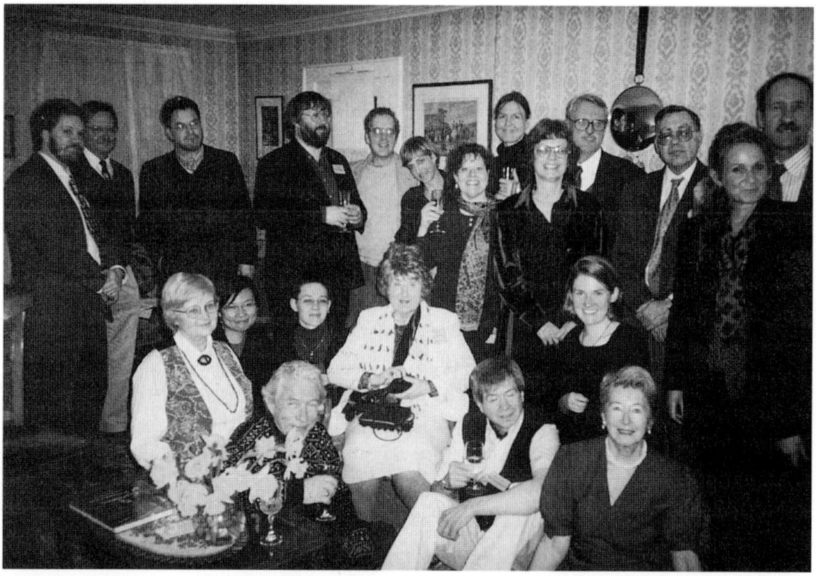

A group of participants at the reception at the World Phenomenology Institute.

HANNA SCOLNICOV

THEATRUM MUNDI IN THE THEATRE: SHAKESPEARE AND CALDERÓN

When the Theatrum Mundi theme is brought into the theatre, one of the terms of this philosophical metaphor becomes concretized, providing a physical framework within which the metaphor is then worked out. In this paper, I shall discuss the special use of the Theatrum Mundi metaphor in drama and theatre.

The Theatrum Mundi metaphor is doubly powerful when used on the stage: If theatre is understood to be a mirror of the world, and the world itself is seen in terms of a theatre, then theatre is of the essence of reality and is raised above all other mimetic arts. Viewing life as a production of a conventional, set scenario provides theatre with a metaphysical dimension and endows it with a general philosophical validity.

I shall examine the intensive use of the metaphor by two of the greatest and most theatrical of playwrights, Pedro Calderón de la Barca (1600–1681) and William Shakespeare (1564–1616). I shall argue that each of them pursued a different strain of thought implied by the Theatrum Mundi tradition. These differences depend on whether the metaphor is seen from an internal or an external point of view, from within or from without, in relation to man or to God. These options were already unravelled by two of this metaphor's ancient proponents, Democritus and Epictetus, whose formulations I shall analyze briefly at the start. I shall argue further that seeing the metaphor from the point of view of the actor leads to a secular and Humanist interpretation, whereas hypothesizing an external spectator who supervises and watches man's performance results in a religious and theocentric interpretation. My approach to the texts will be literary, dramatic and theatrical.

Shakespeare and Calderón are natural choices: Calderón not only used the metaphor as the title for one of his plays, *El gran teatro del mundo*, but also based that play, as well as some of his others, on the many philosophical and theological treatments of the *topos*. Shakespeare gave the metaphor its most famous formulation, "All the world's a stage", and made frequent and wide ranging use of it in his plays. He both wrote within and promoted a theatre culture dominated by the metaphor. The theatre building for which he wrote most of his works was called the Globe, and over its entrance was inscribed the motto: "Totus mundus agit histrionem".[1]

A.-T. Tymieniecka (ed), Analecta Husserliana LXXIII, 3–14.

I

What is this idea of the theatre of the world? It is a metaphor in which the theatre serves as a vehicle for characterizing the tenor, which is the world. In other words, it is an attempt to come to terms with the world or reality, to use the conceptual framework of theatre as a system of ready-made tools for the analysis of life itself. As an artifact, theatre is more immediately perceived, it is the more concrete and definable term. The two terms are not yoked together arbitrarily: The metaphor makes use of their pre-existing, mimetic relationship, reversing that relationship, talking of life in terms of the theatre, instead of adhering to the logical precedence of reality to its mimetic presentation.

The Theatrum Mundi metaphor is a statement about the relationship between our perception and the world. It offers to discuss life within the framework of theatrical discourse. Experience can be sifted and structured as though it were a play. Our dramatic know-how about plots and characters, acting and scenography, can now be brought to bear on life, formalizing and organizing it into meaningful structures.

Such a structuring is clearly visible in what is possibly the earliest instance of the metaphor, attributed to the Greek philosopher Democritus (born ca. 460 B.C.E.):

> The world is a stage,
> Life is a performance;
> You come, you watch, you go.
> (Democritus, fragment 115 DK)[2]

For Democritus, viewing the world as a stage provides an intellectual exercise in emotional detachment. Training ourselves to view life, i.e. our own life, as though we were uninvolved spectators watching a play can help in lessening the pain caused by the reversals of fortune through a conscious avoidance of emotional attachment to all that is ephemeral and evanescent. Democritus advocates extricating ourselves from the flow of life to become its spectators, thus grounding the option of contemplative life in the Theatrum Mundi metaphor.

The Stoic philosopher Epictetus (55–135 C.E.) regarded the theatre of life from the perspective of the actor rather than the spectator, replacing the detachment of the spectator with the resignation of the actor. In his view, the actor is assigned a role in a play over which he has no control:

> Remember that you are an actor in a play, such as the Playwright chose: if short – short, if long – long; if he wished you to act a beggar, act it out naturally; so too, if the part of a lame man, or a

magistrate, or a private person. For this is your lot: to act well the role assigned you; but to choose the part is the role of Another.
(Epictetus, *The Manual* 17)[3]

In the view of Epictetus, every man has been cast by Fate to act a particular role in the world. Instead of a generalized view of human life, he can therefore introduce a variety of possible roles, which he enumerates: the beggar, the lame man, the magistrate, and the private person.

In order to achieve his theatrical perspective on life, Epictetus hypothesizes a transcendent Playwright, who assigns the human actors their roles. Without this hypothesis, the tenor is deficient in relation to the vehicle. The very use of the metaphor seems to imply a transcendent playwright and onlooker, whose point of view we are straining to adopt.[4] These ideas may have been conveyed to Calderón via Quevedo's verse translation of Epictetus' saying (1635).[5]

Life obviously looks very different from our own everyday point of view, where we encounter pain, suffering, grief – and also joy, so that we don't normally live the life of equanimity towards which the Stoics would have had us train ourselves. It is the necessity of that training, *askesis*, that disproves the easy packaging of the metaphor. It does not come naturally to us to view life as theatre – it is a philosophical and, later, religious position that necessitates a basic willingness to distance ourselves from the immediacy of experience.

The theatre of life is a philosophical simplification – but artistically it offers a convenient way of dealing with that abstraction, "life", for which we keep looking (as in the ages of life, the path of life, the voyage of life, and so on). That is why the metaphors have gained more currency in the arts than in philosophy – they are more easily depicted in art, they can be translated into plot lines, they serve as convenient emblems, and so on. Although they originate in philosophical thinking, these metaphors have become naturalized in the different arts, offering conceptual structures for dealing with the amorphousness of life.

A sustained use of the metaphor, and one that links its philosophical origins with theatrical practice, can be found in the Spanish humanist Juan Luis Vives' tongue-in-cheek *Fabula de homine, A Fable about Man* (1518?). Taking his cue from Pico della Mirandola's *Oration on the Dignity of Man* (1486?), Vives proved the excellence of Man through his innate gift of acting, of impersonating the whole scale of creation, from the plants, through the lowliest of animals, up to the gods and Jupiter himself. In this fable, Jupiter not only created the world as an amphitheatre, with the earth as a stage for the actors, but also directed the plays and "prescribed to the company of actors the entire arrangement and sequence of the plays".[6]

Vives developed the Theatrum Mundi metaphor into a curious amalgam of the Roman gods with the Genesis account of the creation of Man in the image of God and with the Christian dogma of Incarnation.[7] Disparate pagan and Christian elements thus became syncretized within his new variation on the theatre of the world theme.[8]

Another important link between the classical formulations of the metaphor and their Renaissance adaptations for theatre is the poignant lyric written by Sir Walter Raleigh (1552?–1618), "What is our life?". Raleigh's answer to that question is that life is a passion play. The shift of the metaphor's vehicle from the classical theatre to the medieval passion play is, necessarily, also reflected in its tragic intensification of the tenor, in its perception of the meaning of life. The emphasis on the suffering replaces the dispassionate Stoic attitude toward the theatre of life. Imprisoned in the Tower of London, reflecting upon his varied career as statesman, colonist and courtier, Sir Walter Raleigh wrote:

> What is our life? A play of passion.
> And what our mirth but music of division?
> Our mothers' wombs the tiring-houses be
> Where we are drest for this short comedy.
> Heaven the judicious sharp spectator is
> Who sits and marks what here we do amiss.
> The graves that hide us from the searching sun
> Are like drawn curtains when the play is done.
> Thus playing post we to our latest rest,
> And then we die, in earnest, not in jest.
> (Sir Walter Raleigh, *What Is Our Life?*)[9]

For the Greek and Roman Stoics, the theatre metaphor was a means of emotional distancing from the turbulence and incertitude of the world. In Raleigh's poem, on the other hand, life is a play of passion, an *imitatio Christi*. Life is charged with pain and suffering, death is very real, and we are judged by Heaven for the quality of our performance. By the simple trick of substituting "a play of passion" for the neutral theatre of the traditional metaphor, Raleigh transformed the classical means of controlling emotion into a religious and highly emotional vehicle for expressing it.

II

In Vives' *The Fable of Man*, we noticed the effort to Christianize the pagan pantheon and to accommodate the classical metaphor within the Christian faith. Raleigh too viewed the theatre of life in religious terms, as a passion play on the merit of which we shall be judged. With Calderón, the Theatrum Mundi metaphor becomes fully baptized.

After Vives had pursued the various theatrical aspects of the metaphor and developed the relationship between man and God within the theatrical hierarchy, the way was opened for its full-scale introduction into the theatre. The full dramatic potential of the metaphor was realized by Vives' countryman Calderón de la Barca, in his extraordinary play, *El gran teatro del mundo, The Great Theatre of the World* (1633?).[10] Calderón took the philosophical metaphor of the world as stage into the theatre, working its network of correspondences into an elaborate and sustained baroque, devotional and theatrical conceit.

Whereas for Epictetus, the Playwright is the Stoic universal Reason or Fate, in *El gran teatro del mundo*, the Playwright becomes the divine *Autor*. This transition from the logos to divinity, Calderón may have picked up, whether directly or indirectly, from Plotinus (ca. 205–270 C.E.).[11] In his discussion of Providence, Plotinus tackles the question of how divine Providence can be justified in the face of evil. He casts divine Providence as the author who gives the actors their parts, but insists that their characteristics are their own. The actors, both good and bad, "existed before the play and bring their own selves to it". The author writes the dialogue, but the actors are responsible for the quality of their acting.[12]

Calderón's poignant exploration of the parallels between the divine Author and himself-as-playwright is carried out within the constricting framework of an *auto sacramental*. The hierarchy of authorships, the homology of divine and human ideational and creative abilities, establishes Calderón's credentials as the author of the *Autor*'s dramatic speeches.[13]

The *Autor* is no less than the Creator himself, who produces the spectacle of life for his own recreation.[14] The act of creation forms a framing play into which the world as we know it is introduced as a play-within-the-play, acted out in front of the divine *Autor*-turned-spectator (II. 628–637). At the end of the performance, He will mete out rewards and punishments to the players, in relation to the quality of their respective performances.

It is the thematic paradox of a play that presents life as a play that provides Calderón with the structure of the play-within-a-play. In the frame play, the divine Author is responsible for the casting, assigning his actors the roles they will play in the inner play. While the *Autor* assigns the playing parts to humanity, it is up to *Mundo* to provide the stage and supply the costumes and props. *Mundo* is the Great Theatre of the World, the created universe in which mortal man acts out his life. The characters represent people from different walks of life: Rich Man, Poor Man, King, Beauty, Peasant and Discretion, the figure of religious devotion.[15] In the inner play, the play of life, the characters enact their unchanging, allegorical personalities in relation to each other, much in the manner of the late medieval moralities.

The Theatrum Mundi metaphor is here extended into an all-encompassing theological framework and dramatic principle. But the force of the idea carries the metaphor beyond the dramatic structure of characters and events and into the very technology of the theatre. The black theatre curtain represents the original chaos and the two stage lamps stand for the two great lights of creation, audaciously inserting Calderón's version next to the Genesis account of the creation of "the greater light to rule the day, and the lesser light to rule the night" (*Genesis* 1:16).[16]

The two stage doors are distinguished functionally, the one serving for entrances, the other for exits, representing cradle and grave respectively:

> Dos puertas: la una es la cuna
> Y la otra es el sepulcro.
> (II. 241–241)
> The cradle is the first,
> The other is the grave.

Once the doors are assigned their meaning by *Mundo*, the passage through them becomes a theatrical metaphor for birth and death, a metaphorical action. But it is the assigning of meaning with words, the act of signification performed in World's speech, that adds symbolic meaning to the simple movement through the stage doors.

The ideological and doctrinal totality of the religious use of the Theatrum Mundi motif is expressed through working out its inner logic down to the physical details of the stage. Only in one crucial respect does the Great Theatre of the World swerve from theatrical practice: Here, no rehearsals are allowed, and each actor is judged on the merit of a single performance.

What is fascinating about Calderón's use of the Theatrum Mundi metaphor is the way in which it becomes anchored in the concrete experience of the theatre-goer, in the welding of the abstract conceit to the phenomenology of performance. From the dramatic point of view, Calderón turns his gaze on the theatre itself, dramatizing its materials, asserting his control over all aspects of production. By turning the metaphor back upon itself, he succeeds in making the technology of the theatre into a potent vehicle of allegory.

At first, this seems to be a retrogressive step, going back to the late medieval moralities with their allegorical characters, plots and scenic conceptions. It is only when we understand the play as the meticulous working out of the Theatrum Mundi metaphor in the theatre – as a theatrical production depicting the world as a theatre – that we realize the Baroque totality of its conception.[17]

The beauty of viewing life and death in terms of a stage entrance and exit lies in the deduction of theatrical parallels once the basic analogy implied by the metaphor is granted. This is the great attraction of the metaphor: The complexity of life, the mystery of birth and death, the nature of experience, are all neatly compartmentalized into the techniques and conventions of the theatre. The underlying assumption is that if theatre is a mimesis of life, whatever we understand by mimesis, that relationship may be reversed, so that life can be seen in terms of the theatre, or even as mimetic of theatre. Developed on stage, this idea then creates a dazzling puzzle of which of the two comes first, which should be seen in terms of the other.

Calderón's play obviously lacks a dramatic conflict. The dramatic plot is supplanted by the working out of the world-as-theatre hypothesis.[18] The plot is no more than the interaction of the various types thrust together in the theatre of life. Due to the absence of any exciting events, and to its doctrinal nature, the interest is naturally shifted from the inset play of life to the relationship between the inset and frame plays. The precarious balance struck between predestination and free will is translated into the differentiation between the assigning of acting parts and the individual performance abilities exhibited by the actors.

The creative energy has been spent on the setting up of the correspondences between life and theatre, and our interest as spectators is directed to the idea and practice of theatre itself. Today, when almost every work of art is seen as reflexive, the play becomes exceptionally interesting, engaging as it does in the aesthetic question of the relationship of art to reality.[19] However, we should bear in mind that the play's shape was determined by doctrinal, and not by aesthetic, considerations.[20]

III

While Calderón enlisted the metaphor into the service of religious justification of the inscrutability of the ways of God toward man, Shakespeare employed it in the service of the theatrical medium itself. Calderón created the figure of the *Autor* as a divine analogue to himself as playwright. Once he has set up the show, the *Autor* retires to become a spectator and judge the quality of the show. Shakespeare has no similar super-figure, no external, transcendent point of reference. In his plays, the Theatrum Mundi is seen from a totally human and Humanist perspective: It is we who choose to look at life through the medium of the metaphor. Thus the two playwrights developed the metaphor along theocentric and anthropocentric principles respectively.[21]

For Shakespeare, the Theatrum Mundi is a guiding metaphor, the basis of his theatre, a self-reflexive statement about the relationship of the theatre to the world that explains why society requires theatre in order to reflect on its own problems. Shakespeare did not develop the metaphor as extensively as Calderón did, within the confines of any one single play. But the metaphor is widespread in his works. Thus, for example, Macbeth reflects on life:

> Life's but a walking shadow, a poor player,
> That struts and frets his hour upon the stage,
> And then is heard no more. It is a tale
> Told by an idiot, full of sound and fury,
> Signifying nothing.
> (*Macbeth* 5.5.24–28)[22]

Lear's "When we are born we cry that we are come/To this great stage of fools" (*King Lear* 4.6.182–183) comes to mind, as does Cassius' awareness that Caesar's assassination will be "acted over/In states unborn and accents yet unknown!" (*Julius Caesar* 3.1.112–113).

Shakespeare brings the world into the theatre and the theatre into the world as plays within plays. His repeated use of the play-within-the-play convention is clearly associated with his fascination with the mirror-relationship between the theatre and the world. *The Mousetrap* in *Hamlet* provides a clear example of such a mirror of nature. Staging this play provides Hamlet with a reflective distance from the immediacy of experience. As we have seen, training oneself to look at life itself as though one were sitting in the theatre is a Stoic exercise of removing oneself from the directly perceived pain of raw, unmediated existence.

Throughout the play, Hamlet's problem is how to deal with his passions, i.e. his feelings. He compares himself disparagingly to the player, who gives vent to an excessive emotion "but in a fiction, in a dream of passion" (*Hamlet* 2.2.552). Both personally and in his theory of acting, Hamlet pursues a more restrained and dignified means of expressing emotion. But his ideal is the Stoic bearing of his friend Horatio, "that man/That is not passion's slave" (3.2.71–72). This is indeed the Humanist ideal of behaviour: An intellectual acceptance of the vicissitudes of Fortune through a conscious effort to objectivize the subjective. Hamlet finally attains this spiritual goal with his new inner conviction that "The readiness is all" (5.2.222).

It would be futile to attempt here even a bare sketch of Shakespeare's varied employment of the metaphor in his comedies, tragedies and histories. Instead, I would like to consider briefly his famous formulation of the Theatrum Mundi metaphor in *As You Like It*.[23]

Jaques' "All the world's a stage" speech takes its cue from the Duke's musings, prompted by Orlando's tale of Adam's ailment:

> Thou seest we are not alone unhappy:
> This wide and universal theatre
> Presents more woeful pageants than the scene
> Wherein we play in.
> (*As You Like It* 2.7.136–139)

The Duke thus offers his fellow actors a double perspective of suffering: as both actors and spectators. When we ourselves are unhappy, we experience life as actors; when we watch the suffering of others, we become spectators. The realization that "we are not alone unhappy", that we are both actors and spectators simultaneously, can evoke two opposed responses: either Stoic detachment or Humanist compassion.

But the Duke's words are not merely a response to Orlando's story. The mention of "this wide and universal theatre" is surely directed at us, at the audience. It is our world that "presents more woeful pageants than the scene" they play in. From within the performance of the play, the Duke not only echoes our sentiments as audience but also turns the tables on us, turning us into the play he and his exiled Court are watching.

With the actors turned spectators, the spectators become actors. An infinite game of mirrors between theatre and reality is thus set in motion. It is this metatheatrical level in Shakespeare's plays that serves to remind us of the theatre's unique attraction, of its theoretical involvement in epistemology and ontology, of its problematic definition as standing in a mimetic relationship to reality, and of its status as a metaphor for life.

The Duke's introduction of the *topos* is followed by Jaques' famous oration on the Ages of Man with its distinct formulation of the Theatrum Mundi metaphor:

> All the world's a stage,
> And all the men and women merely players;
> They have their exits and their entrances,

> And one man in his time plays many parts,
> His acts being seven ages.
> (2.7.139–143)

Shakespeare's universe, as expressed by Jaques, the would-be satirist, is totally anthropocentric and secular, with no mention of divine control at either end of life. Like Democritus, Jaques is an uninvolved spectator at the theatre of life. His typology of characters is somewhat reminiscent of Epictetus, but nowhere is there an indication of any external, divine predestination or involvement. Jaques asks for the license of a fool so that he can criticize his fellow men with impunity. In chastizing the folly of the world and distancing himself from it, he takes up a familiar Humanist stance, adopted, among others, by Erasmus in his *Praise of Folly* (1509), of observing the world as a *Sottie*, or a fool's play.

Shakespeare escapes the authoritarian totality of predestination implied by the metaphor, by refusing to see individual destiny as a matter of vocation, preferring a more democratic view of the uniformity of human life: All men go through the same seven ages, the same seven acts. Shakespeare presents the vicissitudes of life from Jaques' self-distancing, Stoic perspective. This approach contrasts with the religious acceptance of predetermination expressed by Calderón's play, where the Poor Man must accept without demure his divinely ordained role. The difference between the two views explains, perhaps, why Shakespeare's play, and, even more so, his formulation of the metaphor, are so congenial to our own skeptical age, while Calderón's play, despite its undoubted brilliance and beauty, has fallen into the limbo of rarely performed great classical plays.

Shakespeare had too much compassion for the human condition to be able to identify with an *Autor* who could distance himself enough to judge his own creation. Significantly, unlike his character Jaques, he himself never assumed a detached, satirical stance. Only in *The Tempest*, towards the end of his career, on the verge of breaking his staff, did he project himself into the play as something of a magician, an illusionist. But he preferred to cast himself in the Renaissance figure of the mage rather than in that of the divine playwright. As a dramatist, he was obviously an observer of the world and a writer of works for the theatre, but he never dissociated himself from the great theatre of the world, either emotionally or intellectually.

Tel-Aviv University

NOTES

[1] This motto is derived, apparently, from John of Salisbury's *Policraticus*, reissued in 1595. Cf. Ernst Robert Curtius, *European Literature and the Latin Middle Ages*, tr. Willard Trask (New York: Pantheon Books, 1953), p. 141. On the metaphoric significance of the theatre being called the Globe, see: Harriett Hawkins, "'All the World's a Stage': Some Illustrations of the Theatrum Mundi", *Shakespeare Quarterly* 17 (1966), p. 175.

[2] Translation by Samuel Scolnicov. Cf. Hermann Diels and Walther Kranz, *Die Fragmente der Vorsokratiker*, 3 vols. (12th ed., Dublin/Zurich: Weidmann, 1966), Vol. II, p. 165.

[3] Translation by Samuel Scolnicov. Cf. Epictetus, *Discourses, Manual, Fragments*, 2 vols., ed. W. A. Oldfather, Loeb Classical Library (London: Heinemann, 1926), Vol. II, p. 496.

[4] On the evolution of the stoic idea of *logos* into a *poietes*, a playwright, see Lynda Christian, *Theatrum Mundi: The History of an Idea* (New York: Garland Publishing, 1987), p. 7.

[5] Cf. N. D. Shergold, "Calderón and 'Theatrum Mundi'", in *Arts du spectacle et histoire des idées: Recueil offert en hommage à Jean Jacquot* (Tours, Centre d'Études Supérieures de la Renaissance, 1984), p. 171, note 6, and Christian, *op. cit.*, p. 170.

[6] Juan Luis Vives, "A Fable about Man", tr. Nancy Lenkeith, in *The Renaissance Philosophy of Man*, eds. Ernst Cassirer, Paul Oskar Kristeller and John Herman Randal (Chicago: University of Chicago Press, 1948), p. 388.

[7] Cf. Lenkeith, "Introduction" to her translation of Vives, *op. cit.*, pp. 385–386.

[8] An analogous syncretizing tendency between the Christian idea of heaven and the Roman amphitheatre can be observed in some Renaissance architectural designs. See: Richard Bernheimer, "Theatrum Mundi", *The Art Bulletin* 38 (1956), pp. 225–247.

[9] Gerald Bullet, ed., *Silver Poets of the Sixteenth Century* (London: Dent, 1947), p. 296.

[10] Pedro Calderón de la Barca, *El gran teatro del mundo*, ed. Eugenio Frutos Cortés (Salamanca: Ediciones Anaya, 1958); *The Great Stage of the World*, tr. George Brandt (Manchester: Manchester University Press, 1976). There is also a stage adaptation by Adrian Mitchell from a literal translation by Cecilia Bainton (London: The Medieval Players, 1984). All three editions contain valuable introductions.

[11] Cf. Cortés, "Introduction" to Calderón, *op. cit.*, p. 14.

[12] Cf. Plotinus, *Enneads*, with English translation by A. H. Armstrong (London: William Heinemann, 1967), III. 2.16–17, vol. 3, pp. 99–111, and the translator's introductory note on p. 38.

[13] Cf. Barbara Kurtz, "'No Word without Mystery': Allegories of Sacred Truth in the *Autos Sacramentales* of Calderón de la Barca", *PMLA* 103 (1988), p. 270, and also her "'In Imagined Space': Allegory and the *Auto Sacramental* of Pedro Calderón de la Barca", *Romanic Review* 79 (1988), p. 660.

[14] Unlike its English equivalent, *Autor* came to mean theatrical manager, whether he wrote the plays for his company or not. See Hugo Albert Rennert, *The Spanish Stage in the Time of Lope de Vega* (New York: Dover, 1963; first published by the Hispanic Society of America, 1909), pp. 9, 32, 33, 169–170. Brandt translates "The Director", thus obfuscating the hierarchical relationship between him and Mundo. For a slightly different view of the tension between the two meanings, see Cortés, *op. cit.*, p. 23, note 9.

[15] For a further discussion of the characters, see: Anthony Cascardi, *The Limits of Illusion: A Critical Study of Calderón* (Cambridge: Cambridge University Press, 1984), pp. 5–7.

[16] On the self-referentiality of the play, see Stephen Lipmann, "'Metatheater' and the Criticism of the Comedia", *Modern Language Notes* 91 (1976), p. 241.

[17] On the ideological domination and conservativeness encouraged by allegory, see Nancy Campi de Castro, "Allegory You Are a Woman", in *Allegory Old and New*, eds. Marlies Kronegger and Anna-Teresa Tymieniecka (Dordrecht: Kluwer Academic Publishers, 1994), p. 153.

[18] Cf. Barbara Kurtz, "Defining Allegory, or Troping through Calderón's *Autos*", *Hispanic Review* 58 (1990), p. 230, note 6.

[19] On the use of the Theatrum Mundi in plays within plays, see: Howard Pearce, "A Phenomenological Approach to the *Theatrum Mundi* Metaphor", *PMLA* 95 (1980), p. 44.

[20] Cf. Kurtz, "Defining Allegory", *op. cit.*, p. 232.

[21] On Curtius' distinction between the theocentric concept of life in Spain and the anthropocentric concept of life inherent in French and English drama, see Lipmann, *op. cit.*, p. 237.

[22] All Shakespeare quotations are from *The Riverside Shakespeare*, ed. G. Blakemore Evans (Boston: Houghton Mifflin Company, 1974).

[23] I have analyzed this passage in greater detail in: "Ages of Man, Ages of Woman", *Cahiers Élisabéthains* 57 (2000), pp. 61–78.

MUALLA ERKILIÇ

THE THEATER OF LIFE AND IMAGINATIVE UNIVERSALS IN ARCHITECTURAL SPACE

The ancient idea of "theatrum mundi" has cosmological, political, artistic and psychological levels of meaning. Beyond the ontological and conceptual differences between these levels, the meaning attributed to this idea metaphorically refers to "the theater of the world" or "the theatrical representation of the world" where the object of representation is human life itself. The Greek word "theatron", on the other hand, literally means "the place for seeing", indicating illusory and scenographic representations in a theatrical performance.

It is not unusual in architectural interpretations to consider architectural space as "theatron" or a "stage for theatrical performance" where the architectural objects are perceived as visual spatial elements on a stage fulfilling social and functional needs of people in different life scenarios. This descriptive analogy is very helpful for a better understanding of the scenographic nature of architectonic representations. However, this analogy may remain purely at the level of images so long as the ontological significance of symbolic values of architectural spaces, as well as the becoming reasonings behind them, are not considered while disclosing deeper meanings in architectural representations.

This paper aims to scrutinise particularly the symbolic and expressive nature of architectural space as a metaphor and a representation of the "theatre of life" and "a stage of the world" where architecture is perceived as art and more than "techne" (to build, make, etc.), and where it represents human beings' "will to present" their poetic dwelling in the world. Architectural space permits people to spatialize their ideas and world-views relating to realities of cultural life, through their works.

The focus of the discussion in this paper will be twofold. On one hand, referring to some philosophical ideas (such as Giambattista Vico's concepts of "Poetic Wisdom" and "Imaginative Universals", and Ernst Cassirer's "Philosophy of Symbolic Forms"), the theatrical nature of architectural representation and its imaginative poetical realization process will be discussed in a general sense. Secondly, the above generalizations will be particularised and discussed by demystifying a symbolic form of architecture..For this

A.-T. Tymieniecka (ed), Analecta Husserliana LXXIII, 15–37.

purpose, some expressive and symbolic aspects of a traditional religious building of the Alevi and Bektashi people will be analysed in this context.

The *Cemevi* or *Cem House* and *Bektashi Tekke* are the religious gathering halls for Alevi people who, as a marginalised sect of Islam, have lived in Anatolia for centuries and who have their roots in the Pre-Islamic Anatolian and Central Asian traditions. The two important symbolic features of these halls are the *Tütekli Örtü* (layered Lantern Roof) and *Dünya Ağaci* (World Tree), the meanings of which can be revealed by disclosing the "becoming" reasonings behind them.

ALEVI AND BEKTASHI TRADITIONS

Before going into details, it will be helpful to give a brief explanation of the cultural traditions of the Alevi and Bektashi people and the traditional religious buildings that they use for their ritual ceremonies. It can be noted that Islam embraces two main branches, Sunnism, the majority faith, and Shi'ism, which has become a different branch of Islam led by Ali, the nephew and the son-in-law of Muhammed. Although it is a group within Shi'ism, the Alevi and Bektashi traditions developed essentially in Anatolia (in the thirteenth century), and unlike Iranian Shi'ites for whom Shi'ism became an important socio-political foundation for resistance to the Arab-Ommayad sovereignty of that time, Alevism depends mainly on humanism – love of man and nature – in the love of God, and it has its roots in the Pre-Islamic Anatolian cultures and in the Central Asian traditions (Turkoglu 1995: 22, 41, Melikoff 1993: 49, 1998: 27–100, Birge 1965: 213). Furthermore, the Iranian Shi'ite movement follows "Caferism", another sect of Islam which is closer to Sunnism in many respects (Çakir 1998: 64–65).

Alevism has been practised mostly in the rural areas and has become a living tradition in *Bektashi Tarikati*, which is a part of Alevism led by the Dervish Haci Bektas Veli who played a very important role in the dissemination of Alevism in Anatolia and the Balkans.[1] It must be noted also that during the Ottoman Empire as well as the Turkish Republican period and until recently, because of implicit social and political pressures on the Alevi people, these people went underground and continued their ritual traditions secretly. This was because the Empire and later the Republican State represented the majority of the people in the society who were Sunni. For Melikoff, the religious traditions of Alevis and Bektashis are syncretic as they fused traditional local religious beliefs of Anatolia, Central Asian Shamanism, and Christian ideas within their heterodox system (Melikoff

1998, Akιn 1995, 1996, Birge 1965). Akιn underlines the difference between Orthodox Islamic mysticism and Alevi religious ideas, and finds similarities between the spiritual "becoming" reasonings behind Alevism and Western Gnosticism (Akιn 1996: 29). The mystical philosophy of *Tasavvuf* in Alevi and Bektashi traditions also has its roots in Neo-Platonism. Birge notes that "the conception of an ultimate unknowable Godhead causing the world of differentiated beings to emanate from himself; the appearing first of 'aklι kül' and then of 'nefsi kül', 'universal intelligence' and 'universal soul', are the reflection of Plotinus' trinity of the Absolute, Spirit, or Intelligence, and Soul" (Birge 1965: 214).

The Cem House, or what we call today the Cem Cultural House, means the main gathering hall used for musical-religious ceremonies, including eating activities involved in these ceremonies. This duality which makes up the primary difference between a Cem House and a Mosque, is very significant for understanding the religious-symbolic nature of the Cem House as being both "sacred" and "profane". This ritual ceremony as well as the traditional symbolic forms in the sacred building of the Cem House and Bektashi Tekke – such as the *Tütekli Örtü* (layered Lantern Roof) and *Dünya Ağaci* (World Tree) – are the very products of the imaginative world of the Alevi people which have their roots in the cross-cultural scene of Islamic traditions and the Universal Anatolian culture.

IMAGINATIVE UNIVERSALS AND POETIC WISDOM

The symbolic interpretations of "Lantern Roof" or "World Tree" in the traditions of Cem Houses and Bektashi Tekkes, like in many other similar kinds of buildings, demonstrate the eminence of symbolisation, which is one of the basic needs of man who dwells in the world intellectually as well as spiritually, poetically and aesthetically. Symbolic thinking does not only belong to children or poets, it is consubstantial with human existence. Human beings are able to externalise themselves with the help of what Cassirer calls "the symbolic forms" of language, myth and art which are the basic functioning of the intelligible, mythical, critical, imaginative and creative nature of the human mind (Cassirer 1955). During this symbolisation process, man reveals his understanding of and ideas (self-knowledge) about his life in the world that go beyond the representation of symbolic forms themselves.

Throughout the Cartesian thinking tradition, the attainment of intellectual knowledge and human development have had no connection with the "imaginative thought" of men revealed in the form of literature, rhetoric, arts, history, and ethics in developing cultural experience (Verene 1976: 295–302). It has become the purpose of iconographical as well as iconological readings of artworks in modern times to scrutinise the mythological and allegorical meanings of symbolic forms which belong to the "imaginative world" of man, while disclosing the historical, cultural and ideological significance of these works in particular times and places.

It was the Italian philosopher Giambattista Vico who in the eighteenth century first introduced his "Human Science", which unlike the logical and conceptual relations of ideas and theories in the Natural and Pure Sciences, is directly related to the human's ethical, historical being in the world, and implies an internal knowing-making mechanism (a "verum and factum" formula in Vico's terms) in understanding the cultural world. Vico's theory of "poetic wisdom" as a basis for his Human Science as well as the "imaginative universals" central to this theory, offers a vision for better understanding the symbolic and expressive nature of the human mind and its products (Vico 1740). "Poetic wisdom" indicates a "common mental language" shared by people in different times, whereas "imaginative universals" is the concept formation of poetic mind and is related to the means as well as products of the poetic-mythic expression of the human mind. Vico defines the natural act of communication as poetic, representing the pre-reflective or spontaneous consciousness of the human being (Caponigri, 1968: 167). Referring to Vico, Caponigri claims that the poetic expression of the human being is the "first attempt of the human mind to evoke the world of ideas; that is to render or present to itself the totality, the universality of its own being" (Caponigri, 1968: 172).

For Vico, the first man revealed his ideas about the world not with the help of epistemological thoughts, but with his poetic wisdom and imaginative universals, by means of mythical thoughts and symbolic expressions. Man's intention in the symbolisation process, in the first age, was to explain his knowledge about his being in the world, through metaphorical, imitative, allegorical expressions, in other words, through "imaginative universals". We can call this process the "objectionalisation of the knowledge of man by means of symbolic forms". What is significant in this symbolisation process is that we can call it more than simply a myth-making process: it is rather a form of self-expression or theatrical representation of knowledge (like scientific knowledge) about the reality of life in the mythical and poetical mind of man.[2] Furthermore, Cassirer (1955: 29–36) explains that unlike empirical-scientific thinking and consciousness where objectivisation bases the understanding of systematic relationships of universal rules of unity, causality, logic, substantiality, determination etc., in the mythical thinking and consciousness, objectivisation occurs through pure signification, unity of metamorphoses, non-accidental causality, and indeterminist relations of cosmic wholeness.[3] It must also be noted here that the symbolic representations of human life through symbolic forms of cultural works, like architecture, are not coincidental and arbitrary. They are the conscious choices of man closely related to the empirical, natural and cosmic realities of the world.

Furthermore, Vico's thoughts show that today's theory of knowledge, which belongs to the "intelligible universals" of man, is grounded in the theory of "imaginative universals", for example, in the theory of myth. His elucidation concerning the relationship between "imaginative universals" and "intelligible universals" gives us an insight to better understand the status of poetic wisdom and the nature of the unity of the symbolic cultural world in different developmental stages of the human mind. Vico contrasts "imaginative universals", which are the form of thought of the first of his three ages – of the gods, that is the age in which men thought in terms of gods – and of the second age – of heroes – with "intelligible universals" or "intelligible genera", which are characteristics of the form of thought of the third age – that of men (Verene 1976: 304). However, the difference between the three ages is not the form of thought, but the subject matter of self-knowledge shaped by the "imaginative universals" in the expressions of man. Moreover, the movement from "imaginative" concepts to "intelligible" ones is not a simple alteration and the "imaginative universals" of men do not disappear when man passes on to abstract modes of intelligibility (Verene 1976: 306–310). By this account, Verene pointed out that Vico's theory of "imaginative universals" can be conceived as a transformation of the process of concept formation, and in the development of human culture the formation of human experience in terms of "imaginative universals" gives way to or produces the "intelligible universals" (1976: 205). This also reveals that "self-knowledge" and the expression of the humanity of the first man is obviously not the same as modern man's, yet their thought forms always remain poetic, metaphorical and symbolic as an outcome of "imaginative universals" as much as "intelligible universals".

REPRESENTATION OF MYTHICAL AND THEATRICAL SPACE

Since symbolic forms of cultural works including works of architecture, are primarily a means of self-expression and theatrical representation of knowledge about the reality of life in the mythical and poetic mind of man, and since these forms are conscious choices of man related to the empirical, natural and cosmic realities of the world, we can claim that architectural space is created as a means for revealing this knowledge. Thus, it becomes a stage to represent the "theater of life" of human beings. We can find many examples of buildings in the history of art and architecture which represent various ideas and beliefs that transcend the visual expressions of buildings themselves and give ideas about the nature of the poetic and imaginative mind of man. Most religious buildings, including the Cem House and Bektashi Tekke that we will analyse soon, actually symbolise mythical and religious ideas of man which are represented by and limited to man's knowledge about the cosmic and empirical realities of life.

Vico pointed out that imaginative universals – poetic characters of art – always operate metaphorically through abstract forms of symbols and result in the rich and versatile interpretations of art. Metaphors in art work as a creative and open-ended endeavour to bring the work of art to a dimension that transcends it. In the interpretations of mythical-religious spaces, we can find analogical and metaphorical relations with cosmology or astrology, nature and the human body; in other words, with the realities of the empirical world of man. For example, the reality for man in the Middle Ages was confined to knowledge of the astrology of that time, where the cosmos referred to an organised universe in which the regulations of the world were transcendental. In other words, cosmological knowledge and its regulations played an important role in the construction of the mental cosmos itself. Furthermore, man's self-knowledge relating to the reality of the natural empirical world is imitated in the transcendental symbolic forms by means of his "imaginative universe". Cassirer calls these inner ties the fundamental features of a mythical feeling of space – growing out of astrology – which signify the intuition of space (Cassirer 1955: 93).[4]

In order to explain the particular characteristics of mythical spaces, Cassirer reminds us that mythical space is related to the space of perception and is strictly opposed to the logical space of geometry (Cassirer 1955: 84). According to Cassirer,

> The limits which the mythical consciousness posits and through which it arrives at its spatial and intellectual articulations are not, as in geometry, based on the discovery of a realm of the fixed figures amid the flux of sensory impressions: they are fixed on man's self-limitation in his immediate relation to reality, as a willing and acting subject – on the fact that in confronting this reality he sets up specific barriers to which his feeling and his will attach themselves (Cassirer 1955: 85).

For example, in the spatial expressions of religious buildings (including the Cem House and Bektashi Tekke) we find diverse relations, intrinsically unspatial forms, intuitive directions and lines which may seem strange at first glance, that all have their own symbolic meanings. Symbolic articulation and mythical-religious interpretations of these spaces are spiritually functional and "serve as instruments and organs for an explanation of the world. What happens in this explanation is simply that a merely sensuous content is poured into a spatial mold in which, one might say, it is re-formed, and through which it is apprehended in accordance with the universal laws of geometry" (Cassirer 1955: 88–92).[5]

So far, we have tried to identify the universal character and the role of "poetic wisdom" and "imaginative universals" in the becoming process of symbolic forms. Although the process of symbolisation of the cosmological perception of reality in the mental cosmos of man has a universal character, its interpretations vary with different cultural entities in different times and places. Diverse forms or spaces may symbolise similar ideas, or similar metaphors may be used to represent different ideas depending upon the traditions of cultural formations.

In the next section, I will refer to the particular symbolic aspects of the Cem House and Bektashi Tekke, and try to disclose the relationship between these forms and the meanings attributed to them, through which we will see how ideas related to the theater of life were shaped in the mental cosmos of those people and how they were interpreted metaphorically on the stage of architectural space through the "imaginative universals" and "poetic wisdom" of man. The ritual ceremonies that take place in these halls are very significant in helping to demystify the above relationship between the symbolic forms and the meanings that transcend these forms, revealing the theater of life of the people.

ARCHITECTURAL SPACE AS A STAGE OF THE WORLD

Since Bektashism and Alevism have their roots in both Islamic traditions and the Pre-Islamic Anatolian and Central Asian cultures, the ideas related to a theater of life for the people in these traditions became so rich because they have integrated different cultural scenes from different historical times and places. The cross-cultural nature of these traditions has also reflected upon the interpretations of ritual ceremonies, as well as the symbolic forms of architectural elements.

The Cem House consists of four parts, the *mihman evi* (guest house), *As Evi* (kitchen), *Ekmek Evi* (bakery), and the main ritual Lodge Room (Fig. 1). The main hall that Alevi people use during their ceremonies is called either *Cemevi*, *Kırklar evi*, *Erenler Meydanı*, or simply *Meydan*, as Bektashis call it (Akın 1995: 71, Birge 1965: 175–177). One can enter the main gathering hall, *meydan*, through a lower door with its *esik*, threshold. At the opposite end is a small throne consisting of three steps holding twelve candles which symbolise Twelve İmams, the followers of Ali. During the *Cem* ceremonies, both women and men sit down in a circle and the Alevi *dede*, leader, sits in the centre. There is no main direction in the hall, yet all four directions of the hall have sacred symbolic significance. The Cem House is considered as the microcosmos of the world with the vertical invisible axis at its centre (Akın 1996: 29, 1995: 72).

The ceremony is performed in a theatrical way, poems are read and people dance with live music, and at the end of the ceremony the *lokma* (meal) is served (Eröz 1990: 120–121, 302–304). There are different ritual ceremonies

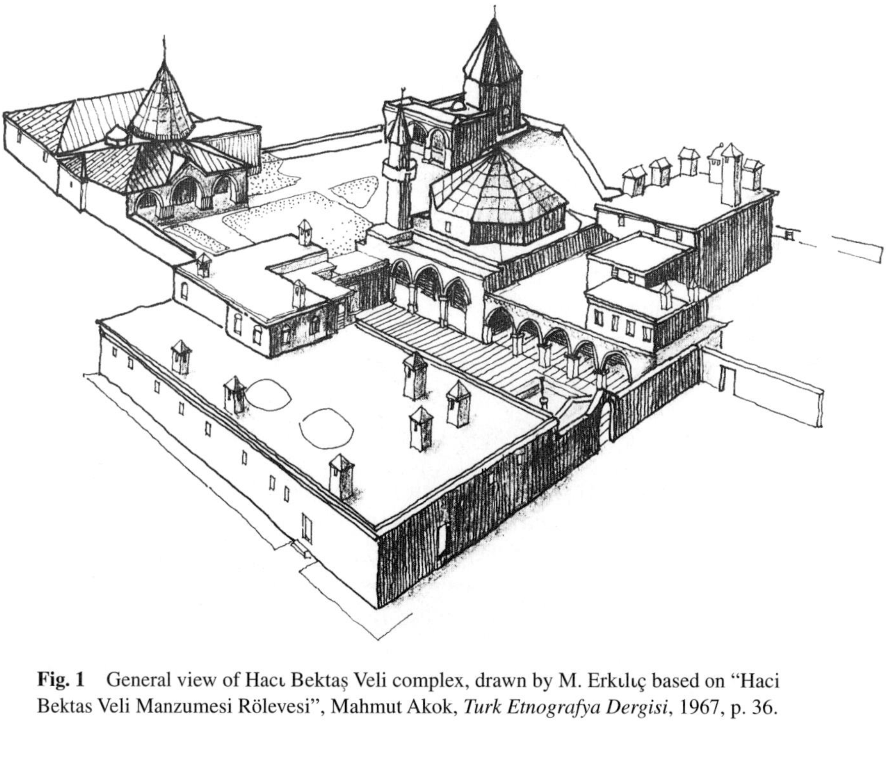

Fig. 1 General view of Hacı Bektaş Veli complex, drawn by M. Erkılıç based on "Haci Bektas Veli Manzumesi Rölevesi", Mahmut Akok, *Turk Etnografya Dergisi*, 1967, p. 36.

representing different beliefs and traditions. The most important among them is the *Ayin-i cem* which is performed to worship Ali and his unification with God. In the *Ayin-i cem*, *Dede* or *Baba*, the leader, gives particular roles to twelve people and each role has transcendental symbolic significance. These roles are: 1) *İmam*, 2) *Ferras* (sweeper), 3) *Hallak* (barber), 4) *Zakir* (who plays and sings), 5) *Sofracı* (who sets the table), 6) *İbrikçı* (who pours water), 7) *Sakii* (cupbearer), 8) *Hadim* (servant), 9) *Gözcü* (guard), 10) *Pervane* (one ready for any service), 11) *Çirağici* (candle-lighter), 12) *Bevvap* (doorkeeper) (Birge 1965: 179–180). Zakir plays and sings *Nefes* in the ceremony and invites those who wish, to rise for the dance *Sema* (or *semah*) which is done in couples of one man and one woman (Birge 1965: 199) (Fig. 2). *Sema* or *Semah*, which means "sky", is the name of a musical and theatrical ritual

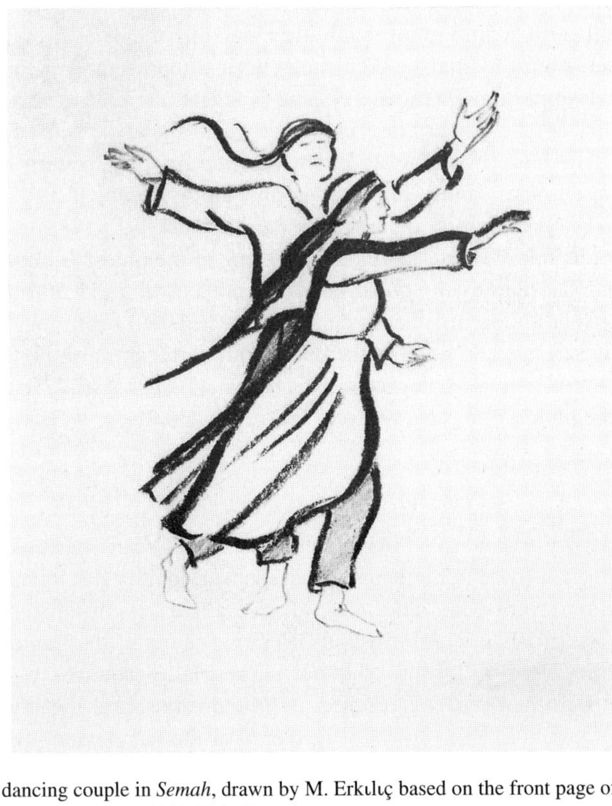

Fig. 2 A dancing couple in *Semah*, drawn by M. Erkılıç based on the front page of *Alevilerde Semah*, Erseven Ilhan Cem, 1990, Ekin Yayınları.

dance during which both men and women dance around a circle symbolising some religious beliefs with different figures. The traditions of Cem ceremonies as well as the *Semah* dances have their roots in Central Asian Shamanism (Birge 1965: 213, Erseven 1990: 98).[6] Birge, referring to Ziya Bey, reminds us that some of the "figures" which were used by Shamans in their religious ceremonies are still used today in many places of Anatolia and are called *Halay* (Birge 1965: 199).

The interpretation of architectural space and the arrangement of particular objects in the Cem House contribute to the theatrical representation of their transcendental meanings in Alevi traditions. Unfortunately, there are not many examples of Cem Houses which can help us to analyse their symbolic features in architecture. As mentioned before, Cem Houses have long been uncultivated because of the social and political pressure placed on Alevi people. Refering to the limited number of representative buildings, we will try to dwell on two important symbolic elements, the Tütekli Örtü and the World Tree, and try to analyse their metaphorical representative characters.

The most important Bektashi Tekke is Hacı Bektaş Veli, which is located in Kırşehir[7] (Figs. 3–4). The Hacı Bektaş Tekke is arranged around a courtyard. There are two main rooms with Tütekli Örtü (Lantern Roof), one for Cem ceremonies, and a big kitchen and dining room. It also includes a mosque and some small rooms for *türbes* and for prayers.

The Merdivenköy Bektashi Tekkesi was one of the oldest Tekkes ever built in Anatolia. The date of its construction is not known, and unfortunately it was demolished in the nineteenth century (Akın 1989). From the documents of Merdivenkoy, we can see that the main symbolic feature of the *tekke* is the central column which symbolises the *Dünya Ağaci* – World Tree – at the centre of the hall.

THE TÜTEKLI ÖRTÜ – LANTERN ROOF

The Tütekli Örtü (Lantern Roof) which covers the *Cemevi* has a geometrical structure formed by either squares or octagons gradually diminishing in size and placed on top of each other at 45 degree angles, rising in steps towards the centre (Akın 1995: 71) (Figs. 5–6–7a–7b). The top of the Lantern Roof is left open to bring light into the hall. The rising squares or octagons symbolize, like the dome of the mosque, the cosmos, and the layers of the lantern roof symbolise the layers of heaven. For the Alevis, the invisible axis passing through the centre of the hall as well as through the hole in the Lantern Roof, reaches the sky and metaphorically symbolises *Dar-ı Mansur*,

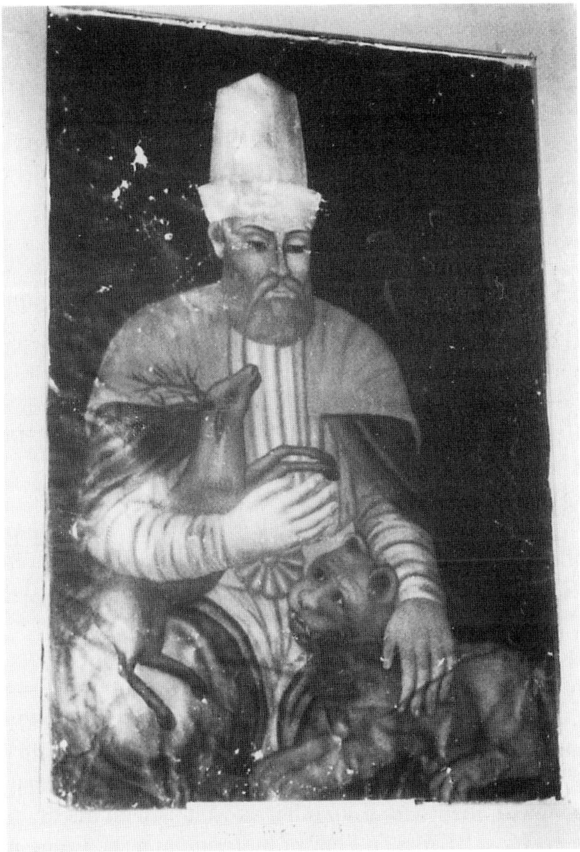

Fig. 3 Portrait of Hacı Bektaş Veli exhibited in the Hacı Bektaş Veli Complex in Kirsehir. Photo: M. Erkılıç 1997.

the *axis mundi*, which refers to the death of Mansur-al Hallaj and his unification with God (Akın 1995: 72).

Generally also in Islam, the vertical axis metaphorically symbolises Muhammad's ascending to Heaven to meet the Divine, and this Celestial Journey called *Miraj* is one of the few miracles attributed to Muhammad (Akin 1995: 72). The Alevis and Bektashis believe that Ali also went on the Celestial Journey with Muhammad, and during the Cem ceremonies following the "Hymn to the Miraj", the participants tell each other that they have seen the *Miraj* (Akın 1995: 73). During a *Semah*, men and women dance

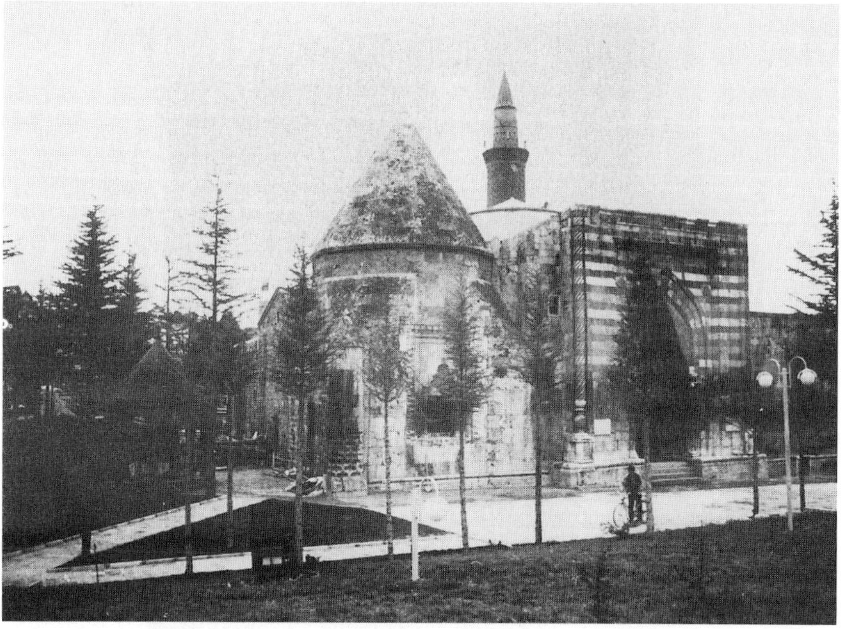

Fig. 4 Hacı Bektaş Veli Complex in Kırşehir. Photo: M. Erkılıç 1997.

around the central vertical axis of *dar* with music, first slowly, then with increased speed in harmonious, rhythmic movements (Birge 1965: 199–200). This part of the ceremony is called the *miraj*, during which dancers believe that they become unified with each other as well as with the divine, making a harmonious cosmos. "This is a kind of hierophany more in the nature of the infusion of the divine into a creature (*hulül*) and the unification of divinity with humanity (*ittihad*)" (Akın 1995: 73).[8]

These interpretations are actually symbolic representations of mythical-religious thinking which have their roots in the Pre-Islamic Asian tradition of Shamanism. We can find similar metaphoric representations in the Shamanist sacred space, according to Eliade, as stated by Akın,

> Sacred space had been the scene of a hierophany and so manifested realities ... that were not of our world, that came from elsewhere and primarily from the sky (Akin 1995: 72).

Akın also points out that,

> As the Shamanist concept of the structure of the universe is formed by three successively traversed planes such as sky, earth, and the underworld, ... the axis mundi passes through these

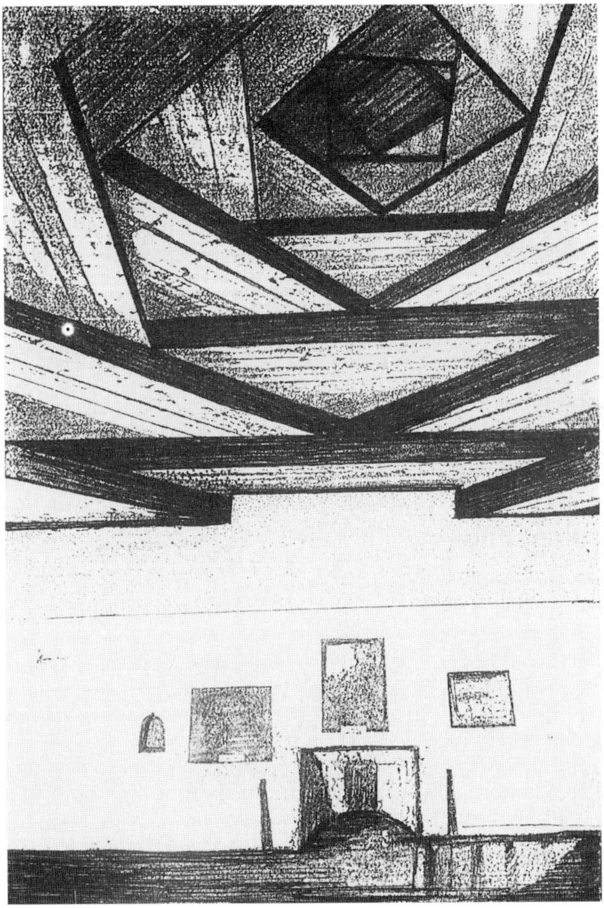

Fig. 5 Tütekli Örtü or Lantern Roof. Interior view from Hacı Bektaş Veli Complex in Kırsehir, drawn by M. Erkılıç based on "Anadolu Cami ve Tarikatlarında Tütekli Örtü", Günkut Akın, Vakıflar Dergisi, no. 22, 1991, p. 347.

holes ... the Shaman climbing through the smoke and light hole in the top of his yurt or semi-subterranean house with a ladder to communicate with heavenly forces (1995: 72).

The Tütekli Örtü was also widely used in the houses designated for ritual ceremonies in Central Anatolian Alevi villages (Fig. 8). According to Akın, the roof symbolises "the cave", man's first dwelling in the world, as well as happiness and love of family life (Akın 1991: 332). We can see the

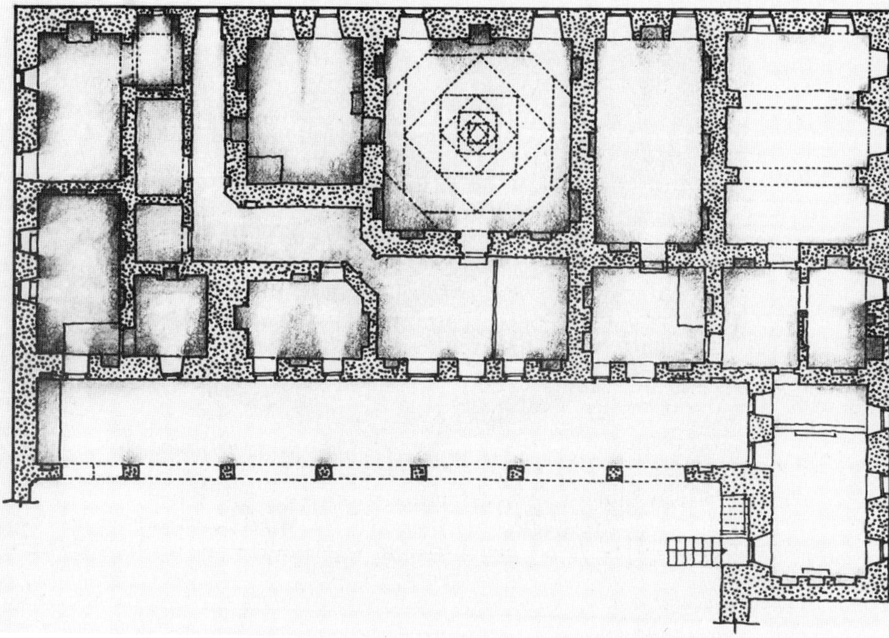

Fig. 6 Partial plan showing *Meydan* in Hacı Bektaş Veli Complex in Kırşehir, drawn by M. Erkılıç based on "Hacı Bektaş Veli Manzumesi Rolevesi", Mahmut Akok, *Turk Etnografya Dergisi*, 1967, p. 41.

Tütekli Örtü, Lantern Roof, in many traditional houses both in and outside of Anatolia (Akın 1991). This form has been used to symbolise various beliefs with moral-practical reasonings behind them. It was widely used, for example, in India, China, Central Asia, the Middle East, and also by the Pueblo Indians in the United States (Akın 1991: 323–354). In Turkey, apart from traditional houses in Central Anatolia, we find many mosques which have a Tütekli Örtü, a Lantern Roof. Cedit Camii – Erzurum, Mollabey Camii – Kütahya, Ispir Camii are some examples of this (Fig. 9).

DÜNYA AGǍCI, THE WORLD TREE

The symbolic characteristic of the Dünya Ağaci, World Tree, which may also be called the "Sacred Tree" in the Merdivenkoy Bektashi Tekke, is an important feature of Alevi Bektashi ideas (Fig. 10). The meanings attributed to the symbolic form of the "tree" vary in different mythical interpretations.

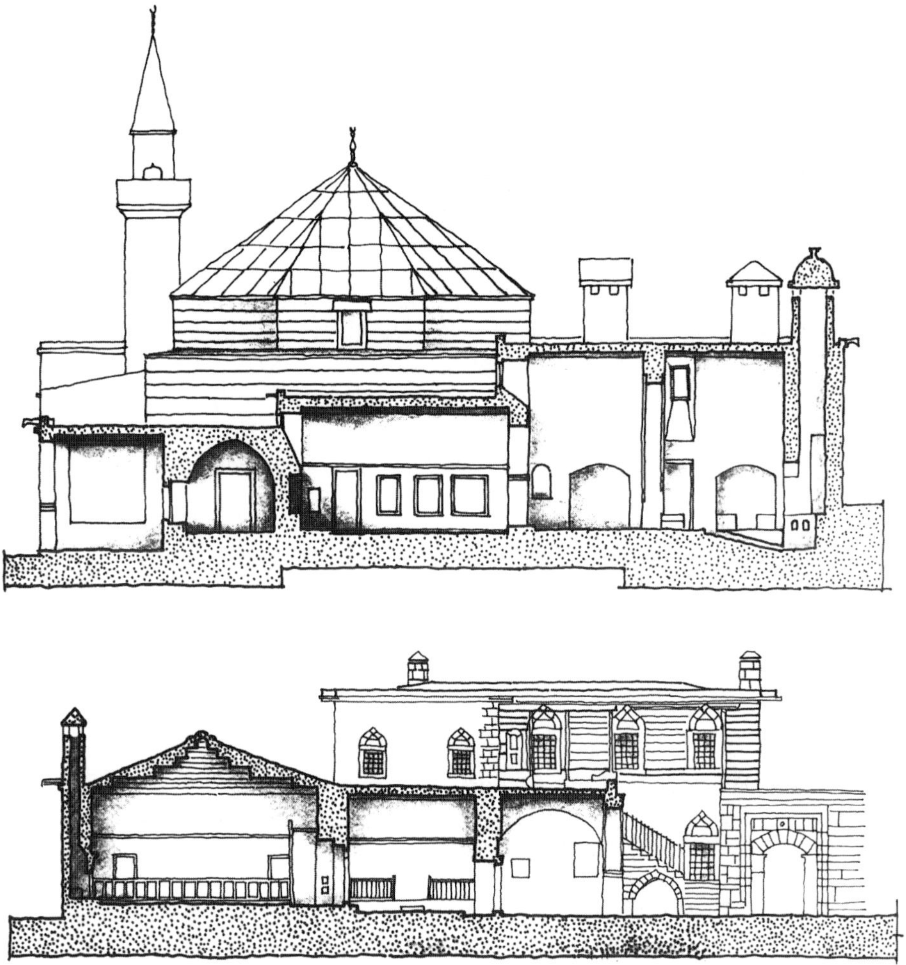

Fig. 7a–7b Elevation and Section of *Meydan* in Hacı Bektaş Veli Complex in Kırşehir, drawn by M. Erkılıç based on "Hacı Bektaş Veli Manzumesi Rolevesi", Mahmut Akok, *Türk Etnografya Dergisi*, 1967, p. 42.

Akın stated three meanings for the sacred tree: the Family Tree, the Gallows and the World Tree. The column at the centre of the sacred hall divides the roof into twelve parts, each of which symbolises the "Twelve İmams" in the Alevi tradition (Akın 1989: 69). According to Akın, the "Twelve İmams" came from the same family, so the twelve-piece structure symbolises a

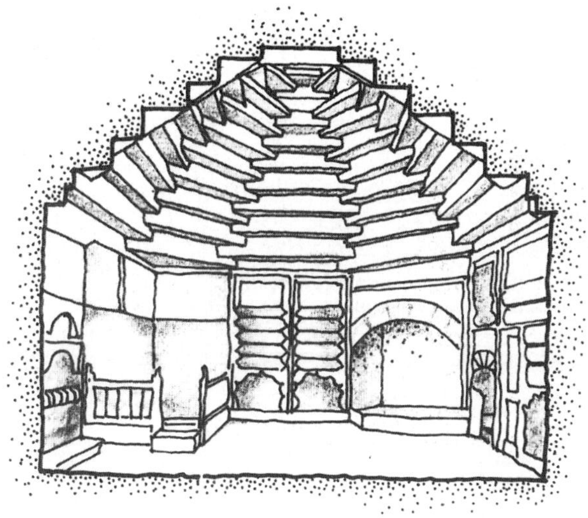

Fig. 8 Somuncuocğlu House, lantern roof in Erzurum, "Anadolu Cami ve Tarikatlarıcnda Tütekli Örtü", Gunkut Akın, *Vakıflar Dergisi*, no. 22, 1991, p. 348.

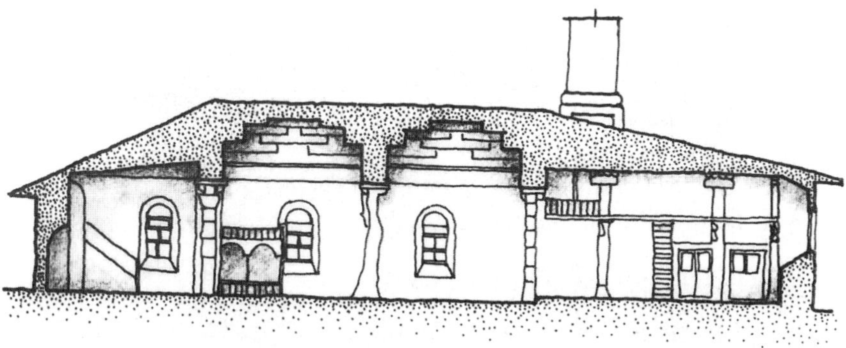

Fig. 9 Ispir Carsi Camisi, section, "Anadolu Cami ve Tarikatlarında Tütekli Örtü", Günkut Akın, *Vakıflar Dergisi*, no. 22, 1991, p. 350.

Family Tree (1989: 69). The tree at the centre of the hall also symbolises the gallows of *Mansur-el-Hallac*, or what we call *Dar-ı-Mansur*, as mentioned above in the vertical axis of Tütekli Örtü.

Islamic philosophers often indicate in their texts that Muhammad achieved the highest part of the *Miraj* where *Sidretu-i-Muntena* (the last tree) is located, and that he was able to see the garden of the Heaven on top of this tree (Peker

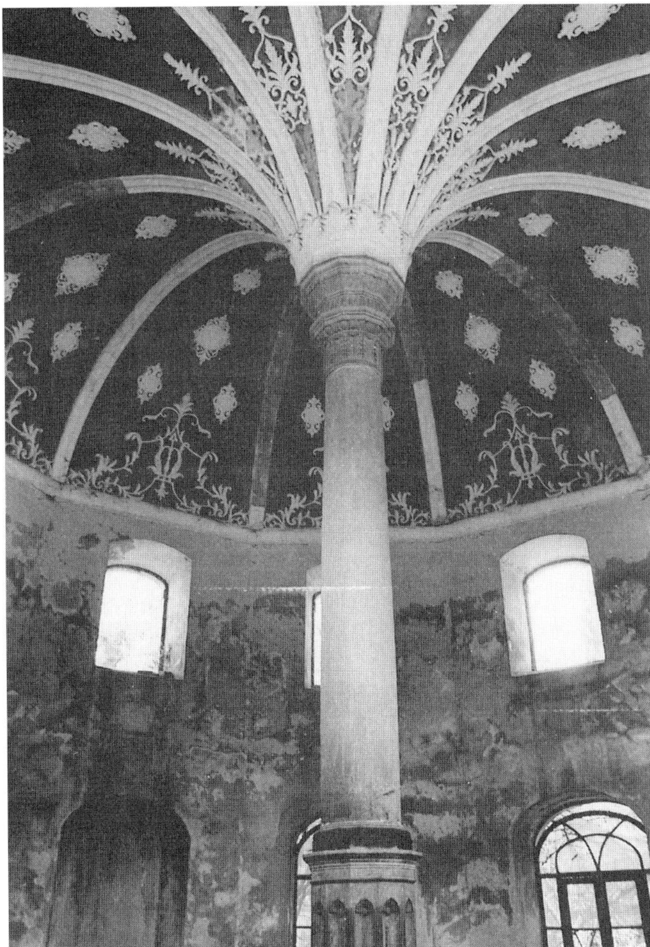

Fig. 10 Merdivenköy Bektaşi Tekkesi "world tree", Günkut Akın, "Merdivenkoy Bektasi Tekkesi'ndeki Dunya Agaci", *Sanat Tarihi Arastirmalari Dergisi*, 1989, p. 69.

1996: 80). In Islamic literature, the name of the Heaven Tree is *Tuba*. In the Koran, the tree of *el-Zakkum* is situated at the very bottom of the earth, which reveals that the world is located between the two cosmic trees (Peker 1996: 81).

Peker reminds us that Ibn al-A'rabi, the thirteenth-century theosophist born in Endulus, sees the sacred tree as the symbol of the last station of the Celestial Journey (*Miraj*) at the end of the seven skies and sensible world

(Peker 1996: xvi): "According to him, its roots symbolise the lower worlds, its branches stand for the sublime worlds, its leaves represent heavenly states and its fruits symbolise sciences and talents governed by these stages" (Peker 1996: xvi). The World Tree in that sense represents the cosmic unity of the world with its different branches, and indicates a harmony between the human world and the divine cosmic world. During the ritual ceremonies of the Alevi people, the hall, with the World Tree at its centre, offers a microcosm to the people who aim to achieve a macrocosm by turning around it in the *Semah*.

For Shamans, the house that they dwell in is also sacred and a microcosm in the centre of which there is a column symbolising the Sacred World Tree. According to Öney, for Shamans the Sacred Tree indicates the centre of the world by means of which one can travel to the sky and the underworld (Öney 1969, Peker 1996: 69).

We can find various symbolic metaphoric expressions of the natural form of the "Tree" such as the Light Tree, Life Tree, Heaven Tree, and so on, all over the world. In many other examples, the "Tree" figure is merged with floral and animal figures representing different symbolic meanings related to particular myths and beliefs. In the Yakut Turk myth called *Er-Ssogotoh*, in the highest part of the sky there is the *nine felek cigri*, under which is the earth with seven levels of sky, with Heaven located on top of the seven levels of sky (Peker 1996: 74). In the "Gilgamish Myth", the Sacred Tree is the Light Tree located in the East, signifying the beginning of life (Peker 1996: 74). Peker also points out that the Light Tree in the Babil tradition is the Palm Tree of Heaven (1996: 74). Altay Tatars believed that at the centre of the earth there is a very big pine tree which reaches the God of the Sky and is called *Bay Ulgen* (1996: 75). The Indian Life Tree is symbolised by a zambu tree.

We can add examples of symbolic representations of trees from many other cultural regions. The cross-cultural dissemination of symbolisation at the level of both physical images and ideas is inevitable, that is to say, the symbolic representations of the "Christmas tree", "Promising tree", or "Selvi tree" in the graveyards today, are not accidental.

EVALUATION: POETIC IMAGINATIVE REPRESENTATIONS ON THE THEATRICAL STAGE OF THE WORLD

It can be surmised from the symbolic meaning beyond the Lantern Roof and the World Tree in Cemevi or Bektashi Tekkes that the symbolic expressive nature of the human mind is imitative, analogical, and metaphorical as well as poetic. While the cosmic unity of the world is represented and imitated in the image of the Lantern Roof, the same unity is found in the metaphorical expression of the Tree, an object from the natural world. This signifies, as Vico stated, a conscious objectivisation of man's self-knowledge of the empirical reality of the world through the "imaginative universals" of symbolic forms.

However, what should be emphasized here is that the figurative choice of the model of imitation (in the case of the tree and in the form of the lantern roof) is not adventitious and sequential. Traditional religious buildings called Cem Houses and Tekke are not simply unconscious representations of people's traditional beliefs; they rather represent human beings, mythical-religious consciousness (limited to their knowledge) and their "will to present the harmony between the human being and the universe" in the theatre of life. Architectural space, accordingly, is a "stage" and the very product of the "poetic wisdom" of creative, imaginative and interpretative human beings. During the process of symbolisation, the mythical and religious thinking is realised both at the level of ideas (intelligibles) and images (sensibles). Although the layered structure of Tütekli Örtü, the Lantern Roof, symbolises the levels and unity of the cosmos with its physical form, in the mythical-religious thinking or in the spiritual world of man, "the inclining form" of the roof symbolises primarily man's understanding and conception of the ideas of, for example, hierarchy, unity, continuity and wholeness. These ideas actually belong to the "imaginative universals" of man formulated through his self-knowledge, which is continuously developed and reorganised in man's cultural world.

Many examples in the history of art from different cultures have referred to the symbolism of the Tree because of its figurative shape, its life cycle, its colour and pattern, and finally because of its relation to man's cultural and natural being in the world. Apart from its figurative form, its natural life cycle is metaphorically associated with many ideas such as fertility, liveliness, newness, freshness, continuity, progress, consciousness, enlightenment, which are only a few examples of the concepts related to the cultural life of man. This process may be called the objectivisation of knowledge and man's self-image in his life, revealed in symbolic forms. Metaphorical representations of the sacred tree, then, symbolise the creative, mythical, poetic, religious and imaginative mind of man before all.

The poetic nature of man's "imaginative universals" is not a characteristic of man in the Middle Ages, and as Vico and Cassirer pointed out clearly, it does not disappear when man passes on to the more abstract and logical mode of thinking, or "intelligible universals". The poetic, metaphorical and symbolic expression of the Tree has also been referred to by some of the critical cultural philosophers in the modern world. Modern man often expresses his reflective and ironic symbolic representations by metaphors in a poetical way. Al-Arabi, a thirteenth-century Islamic theosophist from Endulus, and Cassirer, a critical cultural philosopher of the twentieth century, both refer to the poetic, metaphoric expression of the Tree as a symbol of "enlightenment and consciousness". What they share is the poetic nature of imaginative universals in their expressions; although, however, what they understand about "enlightenment and consciousness" separates their expressions at the level of intellectuals.

METU – Department of Architecture
Ankara, Turkey

NOTES

[1] Alevism is in many ways carried out in the traditions of *Bektashi Tarikatı* which is a part of Alevism led by the Dervish Hacı Bektaş Veli who came from Turkistan (fourteenth century) as a member of Ahmet Yasevi Tarikati in order to disseminate the ideas of Yasevi and Alevism (from Central Asia to Anatolia and the Balkan countries) (Melikoff 1993: 35–36, 155–178). Although Alevism has nothing to do with dervishes, it has similarities with Bektashism, especially its religious ceremonies. Yet, while the Bektashi order had its organisation in towns, and had built *tekkes* (religious places), Alevi people lived in rural communities and carried out their ceremonies mostly in houses (Melikoff 1993: 29–30).

[2] In order to prevent a misunderstanding, it must be pointed out that "myth-making", either as narration or mythification of knowledge, is different from what we call here the mythical mind or thinking as a form of thought of man (Cassirer 1955). The realm of myth, mythical thinking and consciousness, for Cassirer and also for Kant, is concerned not with psychological genesis but with pure being and value (Cassirer 1955: 4). For them, "like knowledge, morality, and art, myth becomes an independent, self-contained world, which may not be measured by outside criteria of value and reality but must be grasped according to its own immanent, structural law" (Cassirer 1955: 4).

[3] Mythical thinking for Cassirer, like the other forms of thought, has its own advancing road in which it has its own problems, turning points, changes (1955: 35–36). Moreover, we cannot perceive a linear and directional relationship between empirical thinking and mythical-religious consciousness as the positivist philosophers suggested it. Cassirer underlines that,

> as formulated especially by Comte, "mankind gradually rises from the primitive phases of consciousness up to the theoretical knowledge and complete spiritual domination of reality. From the fictions, phantasms, and beliefs of these first phases the road leads more and more definitely to the scientific view of reality as a reality of pure facts".

According to Comte,

this progress falls essentially into three stages: the "theological", the "metaphysical", and the "positive". In the first, man transforms his subjective desires and ideas into demons and gods; in the second he transforms them into abstract concepts; it is only in the last phase that he differentiates clearly between "inside" and "outside" and limits himself to the given facts of inner and outer experience. Here, then, the mythical-religious consciousness is gradually overcome by a power alien to it. (Cassirer 1955: 236)

Therefore, in the positivist perception, the goal of the myth and religion is sought outside those in a fundamentally different space (Cassirer 1955: 237). However, as stated before, the mythical-religious spirit has a purely inward dynamic which has its own motion and supreme productions to achieve its own fulfillment, completeness, and continuity (Cassirer 1955: 236–237).

4 Furthermore, the orientation in the intuitive form of spaces, the mythical geography, also plays an important role in the logical formation of mathematical spaces. This relationship is emphasized by Kant, for whom the orientation of spaces begins with a sensuously *felt* distinction – namely left and right – and it rises to the sphere of pure mathematical intuition and ultimately to the orientation of thought as such, of pure reason (Cassirer 1955: 93).

5 Imitation of the cosmic wholeness and the life-cycle of natural objects or the human body becomes intelligible for the mythical consciousness and through the imaginative universals of man. Cassirer reminds us also that "the relation between what a thing 'is' and the place in which it is situated is never purely external and accidental; the place itself is a part of the thing's being, and the place confers very specific inner ties upon the thing" (Cassirer 1955: 92).

6 Birge pointed out the resemblance between Shamanism and Alevi Bektashi ritual ceremonies. These are: 1) that women participated in the formal worship, 2) mystic hymns, the *nefes'es*, came to take the place of the incantations of Shamans, 3) the *Sema*, or ritual dance, resembles the ecstatic dance of Shamans, 4) the sacrifice of the sheep or ram at *ayin-i-cem* is reminiscent of the custom of sacrificing cattle among the Asiatic Türks, 5) the miracles performed by the saints, the metamorphosis of a human into a bird, the flying through the air, etc., are similar to the stories of saints in Chinese Turkestan, 6) other legends of the saints show a type of folklore that is common to Bektashism and to the Buddhist influence which entered Shamanism through Tibet and Chinese Turkestan, 7) sacred places, particularly sacred trees, are common to both (Birge 1965: 213–214).

7 There are only two examples of documented Cem Houses today. These are: Yahyalı Cemevi in Şarkışla (1883–1884) and Baskul Bilalusagi village Cemevi in Elazig (Akin 1989: 71–72). We can get important information from the Bektashi Tekkes which, however rare in number, exhibit the traditional aspects of Alevism in their spatial characteristics of *Cemevi*.

8 Akın also stated that this understanding of unification in Alevi thinking is different from the Orthodox Islamic conception and also from the Sufi concept of unity. He claims that for Ottomans there are three different understandings of the concept of *tawhid* or unity. These are: 1) The Orthodox Islamic concept of unity which considers God as unattainable and arises from the radical distinction between God and all other things (*masiva*); 2) the Sufi concept of unity which considers man, nature, and the universe as a reflection of God and aims at reaching esoteric knowledge through contemplation; 3) the Shamanistic concept of unity which aims at "real" unification (*hulul* and *ittihat*) with the sacred through ecstasy, and reaches catharsis through such a transcendental experience.

REFERENCES

Akın, Günkut (1989), "Merdivenköy Bektasi Tekkesi'ndeki Dünya Ağaci", *Sanat Tarihi Arastirmalari Dergisi.*

Akın, Günkut (1991), "Tütekli Örtü Geleneği: Anadolu Cami ve Tarikat Yapılarında Tüteklikli Ortu", *Vakiflar Dergisi*, no. 22.

Akın, Günkut (1995), "The Müezzin Mahfili and Pool of the Selimiye Mosque in Edirne", *Muqarnas*, vol. 12.

Akın, Günkut (1996), "Cemevi Mimarisi Üzerine Notlar", *Cem Kültür Evi Mimari Proje Yarismasi Sartnamesi.*

Birge, John Kingsley (1965), *The Bektashi Order of Dervishes*, Luzac and Co., London.

Cassirer, Ernst (1955), *The Philosophy of Symbolic Forms*, Volume 2: Mythical Thought, Yale University Press.

Çakır, Ruşen (1998), "Political Alevism verses Political Sunnism: Convergences and Divergences", in *Alevi Identity: Cultural, Religious, and Social Perspectives*, ed. Tord Olsson, Elisabeth Ozdalga, Catherina Rraudvere, Swedish Research Institute in Istanbul, vol. 8, Istanbul.

Caponigri, Robert (1968), *Time and Idea: The Theory of History in Giambattista Vico*, University of Notre Dame Press.

Güler, Ahmet (1991), *Tarikarlar Ansiklopedisi*, Milliyet Yayinlari.

Eröz, Mehmet (1990), *Turkiye'de Alevilik ve Bektasilik*, Kültür Bakanlığı Yayınları.

Erseven, İlhan Cem (1990), *Alevilerde Semah*, Ekin Yayınları.

Melikoff, Irene (1993), *Uyur Idik Uyardilar*, Alevilik Bektasilik Araştırmaları, Cem Yayınevı.

Melikoff, Irene (1998), *Haci Bektas: Efsaneden Gercege*, Cumhuriyet.

Öney, Gönül (1969), "Anadolu Selçuk Sanatında Hayat Ağaci Motifi", TTK Belleten, Ankara.

Peker, Ali Uzay (1996), *Anadolu Selçuklulari nın Anıtsal Mimarisi Üzerine Kozmoloji Temelli bir Anlam Arastirmasi*, Doktora tezi, ITU.

Sezgin, Abdulkadir (1990), *Haci Bektas Veli ve Bektasilik*, Kultur Bakanligi Yayinlari.

Türkoğlu, Orhan (1995), Alevi Bektasi Kimligi: Sosyo Antropolojik Araştırma, Timas Yayinlari, İstanbul.

Verene, Donald Phillip (1976), "Vico's Science of Imaginative Universals and the Philosophy of Symbolic Forms", in *Giambattista Vico's Science of Humanity*, John Hopkins University Press.

Verene, Donald Phillip (1997), *Philosophy and Return to Self-Knowledge*, Yale University Press.

Vico, Giambattista (1740) (1948), *The New Science*, trans. Thomas G. Bergin and Max H. Fish, Ithaca, Cornell University Press.

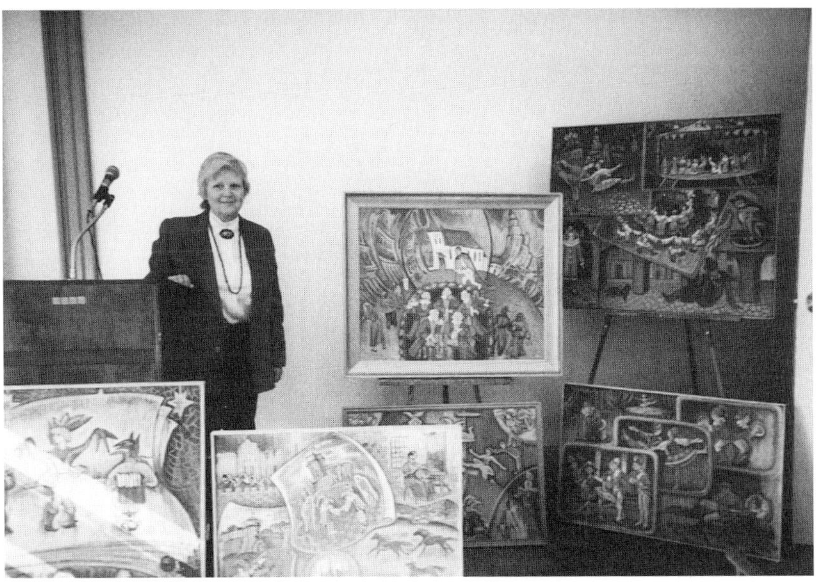

Ludmila Gayvoronski, painter-in-residence at the WPI, interpreting her paintings of the life-theater.

MATT LANDRUS

LEONARDO DA VINCI'S IDEAS OF WORLD HARMONY

Leonardo believed in painting as the only medium to display a stronger visual effect than reality. To achieve such an effect, *pittura*, for Leonardo, signified a unity in effects of the cosmos (macrocosm), human soul (microcosm), world soul (God's creations), and world harmony. I discuss here the way in which this signification of phenomena formed the primary nexus for Leonardo's interest in expressive visual effects and his artistic methodology.

Though not familiar with the modern term, *reality*, his idea of the *real* referred to the "works of nature," the "*meccanismo de natura*." The medieval Latin term, *realitas*, had little use in the Renaissance, where *veritas* sufficed as *truth* or the quality of having an actual existence. Leonardo concentrated on representations of *veritas* of *natura* especially according to their virtual or perceived conditions. Such virtualities come from the thirteenth-century Italian *virtù*, the intrinsic operating influence of a divine being. Whereas Leonardo was not as accustomed to the Medieval Latin *virtualitas*, he was obsessed with studying natural appearances of what the term defines: expressions of the possession of force, essential natural effects of being or potential force.

Drawings which combine symbolic and observed phenomena, like Leonardo's *Water Passing Obstacles* (Figure 1), express more than the originally observed flowing water. Most important for the drawing is its visualization of the movement, form, force, and effect of water upon water. The extent to which Leonardo's drawings and writings represent his visualization of phenomena, beyond that of strictly observed phenomena, indicates his interest in representing the order of a kind of chaotic movement. This intention stems from a belief in the fusion of a kind of psychological (human soul), religious (world soul), musical (world harmony), and physical (cosmos). He would not have used these specific categories. But such categorical concepts do contribute to the organizational elements in his work, which in turn reveal his pursuit of methods for depicting what is more real than reality.

Where observable whirlpools of water convert to projections of turbulent movement, force and effect, Leonardo's study of their forms combines imposed and observed lines. His projected visions of turbulence introduce an intuitive mode of representation in addition to the analytic mode. This intuitive method constitutes his artistic *and* scientific procedure. Thus, to compose turbulence requires good vision, as well as a good sense of what

A.-T. Tymieniecka (ed), Analecta Husserliana LXXIII, 39–50.

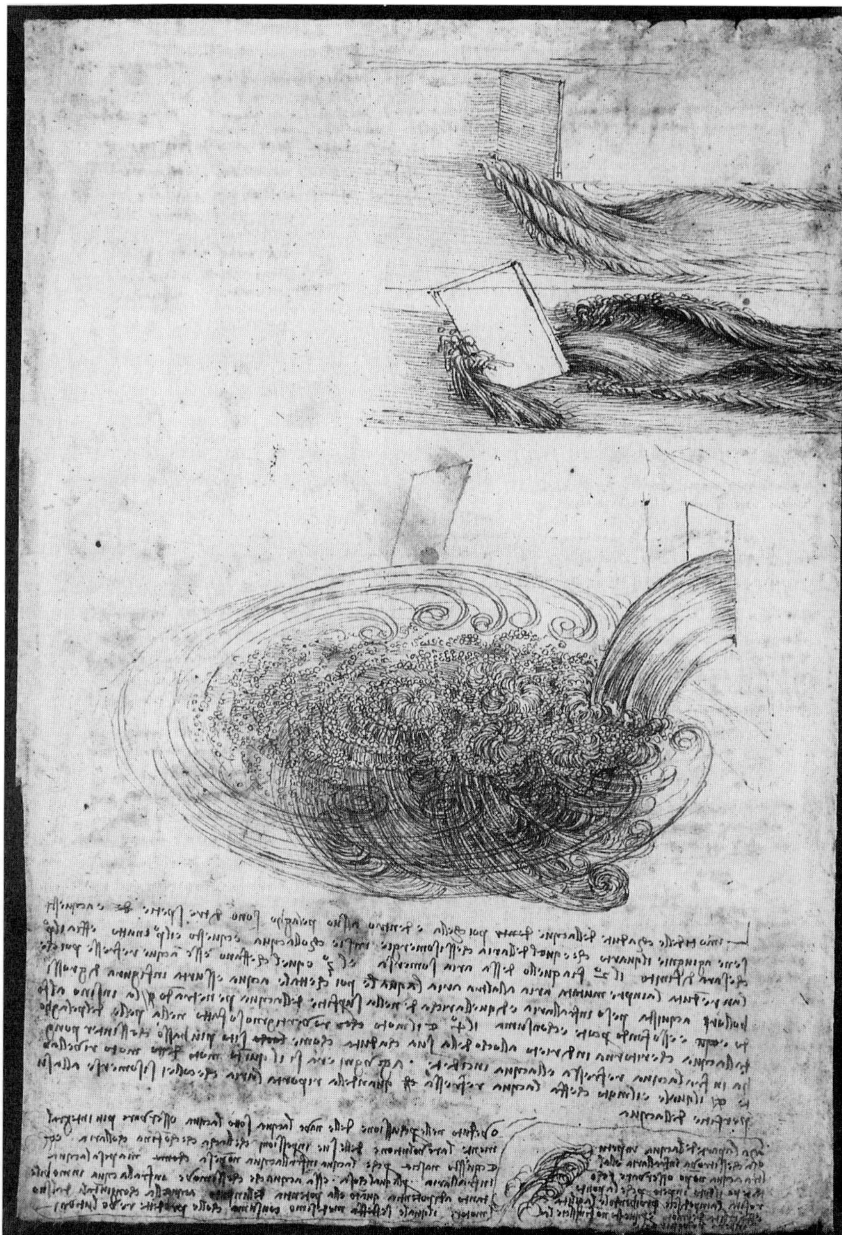

Fig. 1 Leonardo da Vinci, *Water Passing Obstacles* 1507–9, RL 12660v, The Royal Collection ® Her Majesty Queen Elizabeth II.

lines of force *should* be seen. This kind of good sense, or good understanding, basically refers to the intuitive sense of experience. To represent the order of chaotic water, Leonardo based his study on observed experience and on the intuited rules of perceived movement. He states that these intuited rules are based on experience: "Good understanding derives from reason expounded through good rules, and good rules are the daughters of good experience – the common mother of all the sciences and arts."[1]

He often avoids deontological discussions of what is absolutely right or wrong, discussing instead what is "good" or best ontologically. He does not always attempt to unify "the true, the right, and the beautiful," as Ludwig Heydenreich notes.[2] Rather, Leonardo demonstrates that nothing is just "right" in the indeterminant, ever changing forms of nature. He believes in the medieval doctrine that "the medium is best."[3] This medium path constitutes a proportional harmony of mathematical and geometrical laws, as governed by "proper discourse." Through *pittura*, proper discourse appeals to the senses and emotions.

First among Leonardo's priorities for determining the forms and forces of turbulence were his studies of causes and effects, such as his water studies. While he knew the ultimate cause to be God, he did not fully believe in traditional medieval boundaries between the microcosm and macrocosm, between the human soul, world soul, world harmony and cosmos. Erwin Panofsky called this sort of breakdown of barriers in the Renaissance a "decompartmentalization."[4] To Leonardo, a direct, etiological study of nature revealed the complex, interconnected relationships between phenomena.

Central to the context and content of Leonardo's work is his ontological approach to the indissoluble and irrefragable coherence of *pittura*, *verità*, *microcosmo*, *macrocosmo*, *ingegno*, *primum mobile*, *terrore*, *piacere*, *quantità-continua*, *quantità-discontinua*, *linguaggio*, and *fortuna*. Interdependent relationships between these terms, the soul, body, order, causality, sense, and being, by virtue of their *real* natures in the Renaissance, cannot be dissolved into concepts of things. The immense post-Kantian literature on concept-formation (*Begriffsbildung*), including the work of Marx, Cassirer, Wind, and Panofsky, has confused concept-type (*Artbegriff*) with Aristotelian and Leonardian substantial form, that eidetic nature of Being. Kant's focus on the basic form of knowledge devalues the contents of knowledge. True knowledge for Leonardo cannot be the formalized truth-concept (*Wahrheitsbegriff*), rather it would be the matter of directly interpreting the "works of nature." This reality of his reveals itself through an exegesis of the signification of his drawn and written forms.

Without this kind of exegesis, a problem arises, as Martin Kemp notes in his recent book, *Behind the Picture*.

As I came to know more about the proclivities of actual artists, so it was harder to see them as the Renaissance predecessors of Panofsky's and Wind's ways of thinking – not that the great iconographers were overtly claiming that this was the case. The problem with such iconography was, and is, to know exactly where it is located as a mode of explanation… .[5]

Pursuant to the appropriation of hermeneutic areas within original sources, at issue in the present study is the balance of systematic and historical modes of interpreting those initial sources. Such an exegesis involves the study of a number of ordering principles in Leonardo's work, from his general

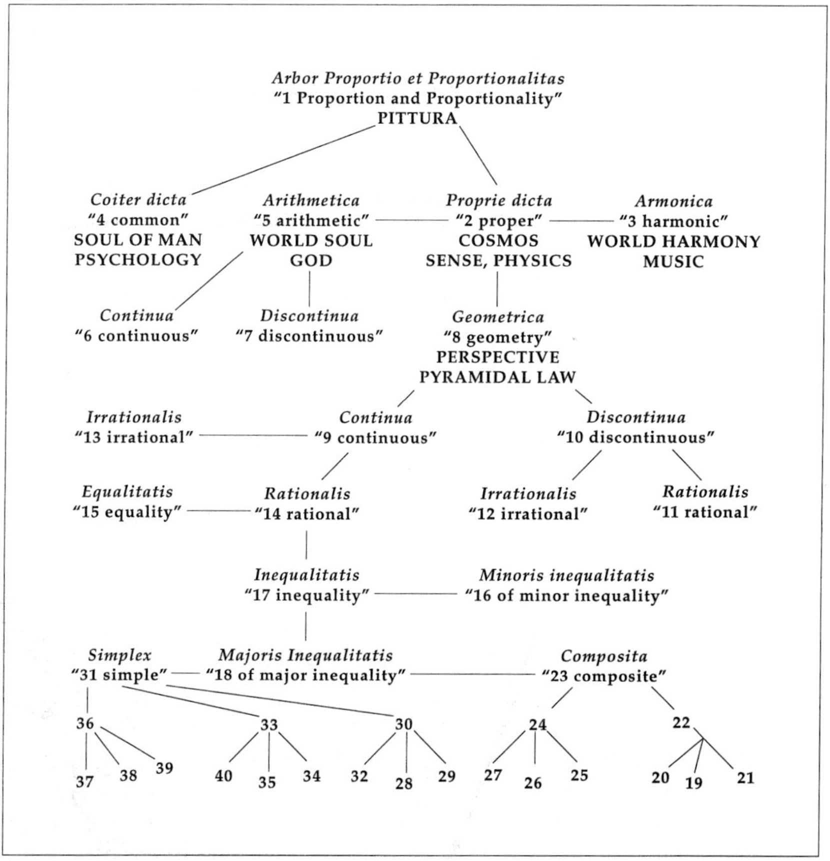

Fig. 2 Author's diagram of Pacioli's 1494 proportion tree (in italics) as it may relate to Leonardo's numbering of its main parts in MS Madrid II 78r. Words in capital letters interpret the symbolic hierarchy.

appreciation of structural systems like Luca Pacioli's tree of "proportion and proportionality," to the role of a *primum mobile* in the etiology of art/science, to the very specific order of the features and functions of sensory perception. These ordering principles express the interdisciplinary, ontological context and development of Leonardo's art and ideas. I combine the ideas of Pacioli and Leonardo in the diagram of Figure 2. Examining these principles further demystifies Leonardo's intuitive method of art *as* science without initially subjecting that method to today's formalist concepts. Mapping trajectories of the creative process, this scientific method provides a framework of systematic and historical modes for interpreting self-similar, reductive, non-reductive, and symbolic forms in Leonardo's draftsmanship.

The tree of proportion first appeared as part of Luca Pacioli's *Summa de arithmetica, geometria, proportioni e proportionalità*, published in Venice in 1494. At the Sforza Court in Milan beginning around 1496, Pacioli and Leonardo shared similar ideas about divine proportion, classification of the sciences, and perspective as a liberal art.[6] Leonardo briefly copied Pacioli's proportion tree around 1504, while also writing about arithmetic, perspective and physical rules. He translates Pacioli's Latin and numbers the sections as follows in this English interpretation: first, "proportion and proportionality"; second, "proper," for "*proprie dicta*" (proper discourse); third, "harmonic," for "*armonica*"; fourth, "common," for "*coiter dicta*" (of commonly encountered discourse); fifth, "arithmetic"; sixth, "continuous"; seventh, "discontinuous"; eighth, "geometry"; ninth, "continuous"; tenth, "discontinuous"; eleventh, "rational"; twelfth, "irrational"; thirteenth, "irrational"; fourteenth, "rational"; fifteenth, "equality"; sixteenth, "of minor inequality"; seventeenth, "inequality"; eighteenth, "of major inequality." After this, he only identifies twenty-three, "composite," and thirty-one, "simple."

If there could be any order to Leonardo's methods, as Claire Farago seems to find as a fractal format of Codex *Madrid* II, Pacioli's proportion tree may have influenced that order.[7] Since Leonardo was most likely interested in Pacioli's study of divine proportion as early as 1496, to what extent had he used the tree format? Various concepts important to Leonardo appear to have an order generally befitting these ideas of divine proportion. As Farago finds, however, the various ideas may not conform to linear patterns; headings often contain nestings of subjects within subjects, at regular intervals of self-similar irregular material. When comparing Codex Madrid II 78r, Codex Atlanticus 151ra, Pacioli's 1494 proportion tree, and Aristotle's *Metaphysics* 1025b25 ff and 1064a17 ff, there would appear to be a direct relationship between these sources, suggesting Leonardo's interest in a teleological network of inseparable phenomena: emotion (psychology), *armonica* (music), cosmos (physics/nature), and God's creations (world soul/*verità*).

Suppose one were to impose Leonardo's ideas in his Treatise on Painting onto Pacioli's proportion tree. Leonardo's first concern, "painting," would occupy the header. "Sense" would occupy the second position as the primary interpreter of facts, the *institutum*, principles, law, "*proprie dicta,*" proper discourse, ethical speech. To this extent, sense encompasses and represents the phenomena of physics or medieval cosmos. "Music," "*armonica,*" takes the third position dealing with world harmony. Leonardo believed music had a nearly lateral hierarchical relationship with *pittura*, well above that of sculpture, language, and possibly politics (in the Aristotelian sense). Completely outside the tree's main structure, though of the same importance as sense and harmony, one might interpret the fourth position as "emotion," "*coiter dicta,*" or commonly encountered discourse, indicating the human soul, psychology, *cultum*, what people want to do, the normative interests, persuasive speech. Essential to interpretations of the senses, the fifth position of arithmetic, and eighth position of geometry, both involve "*quantità continua*" and "*quantità discontinua,*" respectively positions six and seven for arithmetic, and nine and ten for geometry. Possibly more than geometry, arithmetic represents a final test of the scientific-ness of art, since, as Leonardo states, "no human investigation may claim to be true science if it has not passed through mathematical demonstrations."[8] To the highest authority – God – Leonardo attributes this divine truth:

> The eye is the commander of astronomy; it makes cosmography; it guides and rectifies all the human arts; it conducts man to the various regions of the world; it is the prince of mathematics; its sciences are most certain; it has measured the height and size of the stars; it has disclosed the elements and their distributions; it has made predictions of future events by means of the course of the stars; it has generated architecture, perspective and divine painting. Oh excellence above all other things created by God! What manner of praises could match your nobility?[9]

Thus mathematics, according to the proper senses (*proprie dicta*), represents phenomena of God's creations, or the world soul.

Leonardo would have known of the Aristotelian hierarchical system's elevation of contents of sensory knowledge over the influence of poetic inference. Under detailed embryological notes around 1510, he mentions a logical example which claims greater satisfaction with the harmonious effects of art and music than with Neoplatonic poetic rapture.

> When the poet ceases to represent in words what exists in nature, then the poet ceases to be equal to the painter; for if the poet … were to describe the ornate and persuasive words of [someone] … then he becomes an orator … if he speaks of the heavens, he becomes an

astrologer and a philosopher – and a theologian when speaking of things of nature or of God. But if he returns to the representation of any object, he would become equal to the painter, if, with words, he could satisfy the eye … [creating] a harmony to the eye as music to the ear … .[10]

Stating his intention, around 1489, to define the *anima*, the "*cosa divina*," Leonardo produced in 1508 what for him is a satisfactory definition of what the soul causes, with a common medieval explanation of the transformation of animal into man by transmission of the soul.[11]

[Nature] … places within the soul the formative agent of the body, that is, the soul of the mother which first constructs in the womb the shape of man … and the rest of the definition of the soul I leave to the … friars, the fathers of the people, who by inspiration know all the mysteries. Let be the sacred writings, for they are supreme truth.[12]

Leonardo's *Paragone*, his comparison of the arts, gives an example of the way the art/science of super-real painting reproduces much more than just nature's effects:

With justified complaints painting laments that it has been excluded from the number of the liberal arts, since she is the true daughter of nature and acts through the noblest sense. Therefore it was wrong, O writers, to have left her outside the number of the liberal arts, since she embraces not only the works of nature but also an infinite number that nature never created.[13]

Leonardo refers here to the superiority of painting over liberal arts of the trivium and quadrivium, since *pittura* masters rhetorical aspects of the former and mathematical aspects of the latter. By necessity, therefore, *pittura* unifies the human soul, world soul, world harmony and the cosmos. Leonardo also writes about the "divinity" of this rhetorical/mathematical science of painting as a product of knowledge first transmitted through the eye, "the noblest sense" and "window of the soul."[14]

He states his break with the traditional view that "knowledge born of experience is mechanical [and] that knowledge born and ending in the mind is scientific."[15] He believed that *pittura* could appeal to the *ignoranti* as well as the educated by virtue of the experiential scientific knowledge which created the painting. This support of sensory experience over mental analyses attempts not so much to respond to the humanist tradition, as to elevate the status of fine art above its status as technical or mechanical craft. As Andrea Bolland discusses in a recent article, *trecento* views of Boccaccio and Petrarch indicate the believed inaccessibility of the full extent of Giotto's illusionistic effects to the common crowd.[16] She perceptively notes that these views do not call for humanist learning, but for "a wholly new type of expertise: the visual skills necessary to discern and interpret the syntax and the rhetoric of illusionary painting."[17] These visual skills, for Leonardo, utilize arithmetic, geometry and harmony in proper discourse (the physical senses) and thereby relate to the human soul/psychology/emotion through *pittura*.

Leonardo and Pacioli were almost equally fascinated by mathematics and geometry. When the artist and the scholar met 1496, Pacioli's beliefs in interdependent divine, mathematical, and proportional phenomena reinforced Leonardo's earlier ideas for the superiority of *pittura* over the arts, the trivium and quadrivium. This confirmed Leonardo's ontology of *pittura*: to express essential, physical qualities and quantities in a factual manner still cognizant of God and world harmony. He states that painting of this kind "does not speak, but is self-evident through its finished product."[18] In this way, *pittura* becomes the noblest study within what Pacioli would call the *institutum*, an ethical discourse rather than a persuasive or emotional discourse, *proprie dicta* rather than *coiter dicta*, factual discourse rather than normative discourse. Leonardo writes:

> True sciences are those which have penetrated through the senses as a result of experience and thus silencing the tongues of disputants, not feeding investigators on dreams but always proceeding successively from primary truths and established principles, in a proper order towards the conclusion. This may be witnessed in the principles of mathematics, that is to say, number and measure – termed arithmetic and geometry – which deal with discontinuous and continuous quantities, with the utmost truth. … Here all guesswork remains destroyed in eternal silence, and these sciences are enjoyed by their devotees in peace, which is not possible with delusory sciences of a wholly cerebral kind.[19]

These "delusory sciences" include what today may be called psychic sciences: astrology, numerology, tarot, hand-reading, face-reading, and dream analysis. Leonardo's notebooks contain well-known statements against similar unscientific sciences. His "true sciences," like Pacioli's, regard observable examples of arithmetic, geometry, harmony, proportion and God's creations. Leonardo tends to agree with Pacioli's separation of the extensive field of proper concerns from the singular category of normative concerns, as compared in the diagram of Figure 2. Whereas *proprie dicta*, physics and cosmos remain closely tied to arithmetic, geometry and harmony, *coiter dicta* has no direct interdependent link to those phenomena. *Coiter dicta* nonetheless occupies the same high area of importance as its immediate neighbors. To the extent that *coiter dicta* represents commonly encountered discourse, it deals with psychology, the *cultum*, human soul, and what people prefer to do, as opposed to what they aspire to. In a similar sense Leonardo believes that common discourse can degenerate to the following level of people: "These men possess a desire only for material wealth and are entirely devoid of the desire for wisdom, which is the sustenance and truly dependable wealth of the mind."[20] Within the same discussion, he further paraphrases Aristotle's *Metaphysics*: "Good men possess a natural desire to know."[21]

Of similar importance in the neighboring system, *proprie dicta* branches out to harmony, arithmetic and geometry. *Pittura*, rooted in *coiter dicta* and *proprie dicta*, stems from the latter, proper discourse by virtue of the senses, "*Proprie dicta*" could therefore represent the eye, the "noblest sense" of physics and the cosmos for Leonardo. *Armonica*, representing proportional or world harmony, maintains an equal position of importance to arithmetic and geometry. At the same level as concepts of the human soul and the cosmos, world harmony is similarly expressed with proportional harmony.

World soul (God's creation) to Leonardo, reveals itself in the truest of sciences, mathematics. Specifically disinterested in discourse about or around the nature of God, Leonardo bases his *Paragone* on the best ways to represent the works of God. Essential to this representation are the divinely exact mathematical, proportional, arithmetical, geometrical, and perspectival methods necessary for its reproduction. Leonardo claims, "Amongst the great things of mathematics, the certainty of its demonstrations most conspicuously elevates the minds of investigators. Perspective must therefore be preferred to all the human discoveries and disciplines."[22] Instead of a deontological concern for the *right* representation of God's works, which presumes an understanding of God's own concerns, Leonardo seeks the ontologically *best* and *preferred*, observed and reasoned ways to represent visible works of God. Though he did not differentiate between "right" and "best", his comments show his tendency to approach "best" results, rather than "right" results in the absolute sense. He argues:

> You say that a science is correspondingly more noble to the extent that it embraces a more worthy subject, and accordingly, that a spurious speculation about the nature of God is more valuable than one concerned with something less elevated. I reply, we will state that painting, which embraces only the works of God, is more worthy than poetry, which only embraces the lying fictions of the works of man.[23]

Pittura therefore claims a higher position than the Aristotelian *episteme poietike*. With greater satisfaction than poetry, painting hereby masters the *theoria* which leads to *verità*. *Arithmetica* represents the key to reproducing the self-evident *theoria* and *verità* of God's works, naturally manifesting the world soul. Leonardo confirms:

> Painting presents its essence to you in one moment through the faculty of vision by the same means as the *imprensiva* [receptor of impressions] receives the objects of nature, and thus it simultaneously conveys the proportional harmony of which the parts of the whole are composed, and delights the senses. Poetry presents the same thing but by a less noble means than the eye,

conveying it more confusedly to the *imprensiva* and describing the configurations of the particular objects more slowly than is accomplished by the eye.[24]

Possibilities of super-expressive reality in *pittura* necessarily portray a harmonized unity of macrocosm and microcosm. For Leonardo, appealing to the senses in this super-real manner requires the pursuit of completely factual matters of the senses, of *proprie dicta* or proper discourse, of the *institutum*, physics and cosmos. As for harmony, he finds in music and painting conjunctions of proportional parts, like the scales of notes or the "scale of 20 for 20 braccia" in perspective measurements.[25] But unlike the musician's temporary sounds, he says "the painter makes his work permanent for many years, and of such excellence that it keeps alive the harmony of those proportional parts which nature, for all her powers, cannot manage to preserve."[26]

A necessary link to the physical/sensory cosmos of rhetorical/mathematical laws, to the harmony of musical/proportional scales, to the divinity of optical/sensory perceptions of the world soul, is emotion, an expression of the human soul. Emotional content is so important to Leonardo that he requires that a painting, "be made with great immediacy, exhibiting in the figure great emotion and fervor, otherwise this figure will be deemed twice dead … because it [will be] a depiction, and … [will] not exhibit motion either of the mind or of the body."[27] The strength of an artwork's visual effect on a viewer for this reason depends on its ability to please or terrify, excite or depress. Equally, these emotional effects of the human soul would not exist in a painting without the rhetoric and mathematics describing and measuring the cosmos, without the use of universal musical/proportional scales of world harmony, without the divine optical windows giving perceptive access to the world soul of God's creations. All these methods considered, Leonardo seems to have intended to prove that a painting could be more real than reality.

University of Oxford

NOTES

1 *Codex Atlanticus* 221vd/597iiv; translated in: Martin Kemp, ed., *Leonardo on Painting*, trans. Martin Kemp and Margaret Walker (New Haven: Yale University Press, 1989), p. 10.

2 Ludwig Heydenreich, *Leonardo da Vinci*, Vol. I (New York: Macmillan, 1955), p. 175.

3 Urbinas 207v, Kemp (1989), p. 162.

4 Erwin Panofsky, "Artist, Scientist, Genius: Notes on the 'Renaissance – Dämmerung,'" in Erwin Panofsky, ed., *The Renaissance: Six Essays* (New York: Harper & Row, 1962), p. 128.

[5] Martin Kemp, *Behind the Picture: Art and Evidence in the Italian Renaissance* (New Haven: Yale University Press, 1997), p. 264.
[6] See: Claire J. Farago, *Leonardo da Vinci's Paragone* (Leiden: E. J. Brill, 1992), pp. 298–99.
[7] Claire Farago, "Fractal Geometry in the Organization of Madrid MS. II," *Achademia Leonardi Vinci* 6 (1993), pp. 47–55.
[8] Urbinas 1r-v, Kemp (1989), p. 14.
[9] *Ibid* 15v-16r, Kemp (1989), p. 21.
[10] Windsor RL 1910r; translated in: C. D. O'Malley & J. B. de C. M. Saunders, *Leonardo da Vinci on the Human Body* (New York: Crown, 1982), pp. 480 & 506.
[11] For Leonardo's early interest in defining the "cosa divina" see: RL 19038r or Atlanticus 203 r-a. "One might speak of the influences of the planets and of God," Atlanticus 203 r-a as translated in: Carlo Pedretti, *Literary Works of Leonardo da Vinci: Commentary*, Vol. I (Berkeley & Los Angeles: University of California, 1977), p. 112.
[12] RL 19102r; O'Malley & Saunders, p. 474.
[13] Urbinas 15v, Kemp (1989), p. 21.
[14] *Ibid* 8v, Kemp (1989), p. 46.
[15] Atlanticus 154rb/417r, Kemp (1989), p. 10.
[16] Andrea Bolland, "The Education of the Eye and the Experience of Art in Renaissance Italy," *Analecta Husserliana* 53 Bk. 2 (1998): pp. 325–333.
[17] *Ibid*, p. 326.
[18] Urbinas 28v, Kemp (1989), p. 46.
[19] *Ibid* 19r-v, Kemp (1989), p. 10.
[20] Atlanticus 119va/327v, Kemp (1989), p. 9.
[21] *Ibid*. Also Kemp (1989), p. 281.
[22] *Ibid* 203ra/543r, Kemp (1989), p. 49.
[23] Urbinas 15r, Kemp (1989), pp. 32–3.
[24] B.N. MS Institut de France 2038, 19v, Urbinas 11r-v, Kemp (1989), p. 23.
[25] Urbinas 17r, Kemp (1989), p. 35.
[26] *Ibid* 16v-17r, Kemp (1989), p. 62.
[27] *Ibid* 110r, Kemp (1989), p. 145.

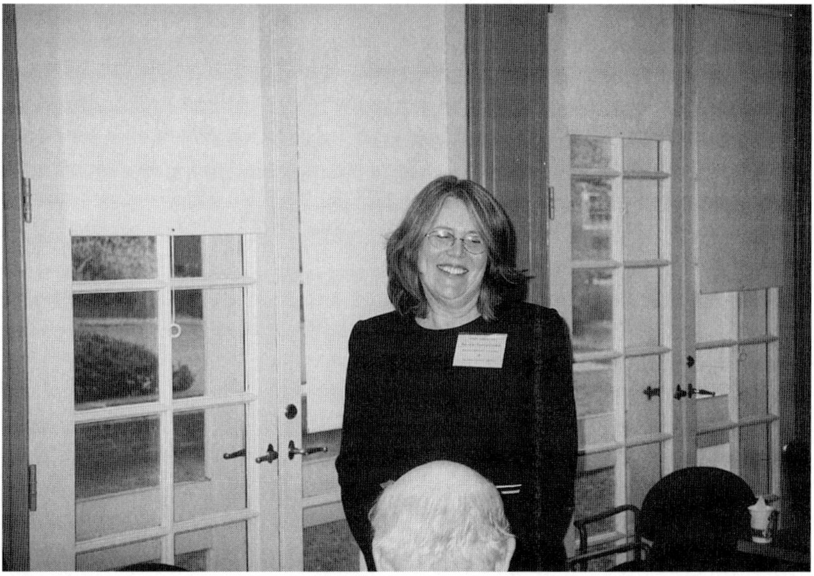

Patricia Trutty-Coohill

PATRICIA TRUTTY-COOHILL

THE RENAISSANCE PAINTER AS DRAMATURGE

For Rev. Richard Meredith, Dramaturge

When Professor Kronegger announced the topic of the 1999 ISPAFA conference, I was much perturbed. What would I talk about? I am a myopic Leonardo scholar. What could I add to the mounds of literature on Leonardo as stage designer? I should prepare a Tymieniecka school study, not a purely art historical one. My reaction does not surprise me; after all, the proposed topic does not seem to be a natural extension of my work. I find, however, that as the title festers in my imagination for a while, and as the meeting date draws near, something hits, and I manage to work within the new parameters. This year I proposed the painter as designer, not of sets, but of the composition in architectural space, as dramaturge. It was pretty. It could be done. But it was somewhat mechanical.

The problem I had with the topic "*Theatrum Mundi*" was that I considered the construct pessimistic. After all, Shakespeare's scenario of life as a stage in *As You Like It* is imprinted on my mind: "All the world's a stage" on which we are but "petty players who strut about and then are heard no more." Added to this is T. S. Eliot's image of the theatre of society as a wasteland. At its worst, the metaphor of life as a stage was a pessimistic cliché. At its best, it was Aristotle's *catharsis*, an artifice by which one could leave daily life behind, put off problems, get away from the pettiness of the world – like the theatre touts who try to lure customers from their shadowy lives in Hiroshige's *Saruwka-cho Theatre*,[1] or Toulouse-Lautrec's imagery of Paris nightlife. Weary escapes *from* reality, cheap tricks, albeit beautifully done.

But not all theatre can be so characterized. A case in point is liturgical drama, especially that of the baptismal rite, which is most dramatically performed on adults during the rite of the Easter Vigil. The third reading, Exodus 14:15–15:1, recounts the parting of the Red Sea by which God saved the Israelites and drowned the Egyptians, their oppressors. The exposition immediately following the reading cautions that the story should be read metaphorically: the sea is the Baptismal waters that have the power not only to save but also to drown and wash away the evil, guilt and care we bear.[2] So "on point" is this image that it has become a cliché: Christ will "wash all your sins away with the tide." However, through art, the reality at the core of the cliché is reawakened.

A.-T. Tymieniecka (ed), Analecta Husserliana LXXIII, 51–60.

Lillian Hellman used a painterly metaphor for the presence of the past: *pentimento*. Max Beckman treated the problem of evil in *The Departure* (1932–33, New York, Museum of Modern Art), a painting he carried out of Nazi Germany into exile. The format is a conscious evocation of the medieval religious triptych. Alfred Barr, director of the Museum of Modern Art, considered it "an allegory of the triumphal voyage of the modern spirit through and beyond the agony of the modern world."[3] While the royal family of the central panel has escaped the tortures of life depicted in the left panel and may be triumphant, Max Beckman does not see any hope for relieving the terrible burdens of guilt. He visualizes the baggage of evil on the right panel which he explains in this way: "On the right wing you can see yourself trying to find your way in the darkness, lighting the hall and staircase with a miserable lamp dragging along tied to you as a part of yourself, the corpse of your memories, of your wrongs, of your failures, the murder everyone commits at some time in his life – you can never free yourself of your past, you have to carry the corpse while Life plays the drum."[4] Only a terrible force could dislodge such a burden. Only a violent action could take its power away, dismember it, and dissolve it completely into the liberating matrix.

This is the purpose of the rite of Baptism.[5] Staged with thought and care, not by rote, it is a powerful drama, most affective in the adult baptism. In the theatre of the ritual, the powerless candidate is immersed full-bodied into water, and the priest forces his head under three times.[6] No catharsis here, no putting off, but a forced drowning and dissolution of the "corpse" we carry. In the final words of the ceremony, the standing baptized man, dripping from his immersions, is released with the words: "Go and walk *like a child* into the light." This theatre overcomes the wasteland; this stage returns the participants unfettered, free, to the life of the world.

It is this sort of theatre that I wish to discuss in terms of the visual arts – not the theatre of illusion or entertainment, but theatre that is involved with the world, with life and our apprehension of it. A theatre that at once hides and reveals, like Christo's wrappings. Christo orchestrates the disappearance of a familiar world by covering the Reichstag in Berlin with a million plus square feet of woven polypropylene fabric with an aluminum surface, held in place by over fifty thousand feet of inch-and-a-quarter-thick blue polypropylene rope to articulate what the drapery has hidden. By blocking and frustrating vision, we are forced to apprehend with our whole body. Christo's is a theatre of ambiguity, a covering up and a revelation, an act of making the familiar strange. After such an experience, our awareness of that piece of the world will be full-bodied and conscious. And in the consciousness there is delight.

THE RENAISSANCE STAGE

The examples above share a common property: their stage is ours. It is not elevated above us, nor is it separated from our space. We participate in its effects without separating ourselves from our life-world. They raise "the gorgeous Playing place in [our] fieldes. ... As they pleased to have it called, a theatre,"[7] and carry us onto their stage. In this way, they are like Medieval and early Renaissance staging, presented not in places separate from the everyday world, but within it. Theaters in the sense we understand were not introduced until the last third of the sixteenth century, although Sebastiano Serlio published Vitruvian stage sets in 1545.[8] The first permanent theatre for the performance of plays in continental Europe was Andrea Palladio's Teatro Olimpico designed for the Vicenza Academia Olimpica.[9] The first free-standing theatre building of the modern age was Vincenzo Scamozzi's design for a Teatro Olimpico in Sabbioneta. It is interesting that the interior of the rectangular hall, though small, is decorated with sculpture, frescoed walls, and columns, to evoke an urban square and a stately court. According to the restorers Anna Di Noto and Francesco Montuori, "the stage is the place where everything is contracted, the theatre and the city identified with one another."[10]

This contraction is an artifice based on the reality of Medieval and early Renaissance pageants and performances. Pictorial examples of theatrical staging in public squares abound, from the procession of various types of carts: emblematic (Piero della Francesca, *Portraits of Federigo da Montefeltro, Duke of Urbino, and His Wife, Battista Sforza*, Florence, Uffizi); musical (Anonymous, *cassone* panel, *Story of Solomon and the Queen of Sheba*, third quarter of the fifteenth century, New York, Metropolitan Museum);[11] or mythological. One cannot help but think of Raphael's *Galatea* from the Villa Farnesina (Rome, 1509–11) as an elaborated theatrical tableau. In addition to pageant carts, there were theatrically staged public processions, e.g. Gentile Bellini's *Procession of the Relic of the True Cross* (on the feast day of St. Mark through the Piazza San Marco, 1496, Venice Academia, commissioned by the Scuola di S. Giovanni Evangelista for their confraternity headquarters). Even executions were "on stage" as is shown in the anonymous *Execution of Savonarola* (1498 or after, Florence, San Marco). In the Middle Ages and early Renaissance, the world of men, the city, is the stage upon which the citizens would "act."

RELIGIOUS DRAMA

It was in the churches that the theatrical events of the liturgy were performed. We may consider the numerous chapels within them, e.g. those at Sta. Croce in Florence, as mini-theaters dedicated to a saint or patron. Such an idea was understood in the seventeenth century, as Gianlorenzo Bernini's funerary chapel of Cardinal Federico Cornaro in the church of Santa Maria della Vittoria demonstrates.[12] Here Bernini gives the sculpted Cornaro family, shown in balconies on either side of the chapel, box seats to the dramatic encounter between St. Teresa and the angel (Anonymous, *Cornaro Chapel*, oil on canvas, c. 1644, Schwerin, Staatliches Museum).

In the Renaissance, however, chapel decorations did not follow such a classical unity of time and place. Rather, their decorations followed the Medieval and Renaissance manner of the emblematic and symbolic conventions of simultaneous staging, as a 1547 illustration of a French theatre shows: Hubert Cailleau, *Staging for the Passion Play at Valenciennes* (Paris, Bibliothèque Nationale).[13] Renaissance chapels were decorated with cycles that broke the wall into sections, into individual sets from the stories of the lives of the saints, all shown together simultaneously, as do the medieval mosaics of the *Last Judgment, Ranks of Angels,* and *Scenes from the Lives of Christ and St. John the Baptist*, in the dome of the Baptistery in Florence (second half of the thirteenth century, central figure of Christ in the *Last Judgment* attributed to Coppo di Marcovaldo).

We shall see that the Renaissance fresco cycles discussed below are also simultaneous stagings, and can even be considered our visual accompaniments to the sung Mass, as Laura Jacobus has recently pointed out in her discussion of Giotto's Arena Chapel frescoes of 1305.

> One might even put this a different way, for in effect frescoes "performed" to musical accompaniment, their famous dialogue being chanted by the choristers below. This may seem a bizarre notion to modern views of the frescoes, but was in itself quite unexceptional to contemporary churchgoers, who would have been used to performing works of art.[14]

It should be remembered that before the era of the proscenium/picture stage, dramas were presentational, in the sense that the audience clearly understood the artifice and did not expect to be drawn into the "picture." Eugenio Battisti, however, emphasizes the importance of the representational qualities of Giotto's naturalism. It allowed the depicted events to "unfold on a psychological plane parallel to that of daily life.... It might even be said that the spirit of the Padua frescoes is peculiarly suited to the devotional outlook of the Latin peoples ... by nature debarred from attaining to the abstract idea of God."[15]

Three layers of individual scenes cover the walls, presenting us at one time with the narrative cycles of the life of the Virgin and the life and passion of

Christ. For scenes taking place in town, architecture forms the stage sets, as it does in the *mise en scène* of the *Last Supper*. We see the event through a framework that acts symbolically to include, but also to exclude us. In the *Apparition to St. Anne* from the upper level, the set is informative. While Anne is told she will conceive Mary in the enclosed room at the right, on the smaller and more open porch, a maid-servant sits spinning. At the obvious level, she performs a duty of daily life. At a symbolic level, her duty is greater: she is one of the Three Fates, a youthful Clotho, the spinner of life. Her gesture, holding the drop spindle,[16] leads to the stairway and therefore to the open "stage" above. To what extent did Giotto draw on this drama to show the divine in terms of everyday life? We know that in the churchyard of the Arena Chapel (built on a Roman arena), the mystery of the Annunciation was acted each year. These were elementary religious spectacles, in the vernacular, just as Giotto's paintings are.[17]

Nearly two hundred years later, we find theatrical staging in Ghirlandaio's Sassetti Chapel cycle (Florence, Sta. Trinita, 1480–85). The fictive scenes are of two types, to suit the two dedications of the chapel: to St. Francis, Sassetti's name saint, and the nativity, in gratitude for the birth of his son Teodoro in 1478–79, after an earlier son of the same name had died. The panel painting depicts the *Adoration of the Shepherds*, while a narrative cycle tells *The Story of St. Francis*. Richard Turner[18] explains the iconography with its numerous portraits of members of the Sassetti and Medici families. The central message celebrates Sasetti's good fortune in living under the protection of Lorenzo de' Medici (pictured at the right). Turner warns, however, that "it would be an error to understand the chapel as a secularization of sacred matters, for the Florentines' sense of their religion lay in enacting it as devout and empathetic participants, whether in procession, mystery plays, or presence in the narrative of a fresco. Any viewer of this private chapel in a fully public place would have understood this without giving it a second thought." And so we do not think it odd to find theatrical devices here – at least in a play with them: the play of *St. Francis Resuscitating the Roman Notary's Son* (set in the Piazza San Marco in Rome) is located in the Piazza Santa Trinita in Florence, with the reference to the live son sitting on a bier obviously related to the Sassetti family history. The street rises in perspective just as the sets of Palladio will, the major players assembled in the foreground, as if on a proscenium. In the upper scene, entitled *The Confirmation of the Franciscan Order by Pope Honorius III,* the major action again takes place on a proscenium, this time emphasized by the low wall draped in red. And before all, and up through the stage floor rise

Poliziano and the Medici sons – as in Leonardo's stage set for Poliziano's *Orpheo*.[19] The children are orchestrated cinematographically: the blonde boy at the left concentrates and looks ahead, the second looks at someone in the audience, and the youngest looks directly at us, emphasizing the oddity and theatricality of the scene. His mediation invites us into the play.

AUDIENCE AND STAGE

Renaissance audiences did not watch spectacles through a proscenium arch. The fictive, the sacred was a part of their own world. Indeed, the divine mysteries were projected into it, as Andrea del Castagno did with the 1447 *Last Supper*, on the rectory wall of the cloistered convent of Sant' Apollonia in Florence. "The theme of the Last Supper was regularly painted on monastic refectories to remind the religious at every meal of Christ's sacrificial self-perpetuation in the form of bread and wine."[20] Castagno projected the "upper room" into the refectory, the action taking place on the proscenium of the projected stage. Castagno's *Last Supper* was probably not seen by anyone but the cloistered nuns, until the convent was expropriated by the Kingdom of Italy in the late nineteenth century. So one cannot judge how much effect it had on what has become the standard image of the Last Supper, Leonardo da Vinci's mural in the Dominican Convent of Santa Maria delle Grazie in Milan.[21]

The stage-like space of Leonardo's painting is made clear in a computer reconstruction that shows what the scene would have looked like from the opposite perspective; indeed it shows how the audience would appear from the players' point of view.[22] Examining later Renaissance stage constructions based on Vitruvius gives us a handle on Leonardo's thought.[23] In Palladio's Teatro Olimpico, the floor of the fictive space behind the proscenium rises quickly in perspective. The seam between the proscenium and the background would be barely evident when the actors were on stage. The recession of such a space is emphasized in the floor grid of Serlio's graphic reconstruction of Vitruvius's stage. Leonardo's theatre, too, is set against an accelerated or forced perspective, as Pedretti's reconstruction shows.[24] The large scale of the figures and the prominence of the table and its cloth cover divert attention from the floor. The apostles at the ends of the table disguise the acceleration of the perspective of the walls.[25] The seam between the ceiling of the refectory in the real world and the stage of the "upper room" is disguised by the architectural valance of the lunettes which carry emblems of the Sforza who sponsored the decoration of the refectory. Leonardo has manipulated the "set" to force our eyes more quickly than we would expect toward the focal point, the face of Christ.[26] The apostles react to Christ's announcements that he will be betrayed and that he will institute the Eucharist. A wave of reaction spreads out and sweeps back to the central figure of Christ, who reacts to their confusion, lowering his eyes to seek the godhead within.[27]

THEATRUM MUNDI

Thanks to the theme of the conference, *Theatrum Mundi*, there is an opportunity to note a detail taken for granted until the recent cleaning of the mural: the doorways between the tapestries. Although they did not seem relevant to the dirty mural, we can now see that Leonardo took care to make them read as articulated spaces that lead out of the room – note the doorways within the hallway; their golden color calls attention to the space within. We may, then, consider them wings of Leonardo's fictive stage set. As such they imply the possibility of horizontal movement in the background that slows the drive of the accelerated perspective of the walls. Their lintels iterate the level of the heads of the apostles and so help us concentrate on the foreground actions.

By taking Christ's view (Fig. 1), we can see, in virtual reality, how the drama includes the whole refectory. From the ledge of a proscenium, about eight feet above the floor, Christ sees the future covering the whole of the opposite wall: his crucifixion.[28] He sees the crowds of faithful, including the kneeling Ludovico Sforza and his wife Beatrice. To reach that wall, because the uninterrupted perspective of the floor is so compelling, he scans past the monks seated along the walls, and over the prior, seated opposite Leonardo's Christ, his back toward the crucifixion. Christ's action, that is, his gaze, passes into the world which "knows him not," to a vision of his sacrifice for that very world. Returning to focus inward, his eyes sweep back through the present, taking it with him. Lowering his eyes in resignation, he accepts all – even "the corpse of your memories, of your wrongs, of your faults, the murder everyone commits at some time in his life."[28] Leonardo has turned the theater of the refectory's multiple stages into a sacred drama that slips us, like children, into the light.

Siena College[29]

NOTES

1 Images of Hiroshige's *Saruwka-cho Theatre* (p. 51) and Christo's wrapping of the Reichstag (p. 52–53) can be found in a Laurie Schneider Adams, *Art Across Time*, New York: McGraw Hill, 1999.

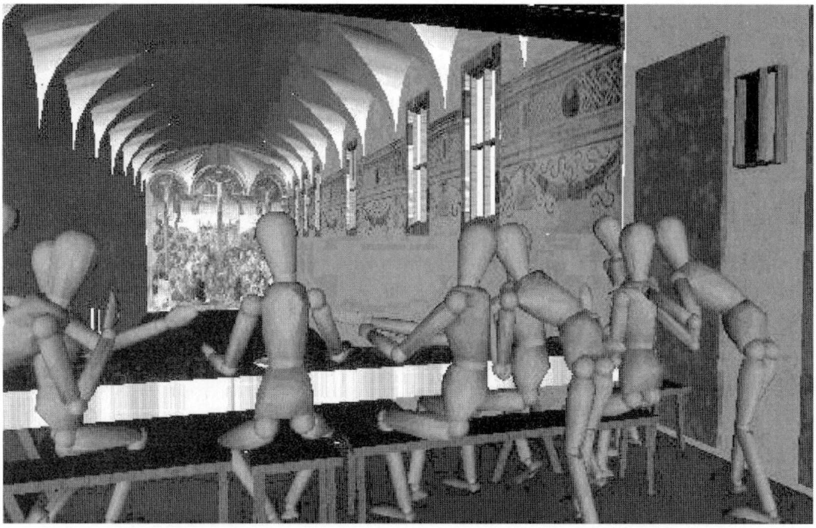

Fig. 1 Digital reconstruction of the interior of the refectory of Santa Maria delle Grazie as seen from the "stage" behind Christ and the Apostles. Donato Montorfano's *Crucifixion* (1495) faces Christ and the Apostles. Reconstruction by Franz Fischnaller, Daniele Marini, Lorenzo Forges Davanzati with the collaboration of Paola Trapani, Marco Belloni, Massimo Pighetti, Mauro Guizzo, Maresa Bertolo, Massimo Nasella, Mauro Marini and Paolo Ferretti; with the contribution of Alessandro Polistina.

[2] "The Red Sea is a symbol of our baptism, and the nation you freed from slavery is a sign of your Christian people."

[3] As quoted in H. H. Aranson and Marla F. Prather, *History of Modern Art*, Englewood Cliffs: Prentice Hall, 1998, p. 290.

[4] Ibid.

[5] The water is blessed with the following incantation: "The waters of the great flood you made a sign of the waters of baptism that make an end of sin and a new beginning of goodness. Through the waters of the Red Sea (you) set (God's holy people) free from sin by baptism.... (May the baptized) be cleansed from sin and rise to a new birth of innocence by water and the Holy Spirit."

[6] The blessing of waters concludes with the phrase: "May all who are buried with Christ in the death of baptism rise also with him to newness of life." It is repeated for all the faithful present in the "Renewal of Baptismal Promises."

[7] J. Stockwood, *Sermon on Paul's Cross*, 1578, p. 134, as cited in the OED.

[8] Serlio, *Il Primo libro d'architectura*, Paris, 1545; his perspectives probably were devised by Baldassarre Peruzzi, architect and painter, cf. his perspective for the Farnesina, a villa he built for Agostino Chigi, 1508–11. The *Salone della Prospettive* creates the illusion that the room is

opened to the outside, through a perspective colonnade with views of Rome. Carlo Pedretti makes a case that a sketch by Leonardo on Codex Atlanticus 286 r-a, 1497–1500, is a sketch of an amphitheater, probably temporary, for a festival or an anatomical theatre; see *Leonardo da Vinci: The Royal Palace at Romorantin*, Cambridge, Mass: Belknap Press of Harvard University Press, 1972, pp. 318–19.

9 Unlike the later Sabbioneta and Aleotti Teatro Farnese at Parma (1619), the Teatro Olimpico had no effect on subsequent theatre architecture. It was designed for the performance of classical plays and was a reaction to late Italian Renaissance theatre. See J. Thomas Oosting, *Andrea Palladio's Teatro Olimpico*, Ann Arbor: UMI, 1981. See also Ludwig Heydenreich and Wolfgang Lotz, *Architecture in Italy, 1400–1600*, Middlesex, England: Harmondsworth, 1974, p. 318.

10 See Di Noto and Montuori, *Restoration of the Sabbioneta Theatre*, text by Maria Argenti, http://www.area.progetto-ed.it/ar31/h/hte.txt. "The theatre embodies an illusory, and at the same time realistic, metaphor. Their project evokes the panels of Urbino, Berlin and Baltimore, 'programmatic manifestos' of an ideal city. [Vespiano Gonzaga wanted Sabbioneta to become an ideal city.] But the process has been inverted, for the city really exists. The stage only transforms it, adapting it to the needs of theatre which is part of a stage. Fiction and reality merge, elevated to the category of urban archetypes."

11 See John Pope-Hennessy and Keith Christiansen, *Secular Painting in 15th Century Tuscany: Birth Trays, Cassone, Panels, and Portraits*, New York: The Metropolitan Museum of Art Bulletin, 1980, pp. 42–43.

12 See Mark S. Weil, "The Relationship of the Cornaro Chapel to Mystery Plays and Italian Court Theater," in "*All the World's a Stage...*": *Art and Pageantry in the Renaissance and the Baroque*, ed. Barbara Wisch and Susan Munshower, University Park, PA: Pennsylvania State University Press, 1990.

13 In Italy, villa design accommodated theatricals, e.g. the north facade of the Villa Farnesina of 1509–11 by Baldessare Peruzzi, where five arches on the facade are flanked by projections. The effect of such staging is best understood by looking at fresco cycles. The Cailleau sketch can be seen in colour on www.arts.usf.edu/theater/cyberhis/pix4.htm. The Sydney Higgins website for theater links is helpful: www.leeds.ac.uk/theatre/emd/linksp1.htm#b.set.

14 "Giotto's *Annunciation* in the Arena Chapel, Padua," *The Art Bulletin*, Vol. LXXXIII, March 1999, pp. 93–107. Jacobus argues that only the annunciation is truly a representation of liturgical drama, and uses the *Apparition to St. Anne* discussed below as an instance of drama "influencing pictorial representation, or of common solutions arising within a common visual culture. [Compared with the *Annunciation*,] it is insufficiently particular about the stage setting to depict a dramatic performance per se" (p. 96). It is not my intention to contend that the illustrations discussed below, especially Leonardo's *Last Supper*, are visual reconstructions of actual performances, but to consider them as imaged stagings. The affective power of theater is very different from that of the more abstract action of looking at pictures on a wall.

15 In *Giotto*, translation James Emmons, Lausanne: Skira, 1960, p. 92.

16 My thanks to my colleague, weaver Jacqui Lubbers, who identified the spindle type.

17 Ibid., p. 96. Jacobus (op. cit., p. 93) describes the procession. Well before 1278, a cult of the Annunciation had used the site for a scene of the *sacra rappresentazione* of the Annunciation.

18 *Renaissance Florence: The Invention of a New Art*, New York: Abrams, 1997, pp. 146–150. Eve Borsook and Johannes Offerhaus, *Francesco Sassetti and Ghirlandaio at Santa Trinita, Florence: History and Legend in a Renaissance Chapel*, Doornspijk: Davaco, 1981.

19 Based on Codex Arundel, fol. 231 and 224 r and Codex Atlantico, fol. 131 V-A; see Carlo Pedretti, *Leonardo architetto*, Milan: Electa, 1978, p. 295.

[20] Frederick Hartt and David Wilkins, *History of Italian Renaissance Art*, Englewood Cliffs: Prentice Hall, 1994, p. 270.

[21] This is the view of Hartt and Wilkins, as above. Ludwig Heydenreich, however, finds similarities between the Castagno's apostles with raised hands and the gesture of Andrew in Leonardo's painting (*Leonardo: The Last Supper*, London: Allen Lane, 1974, pp. 28–29).

[22] The author's aim is to allow the visitor to "enter" the painted "virtual room" where the Ultima Cena is taking place. It is a "3D" ambience environment. The visitor can "walk" inside the room where Christ and the Apostles were having dinner. He can visit the "inside" of the painting, can go around the room, interact, look at the Refettorio from the point of view of Jesus Christ, have a real feeling of how the Refettorio could appear from "inside" the painting, and look at the opposite Crucifixion fresco. See: http://hpux.dsi.unimi.it/imaging/LAST_SUPPER/lastsupper.html, "L'ULTIMA CENA" INTERACTIVE (a work in progress). Authors: Franz Fischnaller, Daniele Marini, Lorenzo Forges Davanzati with the collaboration of Paola Trapani, Marco Belloni, Massimo Pighetti, Mauro Guizzo, Maresa Bertolo, Massimo Nasella, Mauro Marini, and Paolo Ferretti and the contribution of Alessandro Polistina. Carlo Pedretti included Montorfano's *Crucifixion* on p. 71, *Leonardo*, op. cit, p. 70. He emphasized the iconographical rapport between the paintings, comparing the room to Castagno's and to Cosimo Rosselli's in the Sistine Chapel (his *Leonardo*, Bologna: Casa Editrice Capitol, 1979, p. 38).

[23] Vitruvius on stage sets common enough in the fifteenth century: Ghiberti ends his Commentaries, c. 1450 with an account of them (see Kemp, *Marvelous*, p. 35). Evidence of Leonardo's knowledge of Vitruvius before he painted the *Last Supper* is abundant. He drew Vitruvius's scheme of proportions and symmetry in the famous sketch at Venice, dated c. 1487. Kemp cites the following: on July 24, 1490 (Mss. C, Paris Institute, 15v) Leonardo had supper with his friend, Giacomo Andrea da Ferrara, an authority on Vitruvius. Leonardo's explorations of sound were probably encouraged by Vitruvius's account of acoustics in theatrical design; his lists of classical devices were based on sources like Vitruvius (*Marvelous*, pp. 105, 131, 176).

[24] See his *Leonardo*, London: Thames and Hudson, 1973, p. 71, and *Leonardo architetto*, Milan: Electa, 1978, pp. 286–89.

[25] This is not the case in Domenico Ghirlandaio's 1480 *Last Supper* at the Ognissanti in Florence. The pilasters hide the fast recession in Domenico and Davide Ghirlandaio's c. 1470–79 version for the monastery at Passignano. See Heydenrich, op. cit., pp. 28–29.

[26] The proportions of the mural are not all related to the rules of one-point perspective. Luca Pacioli noted this in the third chapter of his *De Divina Proportione*, Venice, 1509. Thomas Bratchert suggests that Leonardo observed the ratio of 12:6:4:3, identical with the Pythagorean intervals related to him by his fellow Sforza courtier, Franchino Gafurio, in his *Theorica musice*, 1492, and later in his *Angelicum* of 1508 ("A Musical Cannon of Proportion of Leonardo da Vinci's *Last Supper*," *Art Bulletin*, vol. LIII/4, 1971, pp. 461–67).

[27] See my "Listening to Leonardo," *Analecta Husserliana* Vol. LXVIII, pp. 301–15

[28] see Max Beckman's statement, p. 52

[29] I wish to thank my colleagues for their help: Nancy Goldfarb for our discussions on her forthcoming paper, "Baptism Averted: Bodily Waters in the *Wasteland*"; James Flynn, medievalist/actor, for enlightening me about medieval staging; and Margaret Barlow for improving the readability of the text.

SECTION II

life on the planet, therefore affecting us as well. But actually this mundane movement reflects the most basic macrocosmic connection placing us within the influence of the planetary movement. This motion never stops but continues while the cultures come to life and collapse, not to mention individuals whose bodies disappear into the earth, even though all their lives they look up – into the sky.

NOCTURNAL SYMPATHY

In many mythological beliefs night was privileged over day; it had potential and a secret. Night was not only the time of waiting for day, but rather, day was a waiting time for the overwhelming experience of night. Therefore only night is promising, as it offers the open window to infinity.

Astral devotion used to be rather common and manifested itself in the past in many different forms. Deification of the astral bodies was observed as a kind of solar monotheism in Egypt but also in the times of the Roman Empire, where it gained a great importance. But as we know from archeological evidence, astral worship achieved the biggest importance during late Babylonian culture.[2] Sun and Moon, as well as other visible celestial bodies, played the role of the omnipotent cause of everything. Although the celestial bodies were not personified, typically for mythological thinking, they reflected eternal and unavoidable laws which subjugated everyone and everything. The result was that astrology became a religion and reading the language of the stars was the noblest occupation.

The Greek tradition borrowed and transformed many oriental ideas,[3] and one of them is undoubtedly the notion of infinity. As Eco claims: "The Greek world is continuously attracted by Apeiron (infinity). Infinity is that which has no modus. It escapes the norm. Fascinated by infinity, Greek civilization, alongside the concept of identity and noncontradiction, constructs the idea of continuous metamorphosis, symbolized by Hermes."[4] The Greeks then, not without fascination, shared the knowledge about those old "barbarians' priests," who "knew the secret links that connected the spiritual world to the astral world and the latter to the sublunar world, which meant that by acting on a plant it was possible to influence the course of the stars affecting the fate of the terrestrial beings. ..."[5] The strong connection of a human being to the cosmos had a spiritual character, but often it was maintained by relations of sympathy, so that a human body reflected the macrocosm in the micro-scale. Jaroslaw Kozakiewicz follows this path in his artwork entitled *On the Boundaries of Bodies*. A Copernican reference is clear, but underneath it there is an astrological hidden layer – of a hermetic kind. The work consists of two sets of life-size lead casts of the artist's body orifices marked with a gold layer on the crucial points. One set of casts is made in the positive, the other as a negative. These small objects are presented in two metal cases lined with velvet. Their positioning in space is crucial – they are situated on a line reflecting the artist's body projected on the scheme of the solar system: nine planets and nine body orifices with the possibility of a tenth, which so far remains a mystery. So "he is no longer an artist, he has become a work of art..."[6] – we could state after Nietzsche.

This artwork can be also associated with the Dionysian aspect of art. It doesn't soothe, but brings satisfaction through transformation, action, motion, and orgy. Kozakiewicz refers to the body orifices as to the subversive realm. "Dionysian power" appears to be a power of the night, revealing itself as a drive for the formless and *nonimagistic* which comes from the darkness and danger of body fluids. All those mentioned features led to disorder which evoked horror, but horror came together with the sublime. Dionysian festivals and the mysteries of Eleusis were older than Greek culture and had much in common with Babylonian festivals welcoming the new year, and also with other traditions as some speculate.[7] But there is no doubt that they were all a path into the horror of darkness, into the sublime located in the starry night. Later, European studies on the sublime,[8] e.g. those undertaken by Edmund Burke, linked it with celestial infinity:

> Magnificence is likewise a source of the sublime. A great profusion of things which are splendid or valuable in themselves, is magnificent. The starry heaven, though it occurs so very frequently to our view, never fails to excite an idea of grandeur. This cannot be owing to any thing in the stars themselves, separately considered. The number is certainly the cause. The apparent disorder augments the grandeur, for the appearance of care is highly contrary to our ideas of magnificence. Besides, the stars lie in such apparent confusion, as makes it impossible on ordinary occasion to reckon them. This gives them the advantage of a sort of infinity.[9]

This is how the European notion of infinity finally lost its privileged position as a locus of the universal Soul sought in antiquity, and infinity became sublime, and only sublime.

But Kozakiewicz's work, as I already pointed out, is not just an intellectual or spiritual game. It is more tactile than visual. The lead casts are small objects which can be easily held in the hand. They invite touch. One can have access to them only when the cases are opened – only for a limited time. It is also a very private portrait of the artist's body – it reviles the secret known only to himself or his lover. Therefore it is the secret of universal connection as well as a secret of intimacy.

THE SECRET AND THE NIGHT ON THE BEACH

The secret syndrome gives the impression that someone possessing it has gained some power – that he/she can make use of it in order to achieve some unknown goal. But just like the hermetic theory of the second century, we do not know the truth. The truth of oneself, one's own soul and body, and their "proper place" remains secret, which encourages us to believe, maybe intuitively, that each body[10] hides part of it, or at least a key to it. The body plays the same role as books, maps and recipes for Gnostics. Moreover, similarly to the syncretism of the long-gone epoch, the existence of a body is possible even though the body itself is full of contradictions.[11] So this peculiar kind of model which Kozakiewicz built with his objects carries a secret message set up on the alchemical context of lead and gold as spirituality and transformation. The most important points are marked with gold: these are actually the points of passing from inside the body/microcosm to the outside/macrocosm. Traditionally this is the dangerous direction, which can cause pollution. This is how the "proper place" is being lost. Ironically though, multiplying the possibilities of transformation can also be limiting. In the hermetic mode, suspiciousness and strangeness proliferate very quickly, but do not promise any way of escape. Meanings are forgotten, or we may forget that there ever were any meanings.

As the "real Reader is the one who understands that the secret of the text is its emptiness,"[12] in the same way the viewer of the work of Kozakiewicz could come to his conclusions. Being an anti-essentialist, what else is he/she to choose if not to involve himself/herself in the macro and micro sympathy of everything with everything, which actually means a complete dissolution in the endless possibilities of connections. But one can also develop a taste for the proposition given by the artist and treat it as one of many possibilities for the interpretative association of his own body with the starry sky. The viewer may treat it as a form of fantasizing, which is worthwhile on a warm and cloudless night at the beach.

THE DAY OF THE BODY OR *LANDSCAPES*

Unlike the almost intimate objects of what I just called the nocturnal work of Kozakiewicz, his project *Landscapes* has a grand scale and public character. Here Sun plays a key role not only as a celestial body central to the solar system, but also as an overwhelmingly powerful emanation of light. Sunlight penetrates the body orifices in the most visible way, although lights of the night do the same thing in the most discrete way. As opposed to the nocturnal contemplative ambiance, daylight brings overwhelming activity illustrated by I. Calvino's story on Thekla – one of the "invisible cities" – where everything is in process, or rather under constant construction in order to prevent the city from "crumbling and falling to pieces," and "not only the city."[13]

Kozakiewicz's idea is to map the European continent, and again, both his body and the solar system serve as a measure. This time the order of human orifices reflecting the celestial bodies' system is superimposed on Europe in its great continental scale. The body orifices, including the very organ which they serve, are designed as grand landscape parks hiding some kind of simple astronomical observatory underground (in an orifice). Similarly to the previous project, Kozakiewicz attributes one body orifice to each planet of the Solar System. The locations are precisely indicated and both the size of each land art project and the real distances between them are not random, but kept on some scale. The left ear coinciding with Mars is supposed to be located in Gdansk, the right ear – Mercury in Elblag, the right nostril is Venus on the Gdansk Coast, the left nostril is Earth in Frombork, the mouth is Jupiter in Warsaw, the anus is Saturn in Berlin, the right eye is Uranus in Vienna, the left eye is Neptune in Brussels, the vagina/penis[14] is Pluto in Rome, and the naval is a possible (but still mystery) planet X in Athens – the place from where Europeans take their cultural origin. In this setting, the size of a planet is represented just by an orifice, for example in the case of Mercury, the whole itself is about 12 meters long (scale of 1: 400,000) whereas the entire ear is a land structure about 200 metres long. The sculpture-park architecture is shaped as an organ of the orifice.

So far this artwork exists only as a project and it may end up only being a utopian project. But this project itself is an artwork consisting of 10 meticulously built independent elements which can be exhibited in a gallery space. They are accompanied by drawings providing precise instructions for potential constructors.

BODY AND/OF EARTH TOPOGRAPHY

A deep human need is to turn an anonymous space into a place of significance. Distinctive space[15] has human-made, and sometimes humanlike, marks. Archeology and astroarcheology[16] have documented the archetypal character of such a need to make one's own environment more distinct, and therefore more telling. That way the environment is not only less hostile, but it also focuses human attention, sometimes only to let it drift away later. It serves a purpose as a point of departure for the imagination. It is the same with *Landscapes* which, if never built on their full scale, may remain a utopian project of fantastic architecture.

The marks of significance of the place may be very public, and easy and obvious visual access to them only strengthens their importance, but often the significant places are actually very private and access to them is strictly restricted. Their significance comes either from a cultural taboo or from a private experience and its memory, that is, from private narrative. Within his project Kozakiewicz embraces both the public element (a work of art exhibited publicly) and the privacy or even intimacy of body orifices, as well as a private belief in a cosmic sympathy between human bodies (his body) and the celestial bodies.

As Lippard states: "The restless artist's preoccupation with travel, navigation, and mapping is often an attempt to address and reconcile the mythic relationship between the daily round and the road to spiritual achievement."[17] Kozakiewicz seems to follow this path, although not directly. His mapping of the ground (Earth) goes unavoidably through a mapping of his own body. *Landscapes*, indeed, rely on the obvious fact that "the body is meaningful, has significance"[18] because it is always heavily invested with libidinal energy and intimate memory. According to a psychoanalytical paradigm, the ego is actually a projection, or a map of the surface of the body. It is not that we have our bodies, but we are our bodies, and this is how "the road to spiritual achievement" – mentioned by Lippard – leads through one's own body. And if the human body serves as a picture of the cosmos[19] in its micro-scale version, *Landscapes* are also another attempt at a primary act of uniting (marriage) Heaven and Earth. The surface of the earth seems to work as a surface of the human body. And then the human body and the earthly body become symbolically the same thing. Celestial mapping is then a primary and ritual assurance of the internal unity of the subject (ego), and of the unity of the Universe.

BODY PERFORATIONS-ORIFICES

Choosing body orifices to build a human image is a brave task even nowadays. Kozakiewicz not only copies (casts) body orifices – as in his previously described work – but in *Landscapes* he actually elaborates on them, making them enormous in size, therefore sublime. *Landscapes* cannot be fully appreciated from the ground – they are too colossal to be visually deciphered as body parts. As mysterious Nazca lines, their detailed and elaborated beauty can be noticed only from the air. (An eye from the sky?)

But the land structures invite one to enter the orifices because the crucial position for astronomical observation is actually located underground. So it is from within the human body, or the body of Earth, that we come out to the macrocosm of celestial space.

As self-made corporeal inscriptions, "make the flesh into a particular type of body – pagan, primitive, medieval...,"[20] the inscriptions on the land turn it into a place with its culture's story consciously or unconsciously exposed. The prehistory, history and posthistory are inscribed here at the same time. The perforation of the left nostril is to be located in Frombork – the city of Copernicus, and the mysterious but possibly existing tenth planet of the solar system leads to the inscription of a navel, which is located in Athens, a very significant place for Europeans whose historical vicissitudes often revolved around the umbilical cord connecting them with this city and its cultural heritage.

The privileged role of the erotogenic zones of the body comes from the evidence that they are the most sensitive, and, in our individual history, they are the very first receptors of pleasure and pain. In the same way the inscriptions on the European land mark the significance of specific zones and give away a plot. This narrative is as private and intimate as the corporeal inscriptions can be. It is, finally, a play with the unlimited powers of man who is able to bring to life a Golem – the man of clay who precedes all men. This is how we start our spiritual journey, and the body is our vehicle.

THE BLUEPRINT

Contemporary interest in macrocosm and microcosm relations obviously has its specificity, although it is a continuation of a timeless, because mythical, tradition. As A. Danto claims, "We must still relate to them [past forms] in our own way. The way we relate to those forms is part of what defines our period."[21] Kozakiewicz's approach combines two traditions which were not very visible in mainstream art history – firstly it is a Gnostic tradition of cosmic sympathy with its condemnation of body, and secondly it is a tradition of representing body as a substance (element) which is in a constant process of change, as described by M. Bachtin. Such a body is not a sealed and idealized vessel of a beautiful form, despite the astral connection, but is rather a leaking body whose body orifices can be subversive. This proposition situates a human being in the important position of unifying microcosm and macrocosm which nowadays may be also described as a paradoxical attempt to combine the photocentric (phallogocentric) tradition with chthonian natural forces of the night.

The reappearing question, already asked once by I. Calvino in relation to Thekla, is back, but now, in our particular context it concerns the body rather than a city: "'What is the aim of a … [body] under construction unless it is a … [body]? Where is the plan you are following, the blueprint?' 'We will show it to you as soon as the working day is over; we cannot interrupt our work now.' … Work stops at sunset. Darkness falls over the building site. The sky is filled with stars. 'There is the blueprint,' they say."[22]

Uniwersytet im. A. Mickiewicza, Poznan, Poland

NOTES

1 Lucy Lippard, *Overlay: Contemporary Art and the Art of Prehistory*, the New Press, New York, p. 86.

2 See Hans Jonas, *The Gnostic Religion: the Message of the Alien God and the Beginnings of Christianity*, Boston: Beacon P, 1970; *Poems of Heaven and Hell from Ancient Mesopotamia*, translation and introduction by N. K. Sandras, Penguin Books, Australia 1971, p. 53; *The Reports of the Magicians and Astrologers of Nineveh and Babylon in the British Museum*, translation and introduction by R. Campbell Thompson, vol. II, AMS Press, New York, 1977.

3 "There was no lack of opportunity for Mesopotamian ideas to reach the Aegean, either through Mycenaean and Greek colonies like Ras Shamra in the fourteenth and thirteenth centuries, or through the Hittite Empire in central Anatolia, or later again through the Phrygian and Lydian kingdoms; but the extent and nature of the contacts is a matter of high controversy. The mythological cosmogonies of the poets, and the first of the philosophical cosmogonies from the Ionian Greek cities in Asia Minor, contain extraordinary similarities. Much of this is too diffused and too generalized to require any direct borrowing or imitation, but some points seem so close that any other explanation would be difficult. There are, also, the persistent traditions of oriental and 'Phoenician' secret books and teaching, which are said to have inspired the earliest Ionian philosophers." – N. K. Sandras, in the introduction to her own translation of *Poems of Heaven and Hell from Ancient Mesopotamia*, *op. cit.* p. 53.

4 *Interpretation and Overinterpretation*, Umberto Eco with Richard Rorty, Jonathan Culler, Christine Brooke-Rose, Cambridge, New York 1992, p. 29.

5 U. Eco, *Interpretation … op. cit.*, p. 31.

6 Friedrich Nietzsche, *The Birth of Tragedy*, translated by W. Kaufmann. Vintage Books, New York 1967, p. 38.

7 Especially Egyptian, but possibly Chinese and Indian.

8 Samuel H. Monk comments: "The sublime from its beginning had been connected with the natural world. Longinus, in crying up man's inherent nobility as witnessed by his inborn love of the great, had cast a hurried glance at natural scenery." *The Sublime. A Study of Critical Theories in XVIII-Century England*, The University of Michigan Press, 1960.

9 Edmund Burke, *A Philosophical Enquiry into the Origin of Ideas of the Sublime and Beautiful*. Basil Blackwell, Oxford 1987, p. 78.

10 L. Lippard states: "Each astrological sign is associated with a specific body part (as an astrological 'house' is a body symbol too), and ancient healers found celestial/biological conjunctions in their magical anatomy studies." *Op. cit.*, p. 86.

[11] See ibid.

[12] Ibid. p. 30.

[13] Italo Calvino, *Invisible Cities*, translated from the Italian by W. Weaver, A Harvest/HBJ Books, San Diego, New York, London, 1974, p. 127.

[14] Kozakiewicz in both works described here follows the old pattern of belief that male and female genitals are actually the same; see: T. Laqueur who in his book *Making Sex*, shows evidence of "how the one-sex body was imagined; to stake out a claim that the one-sex/one-flesh model dominated thinking about sexual difference from classical antiquity to the end of the seventeenth century." Thomas Laqueur, *Making Sex, Body and Gender from Greeks to Freud*, Harvard University Press, Cambridge, Massachusetts, and London, England 1990, p. 25.

[15] See, Yi-Fu Tuan, *Space and Place, the Perspective of Experience*, University of Minnesota, 1977.

[16] L. Lippard writes on astroarcheology: "The field is new one, and still somewhat of a mongrel science, bringing together teams of astronomers, anthropologists, archeologists, physicists, photographers, engineers, comparative mythologists, and even artists." She continues, presenting a history of the discipline; see *op. cit.*, p. 112.

[17] L. Lippard, *op. cit.*, p. 121.

[18] E. Grosz, *Volatile Bodies: Towards a Corporeal Feminism.* Allen & Unwin, St Leonards, 1994, p. 32.

[19] The other possibility is that human being is a center of cosmos; see Yi-Fu Tuan, *op. cit.*, p. 117.

[20] See Elizabeth Grosz, *Volatile Bodies, Towards a Corporeal Feminism*, p. 142.

[21] Alphonso Lingis, *Excesses: Eros and Culture*, New York: State University of New York, 1984, p. 34 in E. Grosz, *op. cit.*

[22] A. Danto, *After the End of Art: Contemporary Art and the Pale of History.* Princeton University Press, Princeton, New Jersey, 1997, p. 198.

[23] I. Calvino, *op. cit.*, p. 127.

DAVID BRUBAKER

DWELLING IN NATURE: ETHICS, FORM AND POSTMODERN ARCHITECTURE

Karsten Harries argues that contemporary architecture is uncertain of its direction. The modern style pioneered so confidently in the 1920's by Walter Gropius and Le Corbusier has given way to the decorative forms of Michael Graves' Portland Building, which refers to structures built in the past. This trend signals a dissatisfaction with functionalism and perhaps the rise of an eclecticism approaching that of the nineteenth century. Yet, postmodern architects also regard historical models with a lack of fervor; there is often a playful irony which keeps us connected to the present and the latest scientific technologies. For Harries, this ambiguity at the level of style is a symptom of our current ambivalence toward the modern period, which has come to rely upon the natural sciences to interpret human needs. In short, we do not seem done with the project of modernity, but many architects and theorists are searching for some new orientation.[1]

What is the task of architecture, given our doubts about modernism and the arbitrariness which mars the new historicism? Harries suggests that we retain Sigfried Giedion's modernist premise that the function of architecture is to convey an ethos, a valid way of living in the world. The emergence of a more unified style will depend on our ability to articulate some new communal ethics, but not from the mere application of decoration to dreary functional structures.[2] The search for a new ethos – one that decenters the objective rationality of the natural sciences – raises a most difficult question: What manner of life is appropriate for us now? Harries is committed to the belief that human existence is anchored within natural life; hence, he seeks an ethical principle that would guide architects toward designs that promote genuine dwelling. Because of the feelings of rootlessness and alienation that have accompanied modernization, we need some rectifying measure of value, some center or ground within actual life, that would apply generally and produce a stabilizing sense of community.

But where shall we find an aspect of material life that will be widely accepted as valid for our period? Where can we turn, in order to reassess our cultural assumptions concerning authentic human existence, the proper way to care for ourselves and others, and a communal development that is in balance with our actual needs? It is here that the architect may begin philo-

A.-T. Tymieniecka (ed), Analecta Husserliana LXXIII, 73–83.

sophical labor. Harries offers help, by evaluating traditional assumptions. His aim is to clear away earlier approaches to architecture that might obscure the path toward an ethos or spirit of dwelling that would indicate how we ought to build.

Several useful premises shape Harries' interrogation. We may accept the thesis that each period style of architecture arises from some communal purpose; our task is to develop a new doctrine of nature that will guide us in living as a community. Moreover, the practice of architecture will surely benefit from any investigation that reveals outmoded beliefs. In this regard, Harries begins with a persuasive estimate of the work required: a moral principle which makes us feel at home in nature can be grounded nether in a metaphysics of pure reason, nor in a doctrine that gives primacy to our scientific knowledge of determinate conditions and functions.[3] This means that we must modify the doctrines of morality and nature that we have inherited from the Enlightenment. When Kant turns to pure reason in order to explain autonomy, critical independence and the idea of freedom, he never supplies an ethics that accounts for the feelings of care and empathy that we have for beings who are actually embodied. It is also clear that we must revise our doctrine of nature, if we are to reduce self-alienation and the attitude of determinism that arises from our scientific thinking. Thus, there is merit in Harries' suggestion that our search for a principle of genuine dwelling will lead to some aspect of embodied life that belongs neither to metaphysics nor to our scientific understanding.

However, there is cause for major concern when Harries goes on to argue against two principles that are central to our modern way of life. Our evaluation of his arguments will determine the direction of our search for a material principle capable of guiding designers. Two Enlightenment beliefs are at stake: the principle of aesthetics which asserts the autonomy of art, and the principle of morality which regards autonomous individuals as sites of absolute worth amid fluctuating natural conditions. First, with regard to the autonomy of art, Harries insists that a viable philosophy of dwelling will replace the hegemony of scientific rationalism, only if we abandon the traditional practice of considering architecture aesthetically. He is unequivocal: the philosophy of architecture must free itself from "the aesthetic approach," if it is to develop a moral principle that will help architects address our actual needs.[4] To preserve the thesis that artworks display autonomous contexts that counter objectification, I argue against Harries' claim that the aesthetic approach is an obstacle to the revitalization of our sense of belonging to a community in nature.

Concerning the principle of the autonomy of individual persons, Harries suggests that we will progress toward a common ethic – one based upon some pre-objective context or center – only if we abandon the modernist notion that the human individual is free, self-determining and separate from the domain of nature which is the source of our cares and interests. In this case, Harries risks an imbalance: he affirms that the value of individual autonomy is one that we refuse to surrender, but he also asserts that a society of autonomous splinters can be healed only by a spiritual authority that transcends the individual's freedom.[5] The aim of cultural unity is certainly laudable, but Harries introduces a troubling supposition: if we are to find a principle of natural existence which makes us feel at home in a larger community, we will have to set aside the Enlightenment idea that freedom and autonomy are central to a valid cultural life. By contrast, I contend that a new nature-based principle of dwelling will be deemed valid by individuals living in our period, only if it also points to subjective bases of self-worth and personal agency that confer value on each unique person.

In the final section below, I suggest that architects may develop a principle of genuine dwelling by examining Maurice Merleau-Ponty's principle of subjective incarnation. With his final remarks on the flesh of the body, Merleau-Ponty writes that an awareness of personal existence depends upon private displays of sensuous wholes that are pre-scientific.[6] If such sensuous contexts belonging to one's own flesh are counted as material bases for a practical principle of morality, then we may arrive at an ethos that coheres with the autonomy of art and the Enlightenment belief that each living individual is an end of unconditional worth. By appealing to Merleau-Ponty's principle of subjective incarnation, we will be able to point to samples of sensuous self-existence, toward which we have both feelings and a moral obligation. As a result, it may be possible to develop an ethos of aesthetic existence (which Foucault anticipates)[7] and also imperatives of skill that will guide architects in exemplifying culturally accepted sites of value: the contexts of flesh that constitute the material bases of each personal lifeworld.

THE AESTHETIC APPROACH: DISINTEREST AND DISTANCING

Our first question is whether the aesthetic approach to architecture is compatible with a viable account of dwelling. To assess Harries' arguments against the aesthetic approach, we must first establish how he uses the term "aesthetic." Once we complete that task, we will be in a position to evaluate his claim that architects who promote an aesthetic awareness will never express the ground of a genuine and satisfying way of life.

According to Harries, the inadequacy of the aesthetic approach becomes clear once we analyze what constitutes aesthetic appeal. He offers a fourfold characterization that is supposed to fit any object of aesthetic experience. First, an aesthetic object should not point to a meaning or reality that is absent; instead, it should simply be. Second, such an object should present itself as a self-sufficient whole, not as a fragment that refers to a whole that is external or elsewhere. Third, an object of aesthetic experience demands a disinterestedness that removes us from the cares, desires, fears and hopes that determine our ordinary way of relating to daily life. Finally, aesthetic objects enable us to be one with ourselves, as they disengage our interests in the future utility of things. In this characterization offered by Harries, each artwork produces an autonomous presence and a feeling of pleasure that is far removed from the mundane interests, edifying thoughts and conceptual systems that we customarily associate with religious or moral concerns. Even the burden of time is lifted, as we manage to exist in the present, at least briefly, without thought of the future or the contingency of daily life.[8] By losing ourselves in the object through concentration, we arrive at what Kant describes as a disinterested satisfaction. Harries goes on to include, without much elaboration, the premise that the production of such a disinterestedness also "involves the senses."[9] Therefore, on this reading of the aesthetic approach, the absorption of sensuous presentations produces an attitude of self-sufficiency that is sought for its own sake, without regard to any interest in objects that exist for us through being known.

Given this description of the aesthetic approach, Harries infers that the absorption of sensuous presentations is a means for our escape to a domain of reason that is distant from any material need or the empiricel aspects of life. He cites Kant as a typical proponent of the aesthetic approach;[10] and indeed Kant does claim that the delight connected with aesthetic judgments is associated neither with the real existence of objects nor with the good, because such judgments are independent of all interests and not founded on any concept.[11] Yet, some twentieth-century defenders of the autonomy of art connect aesthetic experience with our emotions, values and interests, in a way that Kant does not. As Harries acknowledges, Paul Weiss clearly suggests that sensuous artworks, which disrupt our ordinary rational understanding of things, are not merely a means for escaping toward a faculty of reason and the pleasure which results from the purposeless play of our faculties of contemplation. Rather, such sensuous presentations may bring us even more thoroughly *into* the sensible world, in order to tear a portion of that *same* realm away from our distinct thoughts about objects, so that we might finally acquire an attitude that is infused with a sense of value.[12] The connection between sensuousness and moral value is also evident in the writings of Clive Bell, who claims that the sensuous form, considered without regard to ordinary human interests, is "an immediate means to the good" that produces an exaltation of ethical importance.[13] What this means, of course, is that the principle of aesthetic distinterestedness may be interpreted in terms of strong and weak notions of distancing: the strong sort holds that the disengagement of our empirical intuition and understanding removes us entirely from the realm of actual, embodied life in nature; whereas limited distancing entails that each of us loses all experience of real things, but not our awareness of the sensuous wholes which serve as the matter for possible intuitions and cognitions. Thus, we have the option of choosing the weak version of distancing: the aesthetic approach produces an attitude that dislocates us from our understanding of the world of existing, common sense objects. But there are signs that Harries assumes the strong sort of distancing, when he rejects the aesthetic approach. For example, he suggests that architects who cultivate aesthetic experience are dissatisfied "with the larger reality, with life," and not merely with some distortion brought by scientific thinking.[15] He also states that "the aesthetic approach to architecture is opposed to every contextualism."[16] Finally, his claim that the theory of aesthetic distancing arises "from the same root" as liberal democracy may be taken to mean that the absorption of sensuous presentations facilitates entry into the same realm of pure reason and intellect that qualifies persons for membership in Kant's kingdom of ends.[17]

With the strong interpretation of distancing in operation, Harries concludes that the aesthetic approach will cause works of architecture to "turn a cold shoulder" to the constraints, demands and necessities of the surrounding world. Any attempt to produce aesthetic awareness will result in a formalism which "renders talk of the requirements of dwelling or function simply irrelevant."[18] This is to say that buildings which produce aesthetic awareness will transport living individuals away from the cares and feelings that belong to everyday life in our communities; hence, such buildings will be unable to express the kind of dwelling that is required to counterbalance the growth of scientific rationality. After linking aesthetic contemplation with a realm beyond actual life, Harries infers that architecture which functions as art can never exemplify a principle of material embodiment that would be necessary for an account of authentic human existence. Again, the key assumption at work is that the absorption of sensuous presentations can never reveal a material end of absolute worth that would anchor a moral principle.

How should we regard the claim that an architecture of genuine dwelling cannot function as art? It is important to note that this argument against the aesthetic approach has political dimension. Since Harries identifies aesthetic experience with the autonomous rationality that belongs to Enlightenment notions of self-worth, his rejection of the autonomy of art casts doubt on the principle of self-determination that is even today the basis for our belief in human rights. Given Harries' own framework, the search for an account of genuine dwelling places us on a path of great risks. He describes the hope for a new period style that expresses our connection with nature as "perhaps a dangerous dream"),[20] because he doubts that such a style will cohere with our modern belief that communities ought to be organized democratically and in accord with the principle of self-determination. Harries tends to uphold Nietzsche's statement that phrases such as "the freedom of the individual" and "equal rights for all" coincide with social disintegration; indeed, the danger leads him to recite Ernst Gombrich's warning that talk of spiritual collectives as realities independent of their manifestation in individuals will weaken our resistance to totalitarianism.[21] My own belief is that "we moderns," as Harries calls us, will not feel genuinely at home, if our ethos of dwelling depends on a spiritual interpretation of corporeality that makes our cultures less democratic.

Can we deflect the foregoing argument against the aesthetic approach? Harries seems to hold that artworks which display their own sensuous character provide no material context whatsoever for a new principle of morality: presentations of sensuousness subjectively absorbed by a particular individual are supposedly incompatible with the search for a stable communal ethos.[22] Yet, this belief – that a valid way of life for a community cannot be revealed through self-evident sensuous wholes – follows necessarily only if Harries presupposes the strong interpretation of aesthetic distancing, where sensuous artworks are merely a means for a pleasurable adherence to the world of intellect and reason. In other words, the aesthetic approach is incompatible with our search for a new ethos of embodiment only after Harries holds that aesthetic distancing removes us completely from the domain of genuine dwelling. Moreover, Harries seems to rely on the modernist doctrine of nature: that the concrete existence or reality of human individuals is to be defined by determinate conditions that we recognize. Evidence of this tendency may be found in Harries' claim that a great style of art, one opposed to what he regards as the shallow, postmodern aesthetic of the "decorated shed," must belong to "an ongoing community, to a particular region and time," not to a rootless individual absorbed in a self-sufficient presence that is detached from situated life.[19] His critique depends on associating the aesthetic approach with the strong interpretation of distancing, and the strong interpretation becomes inevitable, only if we presuppose that it is impossible for the ground of human existence to be a subjective end that is prior to our thoughts of particular things. But it is this very standpoint of reason that causes our ambivalence and the current search for a new ethos.[23]

THE FLESH OF EMBODIMENT: DWELLING AND DESIGN

Our observations indicate that the aesthetic approach might still be compatible with an account of genuine dwelling in nature, if we could find a plausible account of restricted or limited instances of distancing. By turning to Merleau-Ponty's principle of subjective incarnation, we move closer to a philosophy which explains how sensuous embodiment, detached from our thinking about determinate objects, can nevertheless furnish material bases for a principle of value. This avenue may lead to a practice of dwelling that is valid for our period, because it accommodates a certain privatization of the sacred that affirms the worth of individuals, and because it entails that artworks, as presentations of the sensuous, have a role in exemplifying the ground of a new ethos and communal authority.

Merleau-Ponty develops an account of corporeal existence that offers guidance in how to dwell within nature. The principle of flesh meets Harries' requirements for a genuine account of human existence, since it relies neither on the language of the natural sciences, nor on a Kantian metaphysics of morals that emphasizes pure reason.[25] In this new approach, subjectivity is centered upon an embodied *idios kosmos*, or a secret lifeworld peculiar to the individual person. Each lifeworld of this sort is constituted by complete parts of the flesh of the body, that is, by sensuous wholes that are designated by such labels as "the visible" or "the tactile" or "the audible."[26] These sensuous contexts are evidence of subjective existence; for by touching something, or by watching the visible ground between changing color perceptions, I witness an actual pellicule of my own embodiment that gives me an adhesion to myself.[27] I come to regard each pre-temporal and pre-spatial context of flesh (e.g. the visible or the tactile) as an end in itself, since each supplies material evidence that enables me to regard myself as having actual existence.[28] Thus, the flesh of sensuous embodiment belongs to a subjective lifeworld that is central to self-worth and to keeping human intellect in contact with natural life.[14]

Since the flesh of embodiment displays material ends associated with self-worth and personal existence, it is able to serve as the basis for a communal ethos. The sensuous wholes of flesh are a material foundation for a principle of action that may be presumed to be equally valid for each sentient being. To show this, Merleau-Ponty moves explicitly from the subjective awareness of his own visible and tactile flesh to the conclusion that there are other such domains: Should I not ascribe a similar touch to the hand that I shake? Or a similar secret texture of visibility to my neighbor, who claims to perceive something green?[29] Thus, for Merleau-Ponty, it is the incarnate subjectivity of the flesh, not pure reason shorn of all contact with the sensuous, that makes it possible for each of us to regard other beings as sites of absolute worth. So, finally, we arrive at the possibility of a society of *sentient* beings who ought to care for each other as unique ends of unconditional value.

Does Merleau-Ponty's principle of the flesh of the body furnish the sort of spiritual authority that Harries requires for an ethos of natural life? The sensuous wholes of flesh return each of us to a confidence in the world, because personal existence is constituted by a sensuous cosmos that gives each individual the feeling of having a place within natural life. Each sentient being is an *idios kosmos*, a domain that is stable and changeless, at least in comparison with the contingent figures which emerge or disappear as a consequence of perception, thinking and determinate understanding. With Merleau-Ponty's approach, we moderns would return together, as sentient individuals, to an awareness of a basic aspect of our existence in nature, not through the repetition of some behavior associated with a single timeless archetype, but rather through a reversal which throws the matter of becoming back onto the stamps of being. Self-evident samples of sensuous embodiment give the contingent flux of each human intellect a sacred anchorage within natural life.

Does the flesh of embodiment reduce our sense of alienation and cultural disintegration which arises through the movement made possible by modern technologies? The wholes of flesh possessed by the living person serve as an *axis mundi*, an inalienable hinge which holds together heaven and earth. Since each pretemporal context of flesh is an exemplar of a unique individual's embodied presence, incarnate subjectivity becomes an antidote to the rootlessness that occurs within the intellect when modern technologies accelerate perception, increase mobility and emancipate individuals from the accidents of geography and physical space. The uniqueness and non-substitutability of individuals is affirmed; for no matter how many locations I travel through, no matter how many functional roles I perform, the sensuous wholes of flesh revealed by my own body are inalienable roots that prevent me from ever losing my place within nature.

Finally, the sensuous flesh which marks personal dwelling is a context that may be emphasized by architects. Since Merleau-Ponty's account of the flesh of the body implies a material principle of morality, we may anticipate the derivation of imperatives of skill which would guide the practice of building. Thus, an architect of our own period could begin to organize the basic elements of design into a style that affirms the importance of sensuous flesh as our best material evidence of the intrinsic worth of actual human existence. Architects who use elements of design to exemplify subjective incarnation will convey our commitment to a culture that consists of the concrete interaction and interdependency of intrinsically valuable individuals.

It is possible to foresee works of architecture that will begin to satisfy the basic need of living beings for genuine dwelling, that is, the need to have self-existence confirmed through a non-ideal contact with sensuous nature. This subjective need is not fully addressed by the architecture of the twentieth century, which is shaped by the late-modern assumption that the needs of human existence will be adequately addressed by designs which arise from the knowledge supplied by scientific rationalism. We are, then, left with practical questions: What imperatives of skill will architects derive from Merleau-Ponty's principle of subjective incarnation? How should space be organized to express the flesh that is the medium of personal existence and the straights of our actual contact with others within nature?

Harries suggests that the uncertainty of postmodern styles is due to our dissatisfaction with the objective rationalism of the sciences and the retention of an aesthetic approach that distances ornamentation from the representation of actual life. But this explanation for the emergence of postmodernism does not reveal the full complexity of our predicament. Our results suggest that there are several factors which explain the eclecticism that has emerged since the late 1970's. Harries is correct in claiming that the functional style of modern architecture is no longer satisfying, because it expresses an objective doctrine of nature (and a way of life) that treats human beings as resources to be managed. Part of the uncertainty displayed in architecture today stems from the difficulty of searching for a new ethos of material life at this historical moment; hence, we have warrant for the attempt to group current designs into a style that expresses the transitional or postmodern character of our own period.[24] But if postmodern buildings display a lack of unity (or a mixture of functionality and decorativeness reminiscent of the decorated sheds of the late nineteenth century). This is not so much because they display a style of ornamentation that links moral value to a rationality cleared of everything empirical, but rather because architects are in the process of experimentation, with the aim of finding new forms which express a more genuine contact with nature. The arbitrary quality of postmodern styles will end – the struggle between functionality and decorativeness will be resolved – only after architecture begins to express an ethos of human existence that is aligned neither with a modernism that equates actual life in nature with the things that we think about, nor with a metaphysics of personal existence.

NOTES

[1] Karsten Harries, *The Ethical Function of Architecture* (MIT Press, 1997), pp. 2, 10.
[2] *Ibid.*, p. 2.
[3] *Ibid.*, p. 68.
[4] *Ibid.*, p. 24.
[5] *Ibid.*, pp. 65, 290–291.
[6] Maurice Merleau-Ponty, *The Visible and the Invisible*, pp. 14–15, 139.
[7] Michel Foucault, "On the Genealogy of Ethics: An Overview of Work in Progress," in *Ethics, Subjectivity and Truth*, ed. Paul Rabinow (New York: New York Press, 1997), pp. 260–261.
[8] Karsten Harries, *The Ethical Function of Architecture*, pp. 22–23.
[9] *Ibid.*, p. 21.
[10] *Ibid.*, p. 22.
[11] Immanuel Kant, *The Critique of Judgement*, trans. James Creed Meredith (Oxford: Clarendon Press, 1992), pp. 42–43, 46, 53.
[12] Karsten Harries, *The Ethical Function of Architecture*, p. 23.
[13] Clive Bell, "The Aesthetic Hypothesis: Significant Form and Aesthetic Emotion," in *The Philosophy of the Visual Arts*, ed. P. Alperson (Oxford: Oxford University Press, 1992), pp. 122–123, 125–126.
[14] See also Immanuel Kant, *Critique of Pure Reason*, trans. Norman Kemp Smith (New York: St. Martin's Press, 1965), pp. 65–66. Kant's text seems to imply that appearances have a material aspect.
[15] Karsten Harries, *The Ethical Function of Architecture*, p. 51.
[16] *Ibid.*, p. 24.
[17] *Ibid.*, p. 65.
[18] *Ibid.*, p. 24.
[19] *Ibid.*, p. 63.
[20] *Ibid.*, p. 65.
[21] *Ibid.*, pp. 64–65.
[22] *Ibid.*, pp. 63–65.

[23] Even after we reject Kant's notion of an autonomous reason (which no longer works today as a support for the idea of freedom and for the individuation of persons), and even after we reject the hegemony of scientific thinking, we have not completely removed the risk of a philosophy that distorts our most basic contact with nature. For if our talk about dwelling and a situated existence remains merely the expression of our thinking about a pre-scientific encounter with things, then sensuous nature as such (which is revealed to us as the material wholes of subjective incarnation), will go unmentioned; the philosophy of architecture will still hover at the altitudes of intellect (e.g. empirical intuition, perception, and cognition).

[24] Similarly, contemporary philosophy is uncertain as to exactly which notion of a situated subjectivity will succeed in supplanting pure reason as the complement to our doctrine of material things.

[25] Maurice Merleau-Ponty, *The Visible and the Invisible*, pp. 14–15, 139.

[26] *Ibid.*, pp. 143, 147.

[27] *Ibid.*, pp. 255–56.

[28] The atemporal presence of a material element of my own flesh has intrinsic worth, since it is the mark of my existence and well-being, in an actual world that is not reducible to the contingent forms of mere thinking. Self-existence and absolute worth arise positively out of the sensuous present, without regard to future outcomes, and without any thought about temporal relations and our own mortality. Each of us is able to find peace from our thoughts of the passage of time, because the secret flesh of one's own body (unlike the intelligible body of physical knowledge) is not temporally bound; the sensuous evidence of personal existence is prior to any essence of temporal being. Thus, I disagree strongly with Harries' willingness to base an ethos of dwelling on Heidegger's contention that it is mortality that furnishes an awareness of individuality; see Karsten Harries, *The Ethical Function of Architecture*, pp. 160, 262.

[29] Maurice Merleau-Ponty, *The Visible and the Invisible*, 141–142.

TAMMY KNIPP

VIRTUAL ENVIRONMENTS: PSYCHOSOCIAL HAPPENINGS AND THE THEATER OF LIFE

INTRODUCTION

With the impact of technology, it is becoming harder to distinguish between what is virtual, what is fantasy, and what is considered to be real. Is reality in a postmodern society becoming more like theater, a techno-driven happening whereby our perceptions and experiences are continuous states of illusion? In current times, perhaps illusions are redefining our sense of reality.

METAPHOR: AN ALTERNATIVE PERSPECTIVE

In the theatrical arena, the concept of virtual reality (what is illusional) typically references a metaphor of a "world on stage" and the viewing audience (the "world offstage") references the physical reality. However, interesting thoughts arise if the realities are reversed whereby the world on stage references a postmodern techno-driven reality (a reality we have come to know and experience) and the world offstage references a virtual reality.

For example, within this reversed concept, the audience (the world offstage) represents an atypical reality in that their sensory experiences would be based upon a limiting and stationary perspective. The organized, structured rows of seating in the audience dictate a narrow view with restricted interaction. The on-stage world, on the other hand, would represent a reality dominated and controlled by technology and illusion. The immersed performers would come to experience psychosocial and biological effects among the physical, virtual, and psychological realities. In effect, perceptions of what is real, what is illusional and what is virtual reality would blur. Moreover, the biological and emotional faculties of the performers on stage would come to respond *unconditionally* to the mediated stimuli. The results could produce multiple and illusionary identities. Whether or not the stimuli of the environment (a world on stage) would be factual (real) or illusional (virtual) is not the issue. The important element is the mediated perceptions on the part of the performers in that their psychosocial and biological responses to the stimuli communicate a sense of realism (or an implied truth). In other words, the world on stage may be a *truer representation* of a techno-driven reality – a reality from which we cannot escape.

A.-T. Tymieniecka (ed), Analecta Husserliana LXXIII, 85–102.

THE PERSPECTIVE OF AN ELECTRONIC ARTIST

I am an electronic fine artist who uses "new media" to create virtual environments (social happenings). The works I produce investigate the psychosocial effects in the constructs of techno-driven environments. I use technology to initiate a communal exchange of reactions among both viewers and participants, thus providing a simulated clinical "case study" through an observational perspective. This method of collecting data from an observational view and constructing a phenomenological analysis is a practice employed in the field of existentialism, a discipline of psychology whereby the investigator establishes validity by describing and explicating a verbal report from observing the subject's behavior and perceptual experiences.[1]

In this essay, I reference two specific works of art (phenomenological reports): *CASE STUDY 107* and *CASE STUDY 309*. Using the metaphor of a "world on stage," these works provide a visual, conceptual, and concrete model to illustrate *theater as life* and discuss:

1. Methods for blurring, both literally and perceptually, the boundaries of multiple realities (i.e., the physical, the psychological, and the virtual).
2. Theoretical and scientific influences in the constructs of virtual environments whereby perceptions of trust, truth and self-identity(s) are challenged.

DEFINING AND POSITIONING TERMS: VIRTUAL REALITY, TELEPRESENCE, AND NEW MEDIA

In 1989, the term virtual reality (VR) was coined by Jaron Lanier, a chief officer of a research company. Typically, VR (new media) references particular technological devices such as head-mount displays and motion-sensing gloves that produce electronically simulated environments.[2] From a computer scientist's perspective, VR is "to foster in the viewer a sense of presence: the viewer should forget that [he/] she is in fact wearing a computer interface."[3]

The author of the text *Virtual Reality*, Howard Rheingold provides an alternative view of VR: "The heart of VR is an experience – the experience of being in a virtual world or remote location."[4] Sharing a similar view, Jonathan Steuer in his article entitled "Defining Virtual Reality: Dimensions Determining Telepresence" argues the position of VR to be a "particular type of experience rather than [as] a collection of hardware."[5] Steuer further defines VR in terms of telepresence: "Telepresence is defined as the experience of presence in an environment by means of a communication medium. In other words, presence refers to the natural perception of an environment, and telepresence refers to the mediated perception of an environment."[6] These perspectives shift the focus of virtual reality from a particular hardware package to the perceptions of an individual in a mediated environment. It is from this position of mediated perceptions that I draw a parallel to a postmodern, techno-driven happening (a world on stage) whereby the boundaries of the psychological, the physical and the virtual realities merge.

In the introduction I referenced the term *new media*. In the text *Remediation: Understanding New Media*, the authors state: "Introducing new media technology does not mean simply inventing new hardware and software, but rather fashioning (or refashioning) such a network.... They [new digital media] emerge from within cultural contexts, and they refashion other media, which are embedded in the same or similar contexts."[7] By reinvestigating, or refashioning, the media of theater in the context of virtual environments, we discover the reality of the world on stage to be a mediated perception of an environment–referencing telepresence. Additionally, new media refashion the boundary that traditionally separated performers from audience; this separation has now become a perceptual construct. Thus, *new media* challenge us to rethink and reexamine our basic presumptions about reality and the reality that is virtually perceived.

A WORLD ON STAGE: A REFASHIONED REALITY

Theater and VR share a number of similar practices: creating fantasy and illusive environments; altering perceptions; and, in some respects, attempting to immerse the viewer. In the digital arena, new media refashion other media. Older media (i.e., theater) are revisited and this often results in challenging the views wavering between fantasy (virtual) and reality.

One view known to be shared by Sigmund Freud is that "fantasy is an escape from reality and responsibility."[8] American psychologist Abraham Maslow held a somewhat different view, stating that "fantasy is the means by which a determined man masters reality."[9]

I draw a parallel to these counter views (the humanistic model supported by Maslow and the psychoanalytic concepts of Freud) with the task of VR in the context of existentialism and phenomenology. The typical objective of VR is to immerse the viewer and achieve "immediacy" whereby the presence of the medium is *forgotten, denied or hidden*. Phenomenology is "the study of the data of immediate experience."[10] Thus, the common and key trait described in both VR and phenomenology is "immediacy." Moreover, existentialism emphasizes personal experiences as they relate to one's existence in the *world-design*. "World-design" is a term coined by Swiss psychologist Ludwig Binswanger. For Binswanger, "the world-design determines both how people will react in specific situations and what kinds of character traits they will develop."[11] Therefore, virtual, mediated environments share similar phenomenological experiences, traits, and psychosocial happenings to that of a "world-design" (a world on stage).

Revisiting Freud's view, a person's mode of being-in-the-world of VR perhaps is *denying* the reality of one's existence in the world-design. In reference to Maslow's perspective, one could argue that being-in-the-world of VR relishes a *hidden* or *forgotten* reality. Existentialists and humanists alike "emphasize subjective experience as the primary phenomenon in the study of human nature." In this vein, Maslow wrote: "The basic coin in the realm of knowing is direct, intimate, experiential knowing."[12]

In the constructs of mediated environments, the artistry of a magician or performer, for example, immerses the viewer by methods of "divergence," creating illusive perceptions, fantasies, and virtual experiences. In the context of entertainment, our logical intuition perceives occurrences of this nature to be the art of illusion; thus, the altered perceptions can be rationally and logically explained. However, outside the realm of the arts, theater, and entertainment, if no mediation is perceptually detected, would the immersed fantasy or illusive experience be a "real" or "virtual phenomenon"?

I propose this question in support of studies conducted in the area of mental imagery. Experiments have been conducted whereby patients were asked to mentally visualize a described scene. Researchers discovered that the subject's blood flow increased to precisely those areas involved in that of the "actual viewing."[13] These studies imply that when the subjects "mentally imagine a visual scene, [their] eyes and brain act as if [they] were looking at the scene in reality, correlating mental imagery to perceptual process."[14] This same study suggests that the biological effects (enforcing a sense of realism) induced by the power of mental visualization, blur the boundaries between what is virtual and what is real. Perhaps, what once claimed to be a fantasy has now come to be a "refashioned reality" – a world on stage.

Contemporary German philosopher Hans-Georg Gadamer is referenced as stating that a "work of art requires imaginative activity on the part of the observer."[15] I correlate "imaginative activity" with "mental imaging" whereby perceptual and experiential realisms are framed in a similar remote location to that of the cyber-natural. Lisa Blackman, author and lecturer on media and communications, references author and theorist S. Cubitt as coining the term "cybernatural" and defining its discourse as when "virtual space is viewed as a 'third space' – a space existing beyond those divisions and limits that currently position subjects in the social world."[16] I would like to expand and broaden "cybernatural," as outlined by Cubitt, to include a position similar to that of a bird's-eye/mind's eye/mental imaging perspective. I consider this "third space" to be a fourth-dimensional remote location similar to the observational view inherent in the study of phenomenology. The video/kinetic case studies I will be referencing provide these remote, analytical perspectives and entice "imaginative activity" on the part of both observer and participant.

PSYCHOSOCIAL REALITIES: THE LEVEL OF CONSCIOUSNESS

Michael Polany, in his publication *Personal Knowledge*, claims: "What drives the scientist is an increasing sense of contact with reality."[17] An opposing view is found in the writings of Albert Einstein, which state that "one of the strongest motives that lead men to art and science is to escape from everyday life, with its painful crudity and hopeless dreariness."[18]

There are three described realities outlined by Max Velmans, author and reader in psychology: the physical, the psychological, and the virtual.[19] My focus addresses two main areas of interest: the "source/motive/drive" that creates this longing for the "contact with" or "escape from" reality(s), and the level of consciousness (or the lack thereof) regarding the perceived boundary framing these multilevel realities (the physical, the psychological, and the virtual).

With selective types of VR whereby a head-mount is worn, the obvious factor denoting the point of entrance and exit from the simulated, mediated environment is the physicality of the hardware itself. Cumbersome hardware of this nature provides a tangible, physical "thing" for claiming a reality check; the hardware acts as a reminder, a conscious presence for differentiating the two experiential worlds. A similar reality check is appropriated for the theatrical arena. For example, when a member of the audience steps on stage (equating a mediated environment) and assumes a role with the aid of a physical, tangible costume (equating a hardware component) and interacts with stage props (equating simulated 3-D computer-generated objects), several key elements provide a level of consciousness for the points of entrance and exit between the two realities. These metaphorically shared distinctions not only identify the characteristics of the media (VR and theater) but also define the boundaries and parameters framed by the two worlds and the environments each medium creates.

In the above-cited essay, Blackman references virtual reality to be understood as "either providing the means to enable greater freedom and autonomy (a place where one can choose an identity and play away from the material constraints of repressive society) or, in a more dystopian fashion, as dissolving and fragmenting the 'whole person', leading to greater alienation and estrangement from the self and others."[20] The key phrase I wish to address here is: "A place where one can choose an identity and play." In the context of theater, the term "place" I equate with the stage; the term "play" I equate with role-playing, the multiple roles an actor embodies. Philosopher Kendall Walton, who wrote *Mimesis as Make-Believe*, "develops and applies the idea that many art works are what he calls 'props' in a game of make-believe, and he finds that the value of art rests in the value of playing this game" according to Gordon Graham.[21] (Later in this essay, I will address another correlation with the term "play" in the context of psychology and play behavior with mediated environments.)

I expand the perspectives of both Blackman and Walton to include the psychosocial effects that could, or would, occur if the checkpoints between the realities (the physical, the virtual, and the psychological) became inseparable, undefinable, or invisible. In other words, what if the conscious mind could no longer distinguish between which "identity and play" were to be performed in which reality (world-design); and what if the "props" or mediated/unmediated imagery shared the presence of all realities?

According to Binswanger, the task of existential psychology "is to understand the totality of man's experience of himself in all his modes of existence."[22] A clear example can be found in *The Looking Glass*, written by Brazilian novelist Machado de Assis in 1882, which describes a young man who becomes a lieutenant. On return trips home he wears his uniform and receives astonishing praise from his family, friends, and the entire community. As he gradually becomes accustomed to the attention, he begins to feel more like a lieutenant. One day, however, he is left alone and no one is around to feed his ego. The young man looks in the mirror and sees his outline as blurry and confusing. He becomes overwhelmed, afraid of going insane; thus, he clothes himself with his lieutenant's uniform (mediated/unmediated object) and sits in front of the mirror. Suddenly, the image of himself in the mirror reflects clarity and soundness. In effect, "his feelings of sanity and self-respect return."[23]

In his writings, Friedrich Nietzsche has said: "The great man is the play-actor of his ideals."[24] If the actor unconsciously personifies role-playing in the realm of reality as we know it, then the ramification would be the loss of the self-born identity. Thus, the personification of the perceived mediated-self(s) would take precedence. The possibility of such a mediated psychosocial effect mirrors the "dissolving and fragmenting the 'whole person'" view held by Blackman. Perhaps *the theater of life* is the rearing of virtual and *multiple* identities.

CASE STUDIES: MODELS AND METHODS BY WHICH MESSAGES SHAPE MEANING

In the theatrical arena, omitting the perceptual barrier that separates the viewing audience from center stage promotes interplay among the multiple realities and self-identities. A cultural anthropologist might say that without social interaction and human physical contact, we would lack social cognition and loose the content that shapes meaning – meaning that derives from body language and physiognomic characteristics based on *the interpretation of/interaction with others*. (An interesting psychosocial phenomenon to consider would be a world on stage whereby social cognition and meaning derive from the interaction among the multiple, virtual identities of the mediated self.)

Communication expert John L. Locke, author of *The De-voicing of Society: Why We Don't Talk to Each Other Anymore*, is credited with the notion that "human voice and gesture provide constant feedback [immediacy]."[25] Philosopher and author Paul Ekman echoes Locke, stating that "one important function of facial and other gestural expressions of emotion is to communicate one's inner states [multiple identities] to oneself and others."[26]

As an electronic-media artist, I seek a forum whereby technology facilitates or perhaps instigates a "social happening" (a serendipity), encouraging elements of laughter and humor as humanistic models for interactivity and play. My works consist of 3-D video/kinetic sculpture installations entitled *CASE STUDY*, each of which is assigned an identifying number. As previously noted, these works provide a simulated clinical "case study" – an existential analysis.

Each *CASE STUDY* integrates diverse fields of study that draw from models of research in areas such as: cognitive styles of the haptic (touch)/kinetic (motion) learner; motivational, behavioral, and existential practices; perceptions of danger and risk hunger; social/kinetic language; the psychology of humor and play; and research in the area of selective optimum stimuli. *CASE STUDY 107* and *CASE STUDY 309* (see Figure 1) exemplify the employment of these areas of research, and illustrate the interaction among the multiple realities.

CASE STUDY 107

CASE STUDY 107 (Fig. 1) consists of two chairs bolted back to back, positioned directly under a suspended cinder block. With the aid of four pulleys attached to the ceiling, the cinder block is held in place by a 3/4-inch rope. The rope leads to two black boxes that encase 19-inch color video monitors. As illustrated in Figure 2, the objective is to create an optical illusion, making it appear as though the rope passes through to the underside of each black box. The physical weight and gravity of the suspended cinder block creates an illusion (perception) of stress, tension, and virtual danger – conversely, a "true" physical and emotional risk.

An eight-minute video segment displays a similar 3/4-inch rope, which gradually unravels. As each strand of rope breaks (video imagery), the cinder block overhead physically shakes. Simultaneously, the two chairs are jolted with an electronic vibration – a shock.

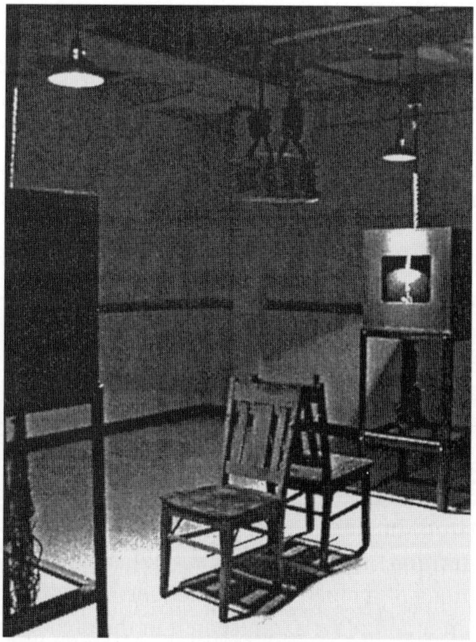

Fig. 1. *CASE STUDY 107*, 1997. Video/kinetic sculpture installation.

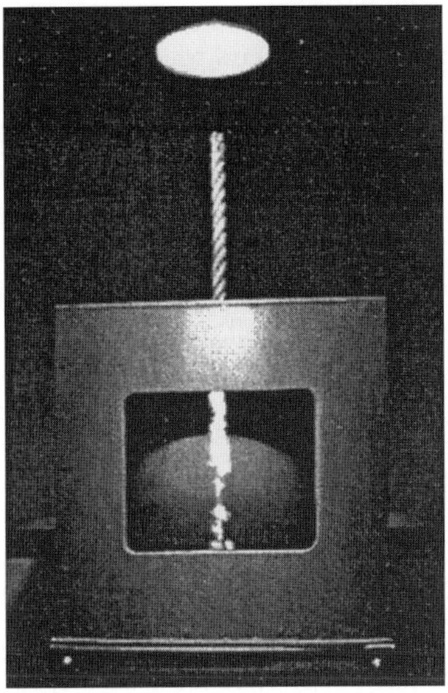

Fig. 2. *CASE STUDY 107* (detail view).

CASE STUDY 107 revealed the following observations:

1. *Diversion: Selective Optimum Stimuli*

The bystanders (the arena of viewers, or audience) appeared to believe that the people in the chairs were jumping out of their seats in fear that the cinder block overhead would fall. In actuality, the jolting chairs caused the unexpected stimulus-response. As in the art of theater, this tactic (optimum selective stimuli of diverting the expected with the unexpected) was used to maintain a level of novelty and arousal, as well as to alter expectations and perceptions of both participants and the audience. Through the combination of real and virtual reality, the immediacy provided each person the opportunity to encounter a peak (immersed) experience.

2. *Merging Multiple Realities: Center Stage*

CASE STUDY 107 merged the perceptual boundaries of the physical, the psychological and the virtual reality. The arena of viewers took on the role of performers. Their participatory response gave additional meaning to the perception and interpretation of the mediated environment. The shared presence of "center stage" (center stage references a central location where the realities interact and merge) encompassed all of the individuals who were present.

If interactivity requires a center stage arena, people are more likely to interact if the invitation accommodates two or more participants (performers). The dialogue becomes, "I'll do it, if you do it." Additionally, participants feel less intimidated and self-conscious if the seating is positioned back-to-back, versus face-to-face. Creating appropriate personal space encourages participation while enhancing the virtual/real experience.

3. *Interfaces: Accessibility, Timing, Attention Span, and Mental Visualization*

Chairs signify an invitation for seating. Metaphorically, the two chairs signify a hardware interface and the two participants in the chairs signify a software interface. The level of interactive accessibility in this piece was intuitively framed in that it did not include the technical complexity common in most VR systems. In other words, the participants were not required to read a set of instructions, point-and-click a hardware device, or be clothed with a cumbersome hardware-wired system.

Pushing the envelope of patience and attention span in linear time, the eight-minute video segment appeared to be the breaking point in this piece. The video ended with a very thin thread virtually appearing to hold the weight of the cinder block. After watching the complete video segment, people proclaimed disappointment that the rope didn't break. It seemed that the slightest possibility of danger was the motive for capturing and maintaining their attention span for the entire length of the "performance." This fascination supports Gadamer's argument that a "work of art requires imaginative activity." Interestingly, the observation of the viewers and participants seemed to include the imaginative activity and physical attraction to the element of apparent danger. In other words, it appeared the viewers were unconsciously drawn to the visual of the cinder block falling on the two participants sitting in the chairs, waiting to see the experiential after-effects. Perhaps this desire of risk with a fascination for danger says something about our society.

The enticing traits of risk and danger employed in *CASE STUDY 107* engaged and merged the psychological, the physical, and the virtual reality, thus challenging a sense of trust, truth, and realism on the part of both the observer and participant.

CASE STUDY 309

Sharing similar characteristics with *CASE STUDY 107, CASE STUDY 309* utilizes a personable language, such as kinesthesia (uniting physical sensation with sound and imagery). The installation consists of two identical structures, each measuring 12 feet high, 4 feet wide, 4 feet deep. Once again, the invitation accommodates viewing for two participants, each having an individual zone. However, the viewing perspective in this installation requires the participants to lie on their backs (a vulnerable position) on creepers, and roll beneath a suspended two-foot-square black box. The boxes (as in *CASE STUDY 107*) encase 19-inch video monitors. The objective is to synchronize the video imagery with that of physical kinetic sensations similar to *CASE STUDY 107*.

One of the towering structures depicts a video image of a cement brick falling in the direction of the reclining viewer. At the moment the brick breaks the glass (simulating the glass of the video screen), the black box physically shakes. Similarly, a video image displays an egg being dropped, then removed by a vacuum cleaner, at which time an actual vacuum cleaner pulls the participant's hair from the headboard of the creeper. Other synchronized visual-audio, kinetic elements are images of machinery parts that correspond to vibrating motions of head-and-body massage units.

CASE STUDY 309 revealed the following observations:

1. *Sensory Adaptation: Timing, Attention Span, Accessibility and Selective Optimum Stimuli*

Sensory adaptation refers to a "decrease in sensory response to a constant or unchanging stimulus."[27] As both stations (structures) were visually identical in construction, each provided different visuals corresponding to different physical sensations. The structures were purposely designed to be identical so as to challenge predictability (sensory adaptation) and to create a momentum of curiosity and interaction between the two stations.

Research indicates that an environment designed to minimize stimulus input is not something humans generally seek out.[28] Dr. Daniel Berlyne, a major figure in the study of motivation, references four valuable traits for research into stimulus selection: novelty, uncertainty, conflict, and complexity.[29] By altering the viewing perspective (a non-traditional approach to viewing a work of art), disorientation was a "stimulus selection" employed in *CASE STUDY 309*, a method of enhancing the realism. Disorientation is defined as overwhelming or conflicting stimuli that complicate the brain's correlation of information. The brain consequently sends false input to the various senses whereby the altered perceptions are, in turn, experienced as reality.[30]

The only way to view and experience *CASE STUDY 309* was to place oneself in a vulnerable position. This disoriented position involved an element of risk with plausible danger of the suspended video monitor falling. The reclining position also included emotional risks of embarrassment and social consciousness. More importantly, the enticing element which encouraged participation appeared to be that of curiosity and the (perceived) physical danger.

The kinetic experience (video segment) lasted 90 seconds. The attention span in a reclining position appeared to maximum between 60 to 90 seconds before the peak experience would have been lost to predictability (sensory adaptation). Timing, attention span, accessibility, and selective optimum stimuli were all important elements of consideration in these interactive electronic works.

2. *Risk Factor: The Demarcation between Virtual Risk and Real Risk*

Galleries displaying *CASE STUDY 309* requested a signed release form from each participant stating the risks involved and declining liability if a mishap occurred. Obviously, this added another dimension to techno-driven virtual environments, and raises many concerns for the electronic artist who creates social interactive works. By raising issues of belief and perceptions of trust, the demarcation between virtual risk and real risk (virtual reality and reality) breaks down.

3. *Merging Multiple Realities: Psychosocial, Multilevel Interactivity*

Similar to *CASE STUDY 107, CASE STUDY 309* displayed multiple levels of psychosocial interactions, blurring the perceptual boundaries of the physical, the psychological, and the virtual realities. Both the on-stage participants and the implied offstage audience shared the presence (telepresence) of center stage.

Referencing Gadamer's philosophy, the "mind" of the art work (*CASE STUDY 107* and *309*) and the "mind" of the audience must be mutually engaged in the creative activity – "the work's creativity needs its audience."[31] This view is shared by author and philosopher R. G. Collingwood: "'Art is not contemplation, it is action' and the function of the audience is 'not a merely receptive one, but collaborative.'"[32] Collingwood rejects any "conception of audience as passive spectator."[33]

SYNOPSIS

In both *CASE STUDY 107* and *309*, the sculptural contraptions took on a life all their own. The structures became instigators for enticing and facilitating a "social happening" – borrowing the term "happening" from Allan Kaprow, artist and author of *Assemblage, Environments and Happenings*. Participants and viewers were unbeknowing "subjects" from an observational perspective, a view whereby human behavior, psychosocial responses, and social interaction could be analyzed in real time. Both case studies provided this analytical view (i.e., the cybernatural/existential analysis and phenomenological report) from which to observe the subjects in a mediated, techno-driven environment, one which also encompassed multicultural diversity.

Each case study invited participants and viewers to exchange roles. The direct experience actually caused participants to become even more knowledgeable viewers. In other words, the haptic-kinetic experience began with a curious viewer, changing into an active participant, then returning to the role of a viewer. This secondary viewing leads to a more sophisticated observational role. It is from this "haptic-experiential perspective" that the most self-learned meaning is constructed. These many perspectives created an experiential embodiment (incarnation) of both body and mind.

In both case studies, the following elements were points of consideration for merging the constructs framed by each reality:

Somesthetic Senses: Somesthetic is a combination of "soma," meaning body, and "esthetic," which means "to feel." Somesthetic senses include the skin (touch), kinesthetic sensors (receptors in the physical body for detecting position and movement), and vestibular senses (receptors in the inner ear for maintaining balance).[34]

Research has discovered that "skin receptors produce at least five different sensations: light touch, pressure, pain, cold, and warmth. Altogether, the skin has about 200,000 nerve endings for temperature, 500,000 for touch and pressure, and 3 million for pain."[35] Because the body has more nerve endings for the sensation of pain, the key stimuli in both case studies was that of (implied) pain, engendered by elements of perceived risk and danger.

The interesting phenomenon is the "sense of truth" and realism that results from creating an illusive virtual pain versus real physical pain. The physical body reacts and responds in a similar biological and physiological fashion, whether or not the pain is physical or perceptually experienced. The brain triggers the release of a chemical called beta-endorphin (similar to morphine) to combat pain.[36] Receptor sites for endorphins are located in the area of the brain associated with pleasure, pain, and emotions.[37] Researchers have concluded that "there is reason to believe that pain and stress cause the release of endorphins. These in turn induce feelings of pleasure or euphoria similar to morphine intoxication."[38] I conclude that the physical and emotional responses resulting from the (perceived/real) stress and tension in both case studies caused these receptor sites to release endorphins, thus producing a "peak experience." A peak experience, a term coined by Abraham Maslow, is defined as the cognition of being, an ecstatic moment, *an awareness of the body*.[39]

In *CASE STUDY 107* and *309*, peak experiences embodied the cognition(s) of the being(s) – the physical, the psychological, and the virtual. The ability to produce peak experiences by enticing the receptor sites was an important factor. The stimulus-response heightened the level of awareness of the body, providing a sense of truth and realism on the part of the participants.

Hapticism: My self-generated definition, or description, of hapticism is akin to the somatic in which we learn principally through the physical body itself. "Haptic" relates to touch; "ism" defines a system, theory, practice, or action. I refer to hapticism as experiential interactive art that utilizes kinetic methods to produce a reality virtually perceived by the haptic learner.

In investigating the haptic learner (that is, one who learns best by experience, utilizing and encompassing as many of the senses as possible), I have discovered the importance of including the element of touch in virtual environments. It is through touch that we define meaning, according to David Katz, a major figure in the study of the psychology of perception. Katz states: "From a perceptual viewpoint, we must give precedence to touch over all other senses because its perceptions have the most compelling character of reality. Touch plays a far greater role than do the other senses in the development of belief in the reality of the external world.... What has been touched is the true 'reality' that leads to perceptions."[40]

Play Behavior: Both *CASE STUDY 107* and *309* challenged psychosocial issues such as social pretentiousness, embarrassment, insecurities, and emotional, guarded boundaries. Therefore, it was important to maintain a light-hearted perspective, engendered by play in these mediated environments. This element of play was induced and promoted by applying both visual and haptic/kinetic absurdities.

According to Sutton-Smith, a researcher in the psychology of behavior, "Play gives the individual the opportunity to engage in social learning without the fear of experiencing repercussions."[41] Joseph Levy, another major figure in the study of play, humor, and laughter, states: "Play is necessary to affirm our lives. When we slip into play, we slip into a self-experience where we can afford to 'let go' and respond to ourselves, to others, and to the environment in an unpredictable, personal way. Living in play means confirming our existence and celebrating life. This means using our senses – eyes, ears, nose, taste buds, skin, even our lungs and heart – in order to experience the world as it really is."[42]

Humor: Just as absurdities promoted play in each case study, it was also important for play to promote the language of humor – the ability to laugh at oneself and with others.

Patricia Keith-Spiegel, a researcher in the psychology of humor, notes that there are four elements deemed by many theorists as necessary (though not sufficient) to appropriate conditions for the experience of humor and laughter: the element of surprise, the element of shock, the element of suddenness, the element of unexpectedness.[43]

With the use of bizarre video imagery, absurd kinetic devices, and tactics that imply risk and danger, I was able to produce these four elements in each case study. These elements enticed (triggered) the release of endorphins, and thus, created peak experiences for both the participant and the audience. It was observed that humor was the communal reaction as a result of the combination of stimulus and response, creating the appropriate conditions for social interactions. This observation of communal laughter reflects examples from anthropology, researched by Jacob Levine, who says: "There is nothing so completely shared as laughter."[44] Konrad Lorenz, one of three recipients to share the Nobel prize for work on behavior, states: "Laughter produces, simultaneously, a strong fellow feeling among participants.... Heartily laughing together at the same thing forms an immediate bond."[45]

Risk Hunger: An additional trait evident in both case studies was the element of risk. Immersed in a technoculture, we wrap ourselves in a cocoon of safety, comfort, and convenience to the degree that we have become bored. Ralph

Keyes, author of *Chancing It: Why We Take Risks*, claims that "we suffer from risk hunger." He suggests that people need to take risks because our nervous systems and our bodies demand the stimulation, that a little danger is good for people. Taking risks brings us back to something primal and original, relieving us from the tedium of everyday life.[46] However, according to Keyes, the concept of risk becomes confusing if we define risk as a fear of loss. If fear is amputated from danger, then it becomes a Disney ride – a leisure activity. Many people aren't satisfied with taking risks unless something of value is at stake.[47]

The definition of risk is subjective, as is peak experience. Paradoxically, what is most revealing may not be the risks people take, but the ones they don't take. For example, for some individuals, the prospect of being rejected by another poses a greater risk than engaging in an activity like bungee jumping. Could virtual environments promote an unemotional, risk-free society whereby the attachment to objects and virtual identities governs human interaction on the physical plane? Because of the emotional and psychological risks involved, will we become a culture that lacks a social/kinetic language which builds on intuition developed from the interpretation of gestures, expressions, and body languages communicated in the physical reality? Locke warns: "'De-voicing' may cause both Westerners and Easterners to become more isolated, distrustful, and unhappy."[48]

If technological devices and trends progressively move us away from social interaction in the physical sense, then we may risk losing touch with our own presence. Then again, our subconscious/conscious attraction to the "world on stage" may be the primary objective in order to avoid the emotional risk of being in touch with our own self-presence.

CONCLUSION

Art critic and author Timothy Druckrey states: "The body is unquestionably the next frontier – the body, and then cognition."[49]

Throughout this essay, I have discussed the invitation of a *theater as life, a world on stage*. I have provided the description of artistic examples with scientific reasoning and applied tactics for merging multilevel realities, while enticing critical and provocative dialogue. Aside from a theoretical perspective of whichever reality we decide to engage in, I suggest that we maintain one important element: our sense of humor – the ability to laugh at our multiple selves and with the multiple selves of others.

Florida Atlantic University

NOTES

[1] Calvin S. Hall and Gardner Lindzey, *Introduction to Theories of Personality* (New York: John Wiley & Sons), 256.
[2] Jonathan Steuer, "Defining Virtual Reality: Dimensions Determining Telepresence," *Journal of Communication* 42, no. 4 (Fall 1992): 73.
[3] David Jay Bolter and Richard Grusin, *Remediation: Understanding New Media* (Cambridge: MIT Press, 1999), 22.
[4] Ibid.
[5] Jonathan Steuer, "Defining Virtual Reality: Dimensions Determining Telepresence," *Journal of Communication* 42, no. 4 (Fall 1992): 74.
[6] Ibid., 76.
[7] David Jay Bolter and Richard Grusin, *Remediation: Understanding New Media* (Cambridge: MIT Press, 1999), 19.
[8] Colin Wilson, *The Essential Colin Wilson* (Berkeley, CA: Celestial Art), 87.
[9] Ibid.
[10] Calvin S. Hall and Gardner Lindzey, *Introduction to Theories of Personality*, 243.
[11] Ibid., 249–250.
[12] Larry A. Hjelle and Daniel J. Ziegler, *Personality Theories* (New York: McGraw-Hill Book Company), 366.
[13] Ann Marie Seward Barry, *Visual Intelligence: Perception, Image, and Manipulation in Visual Communication* (New York: State University of New York Press, 1997), 86.
[14] Ibid.
[15] Gordon Graham, *Philosophy of the Arts: An Introduction to Aesthetics* (London and New York: Routledge, 1998), 15.
[16] Lisa Blackman, "Culture, Technology and Subjectivity," in *The Virtual Embodied*, ed. J. Wood (London and New York: Routledge, 1998), 133.
[17] Colin Wilson, *The Essential Colin Wilson*, 88.
[18] Ibid., 75.
19 Max Velmans, "Physical, Psychological and Visual Realities," in *The Virtual Embodied*, ed. J. Wood (London and New York: Routledge, 1998), 46.
[20] Lisa Blackman, "Culture, Technology and Subjectivity," in *The Virtual Embodied*, ed. J. Wood (London and New York: Routledge, 1998), 132.
[21] Gordon Graham, *Philosophy of the Arts: An Introduction to Aesthetics* (London and New York: Routledge, 1998), 17.
[22] Calvin S. Hall and Gardner Lindzey, *Introduction to Theories of Personality*, 248.
[23] Colin Wilson, *The Essential Colin Wilson*, 82–83.
[24] Ibid., 82.
[25] Jeff Minerd, "The Decline of Conversation," *The Futurist 33*, no. 2 (February 1999): 18.
[26] Stephen Davies, ed., *Art and Its Messages: Meaning, Morality, and Society* (Penn. State Press, 1997), 40.
[27] Dennis Coon, *Essentials of Psychology*, 6th ed. (St. Paul, New York, Los Angeles, San Francisco: West Publishing Company, 1994), 180.
[28] Lyle E. Bourne Jr. and Bruce R. Ekstrand, *Psychology: Its Principles and Meanings*, 2d ed. (Holt, Rinehart and Winston, 1976), 206.
[29] D. E. Berlyne, *Conflict, Arousal, and Curiosity* (New York, Toronto, London: McGraw-Hill Book Co., 1960), 18.

[30] Ronald D. Davis, *The Gift of Dyslexia* (Burlingame, CA: Ability Workshop Press, 1994), 18–19.

[31] Gordon Graham, *Philosophy of the Arts: An Introduction to Aesthetics* (London and New York: Routledge, 1998), 15.

[32] Ibid., 33.

[33] Ibid., 33.

[34] Dennis Coon, *Essentials of Psychology*, 6th ed. (St. Paul, New York, Los Angeles, San Francisco: West Publishing Company, 1994), 177.

[35] Ibid., 178.

[36] Ibid., 183.

[37] Ibid., 183.

[38] Ibid., 183.

[39] See Abraham H. Maslow, *The Farther Reaches of Human Nature* (New York, London, Toronto: Penguin Group, 1971).

[40] Lester Krueger, *The World of Touch* (New Jersey: Lawrence Erlbaum Associates, Publishers, 1989), 240.

[41] Joseph Levy, *Play Behavior* (Toronto: John Wiley & Sons, 1978), 169.

[42] Ibid., 1.

[43] Jeffrey Goldstein and Paul McGhee, *The Psychology of Humor: Theoretical Perspectives & Empirical Issues* (New York, London: Academic Press, 1972), 9.

[44] Jacob Levine, *Motivation in Humor* (New York: Atherton Press, 1969), 12.

[45] Ibid.

[46] Joe Surgarman, "This Thing Called Risk," *College Park* 8, no. 2 (Winter 1997): 16.

[47] Ibid., 17.

[48] Jeff Minerd, "The Decline of Conversation," *The Futurist* 33, no. 2 (February 1999): 18.

[49] Timothy Druckrey, introduction to *Culture on the Brink: Ideologies of Technology*, Grentchen Bender and Timothy Druckrey, eds. (Seattle: Bay Press, 1994), 9.

R. A. KURENKOVA AND O. V. PETROVA

MUSIC ON THE STAGE OF LIFE

I

As we peer into the century ahead of us, we are intrigued by the question of the place that stage music will have in this era.

We would like to speak about the ecology of musical culture. The root of the word "ecology" is the Greek *oikos*, meaning dwelling, home. It is not easy now to answer the question of where the homeplace of music is. Its inhabitants often leave it, and ordinary listeners feel no need for it. Or when they are enraptured by music, it happens only when experiencing it in synthesis with other arts.

Television has become the main drug and teacher of our musical consciousness. You can see everything on TV – the theatre, the ballet, all of the different arts. Technically not being inherently art, music is becoming dependent on the modern means of its transmission and is becoming enslaved by the other arts. As a result, music is tending to lose its identity.

This is not all to the bad. Sometimes this can well prove its value – whenever work that is an integral whole sweeps over an audience. But the music that as a matter of fact we hear on TV at present can usually be called an acoustic prosthetic device.

The most awful thing is that we are ruled by TV. There is an intellectual poverty, even a mental imbalance that comes across to TV viewers in typical music programs. The directors are not in the least interested in raising such questions as what is good and what is bad and what in music is good taste and what is just tasteless. Everything is subordinated to commercial purposes, business, image building, and star development. Commercial and creative phenomena have become almost synonymous. Most musicians are now labeled as either stars or nonstars. There is at work a "popularity $\rightarrow$ money$\rightarrow$ popularity" cycle. We can speak of "copper-bottomed", "new Russian", or "high society" audiences. And where is the real audience? By that we mean listeners who enjoy thinking musically and who silently and rightfully fill the concert halls. This kind of audience and especially our children do not fit into this market and real musical culture. This is happening because our society feels no need for musical culture. And so beauty is not considered necessary for success in the music market.

A.-T. Tymieniecka (ed), Analecta Husserliana LXXIII, 103–107.

Consequently, the music on the screen is not what music is and should be in the historical and theoretical sense of the word. The TV market negates the art of music as a foundation of experience, existential meaning, and identity. Music on TV is various sounds accompanying a fluid picture. As a result, television viewers consider something that flashes before the eyes and ears to be music. What is called music here is not music at all.

We think that the ruling idea of music in the twenty-first century must be that of the *joy* of music. The contemporary social situation calls for this response. Tragic conditions in the social body are transforming and destroying the most light-hearted, elevated and beautiful aspects of life, the beauty of music too. Music is being robbed of its joy. The famous Russian writer Chekhov wrote that while kindness requires the enactment of kindness, beauty does not. But today beauty in general and music as its mysterious manifestation need the demonstration of their beautiful being. Music, the feeling of the joy of music, needs to be defended as we cope with today's social problems and atmosphere of political, national, economic, and religious dissension and even military strife.

Music has no autonomous existence when removed from us. It exists as a feeling, as the joy of music. Looking on the processes eroding social relations, an anthropological crisis, at the increase of iniquity, violence, aggression, abomination, and the demonic, we understand that the space for beauty, for the joy of music too, has shrunk a lot.

We have so many global problems. How to avoid war? How are we to meet energy supply and food needs, respond to population pressures and birth rates, and resolve ethnic and international tensions? All problems now take on a global aspect, but global thinking cannot of itself change humankind's thinking about the prospect for its life in the new millennium. In Krzysztof Penderecki's "Threnody for the Victims of Hiroshima" we can hear expressions of the shocked consciousness of the tragic twentieth century. And we can only have forebodings of what computers, biotechnology, and things we cannot yet imagine will make possible that will shock humankind in the third millennium.

II

Our problem is posed as a question that cannot be answered on the basis of knowledge. This is the question of the content, structure, and causes of the link between the cognitive and aesthetic factors in the musical consciousness of man.

Our research hypothesis derives from the fact that estrangement from or absorption in musical art follows from the personal interests that accumulate in human perception. We see apprehension of the phenomenalism of this personal formation as it tends towards absorption in music as providing the necessary foundation for quickening the processes of musical education.

Our research goals consists in:

- analysis of the classical and modern phenomenological texts;
- defining new categories for the phenomenon of musical consciousness;
- consideration of the phenomenological method of analyzing art and music;
- studying the methods of formation of the phenomenological structures of consciousness in the publications of American scholars and the practical work of American music teachers.

Under the guidance of experienced scholars in the USA, we would like to proceed with the investigation of the problems outlined in a book by the first author of this piece, *Fenomenologiia khudozhestvennogo soznaniia (estetiko-obrazovatel'nye aspekty)* [Introduction to the Phenomenology of Music], (Vladimir: 1996).

Although there are some publications in the Russian philosophical and musicological literature addressing our chosen problems (see the works of V. V. Medushevskii, Iu. I. Kholopov, and T. V. Cherednichenko on the interaction of music and man), the phenomenological approach has never become a force in our native thought.

In this respect the books and papers of the following American researchers are of great interest to us: Anna-Teresa Tymieniecka (her book *Logos and Life*), Lawrence Ferrara (his book *Philosophy and the Analysis of Music*), F. Joseph Smith (his book *The Experiencing of Musical Sound: Prelude to a Phenomenology of Music*).

Phenomenological education is an absolutely new and promising sphere of human knowledge, one especially pertinent in view of the forthcoming reforms in education in this country. There can be no creative thinking without phenomenology.

Perhaps never in human history has there been a greater need for a common language than today. Music is the language that transcends words and may be of use as the language of concord. The phenomenology of music can be musicology's best friend in advancing the universal realization of the feeling of concord in humankind.

The revelations of the founder of phenomenology, Edmund Husserl, have acquired special relevance for the philosophy of music nowadays. In the past, musicologists' theoretical interests were centered on the musical composition as such. Thus, there is for them an unnaturalness in the phenomenological analysis of music for therein the object of thought consists in the immanent world of man's musical consciousness, that is, the acts of musical thought are what constitute the music's content. When the analyst is no more than an objective observer, his musical consciousness is aimed at eliciting pure entities with emotions being shifted to the background. The natural orientation of consciousness is toward the elicitation and investigation of the constitutive elements of musical cognition. The phenomenology of music is based on the phenomenology of musical thought. However, it does not have to do with the empirical musical experience conceived and analyzed in its essential universality. At its base the phenomenological analysis of music focuses on intention, understood as the orientation of man's musical consciousness toward the generalized object. This is the phenomenological conception of the composition's musical integrity. Its purpose is the elicitation of pure entities of musical experiences, the formulation of logical musical notions and their verification.

One can discern three historical models for music's phenomenological interpretation.

The first wave of phenomenological investigation swept over German musicology in the 1920s. Among the phenomenologists involved were such eminent scholars as Hans Mersmann, August Halm, Heinrich Schenker. Here the focus was on the ontological entities of a musical composition.

The second model for the phenomenology of music shifted attention to the exploration of the problems of music's cognition and construction in the spiritual experience of the listener. This was the project of the 1950s and '60s. Representative of this trend are N. Hartmann and H. Kurt in Germany; J.-P. Sartre, M. Dufrenne, V. Jankélévitch, and N. Coriabine, in France; S. Langer and G. Epperson in America; and R. Ingarden in Poland.

The third model for the phenomenological analysis of music belongs to our times. One can tentatively term it "inclusive phenomenology". It consists in a discourse on the phenomenology of man's musical consciousness that borrows elements of other constructive contemporary models of musical analysis. Among the contemporary phenomenologists taking this approach, we should take note of such names as Nicholas Cook, Lawrence Ferrara, Kingsley Price, Alfred Pike, and F. Joseph Smith.

In Russia the phenomenology of music appeared at the close of the nineteenth century in the work of such aestheticians and philosophers of music as M. A. Smirnov, K. Cherkas, and A. Saketti. But the peak of Russian musical phenomenology was reached in the investigations of A. F. Losev, the father of the present-day dialectical-phenomenological method of research in musicology.

While acknowledging our great debt to Husserl and calling him a prodigious thinker, Losev departed from the German thinker's ideas on significant points. Husserl had formulated in clear terms a return to the long forgotten philosophical idea of *eidos*. (In Husserl's phenomenology "form" is represented, even in its highest mental abstraction, quite visually and as independent of thought, thus bringing philosophy in touch with the realities of being again.) Husserl, however, got stuck halfway in sketching his eidetic phenomenology, tying it to a system of schematic-mathematical connections instead of categorical-eidetic ones.

The aesthetic experience of the "Other" cannot be perceived apart from a musical-aesthetic feeling. According to Husserl, the "Other" may be interpreted as the spiritual, personal, and transcendental "Ego", the genuine form of intersubjectivity being bound to the last type. Thanks to intersubjective construction, the "Ego" of the other assumes its form in an ideal musical object alongside the "Ego" of the recipient subject. And vice versa, the "Ego" of the listener – thanks to the intersubjectivity in the strains of the piece of music perceived – has bearing on the composer's or the performer's "Ego". And this is exactly the conductor of the spiritual contact that occurs among people whose communication is limited to music. This form of intersubjectivity in musical culture may be called "musical aesthesis". Its value consists in the fact that it postulates a personal, "involved" attitude toward music. It is impossible to conceive the value of the world of music outside of the medium of intersubjective relations. The new notion of musical aesthesis is meant to denote the ability of a person's musical consciousness to experience personal aesthetic pleasure from music on the basis of perception of its intersubjective humanistic content. *The goal of music education is to* instill in an ever expanding audience the appreciation of beauty and concord in music as well as awareness of the oneness of the world that this beauty and concord provokes.

Vladimir State Teacher Training University

SECTION III

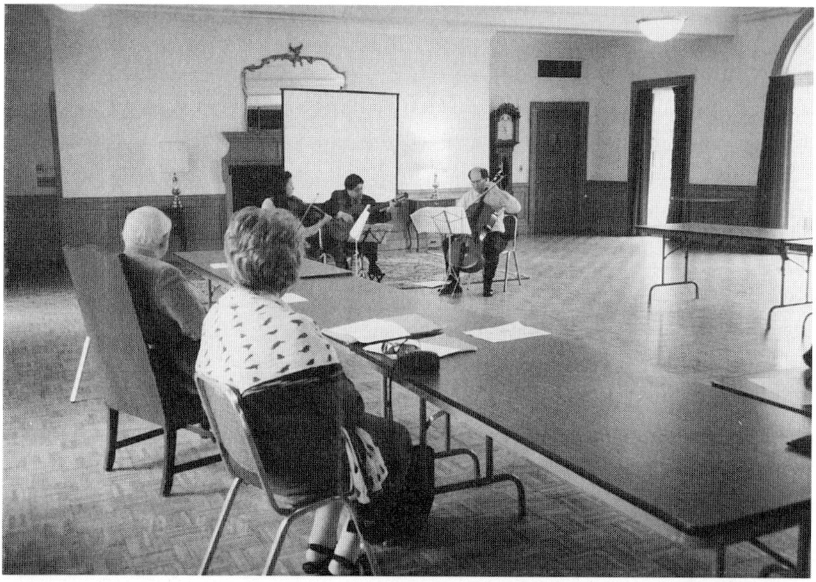

The annual concert

HOWARD PEARCE

ILLUSION AND ESSENCE: HUSSERL'S EPOCHÉ, GADAMER'S "TRANSFORMATION INTO STRUCTURE," AND MAMET'S *THEATRUM MUNDI*

The *theatrum mundi* metaphor is a means of expressing an essential intimation of the world's constituted nature.[1] This ontological metaphor invokes a duality of worlds, in playing back and forth between the familiar, ordinary, reasonable waking world and the images that emerge from the strange, often disturbing worlds of dream or the stage. Experience of the lived world is not always direct and full perception but essentially a manifold of what Edmund Husserl refers to as "living convictions," of preconceptions and beliefs – philosophical, theological, psychological, etc. Heidegger's "fore-structures," similarly, aid us in finding our way and also cover up what must be uncovered. Heidegger's distinction between "earth" and "world" characterizes the world as the constituted setting of human events. My suggestion that the world is "constituted" this way identifies our "fallenness" as a condition in which illusoriness is inherent. In the negotiations between the known and the unknown, the familiar and the strange, the probable and the surprising, the metaphorical play of imagination reiterates a sense of that illusoriness and the potential inversion of the real and unreal dimensions. The idea of play entails, in experience, an essence of ludic possibility. In the dramatic experience, the stage manifests this potential for inversions, in that its inherent unreality, its fictionality before a real audience, asserts itself to be true; and it presents within its dimension the plays of illusion in masks, disguises, character inversions, plays-within-plays.

The Socratic pursuit, a seeking of conversation about another dimension of reality, results not in resolution, closure upon one final version of reality, but in the need for extending the search, for persisting in the commitment to that other world's truth and reality. In his determination to make accessible that ideal dimension, Socrates represents the dimensions as if in conflict, the familiar world affirming its genuineness while revealing its deceptiveness – in the falseness of sense perceptions, when we see as bent a straight object standing in water; when those intimations of the truth we have forgotten must be recalled for us through dialogue or through narratives of, for instance, the myth of Er; when the metaphor under discussion here speaks a truth about our living in a theatre or a dream. In the *Philebus*, Socrates qualifies his

A.-T. Tymieniecka (ed), Analecta Husserliana LXXIII, 111–128.

recalling a "theory that I heard long ago" with the possibility that he "may have dreamed it" (20b), a playful admission of the uncertainty evoked in the *theatrum mundi* metaphor. (These comments on Plato seem like a justifiable reading of the drama of Socrates's search if we take Hans-Georg Gadamer's suggestion that we read Plato mimetically.)[2] Socrates appeals directly to the metaphor of the stage, which in his rhetorical use allows real life to become the stage: "In laments and tragedies and comedies – and not only in those of the stage but in the whole tragicomedy of life" (50b). On the other hand, Socrates can declare with full seriousness that "nothing in mortal life is worthy of great concern" (*Republic* 10, 604c), authentic, serious concern being then for the changeless dimension that is not mortal. We continue, like Socrates, to seek and to imagine genuine, coherent, and enduring sites of reality – the garden, the pastoral, heaven, the ideal body politic, the psycho-social subject, the cosmic setting ranging from the Ptolemaic to chaos theory – both in and out of this world.

In the idea that worlds are imbedded in or discoverable through language, we might again observe Plato's dissatisfaction with words in the *Cratylus*. Beginning by seeking the origins of the true in the origins of words, Socrates ends with the warning that we must keep trying and not become wanderers in Heraclitean versions of a reality that is temporal and Protean. The language accepted in that pursuit might be constrained by boundaries and rules that assure results, that produce for instance the closure of a coherent theory of truth or of a scientific methodology. The theatre-dream metaphor, on the other hand, albeit an imaginative, indirect, playful assertion about the dimensionality of ontological structure, remains a problematic assertion of a version of truth, problematic in the first place because it helps to open the play between alternate or opposed versions of truth, between "possible nonactual worlds" (Spariosu). Rather than a reduction to clarity and the simple Oneness Plato imagines for an adequate language, the metaphor plays and maintains an open structure of relationship between dimensions; it presents an illusion of mimetic relationships that challenge the Egyptian spirit Plato admires in the *Laws* (2.656d–657b).

I suggest that illusion is not an aberration or a deficiency, as Plato would have it, and simply a detriment to our apprehension of the truth; nor is it an unreality constituted in an either-or structure as over against reality. The idea of illusion is not here intended as indicating a state of unreality as opposed to the real, the verifiable, the true. Illusion [*in* + *ludere*] is the play, the activity, that is not yet resolved into the real and an opposed unreal; what seems to be unreal in this play of illusion might be resolved as the real, converting what seems to be real into the unreal. Illusion is the very play, as in the play of light or the *jeu* of fountains (Gadamer, *Truth & Method* 103; *Relevance of the Beautiful* 22), the sleight-of-hand man's manipulation of his shells, and diverse human games, that is always, in human terms, looking within or directly at the present appearance, not at an absent reality behind the false but at what might appear here and now, the yet-to-be-discovered in the ongoing activity. It is an essence of experience. When I suggest that illusion is grounded in and continues through play, this thought seems to be indebted to Husserl's idea of experience as "intentional" – the activity of perception not being reducible to either site, the subjectivity or the objectivity, but remaining as a mobility in the event, in time.

The activity of engagement in experience understood as Husserl's "intentionality" occurs as if between the abstracted antitheses of subject and object. This view of experience as a temporal unitive mediation, as an insight and a negotiation, is not only retained as impetus in the hermeneutic philosophy developing from Husserl's thought, but also echoed in a range of contemporary attempts to illustrate a model of experience. In considering the idea of "possible other worlds," Mihai Spariosu affirms the emergence of the ethical in imaginative literature. The locus of that possibility is in the "liminal" or "ludic," the medial place that is both between worlds and a place in itself (like the stage) that interrelates with the worlds it lies between. The *ludic* (which opens up this place, releases it from the closure of one world's seriousness or "living convictions") and *liminality* (by which the place remains as between worlds, as transitional) are accompanied by a third essence, the *irenic*, which puts out of play the real world's "voluntaristic" or "rationalist mode of thought, feeling, and behavior" and promotes the generation of "various ethical codes" (xiii). This "ludic-irenic view" sees "liminality as a margin … providing a playful opening toward alternative worlds that are incommensurable with ours … [and] act[ing] as a threshold or passageway from one historical world to another."

The phenomenologically based Jungean psychology of James Hillman, too, imagines a "perspective … a viewpoint … [that is] reflective; it mediates events and makes differences between ourselves and everything that happens. Between us and events, between the doer and the deed, there is a reflective moment – and soul-making means differentiating this middle ground" (xvi). The ludic appears in Hillman's thought as a resistance to the "literalism," or the reduction of metaphor to "fact," that builds a world certain of its convictions and inflexible in repudiating alternatives, possibilities. According to Hillman, "An adequate psychology must be one that cannot take itself or any of its ideas literally.... It will have to consist of fictions.... Fictions ... are full of contradictions and 'logical impossibilities.' They can never be taken literally on their own terms, but carry on their calling card the prefix of their class mark, 'as-if'" (150).[3] This imagination of a territory between two worlds has been a recurring metaphor in Western thought. Spariosu finds an instance of it in Odysseus's taking "shelter under a wild olive bush inextricably intertwined with a tame one" (xi). The "intertwined plants ... represent the liminal space or no-man's land between sea and shore, between wilderness and civilization" (xii). Plato's Myth of Er provides another instance. The meadow the "warrior bold, Er," tells about serves as a temporary abode for those returning, who were "encamped there as at a festival."[4] This setting provides them with a ludic freedom to share stories of their lives, and Er is assigned the role of "messenger to mankind to tell them of that other world." Nathaniel Hawthorne's world of moonlight, the site of the romance as opposed to the novel, is a place of transformation and generation of images that mysteriously elude the matter-of-factness of the daylight world.[5] Husserl's intentional activity that takes place between subject and object seems to seek that persistent ludic, liminal site open to play. It entails that quality of experience that is ever *playing toward* the establishment of a substantial truth, which might be *a priori* in a Platonic Idealism or *a posteriori* in a phenomenological reading. In Martin Heidegger's hermeneutic, this negotiation appears as a dissatisfaction, an incompleteness; and it recognizes the not-as-yet in the appearance and in its "withdrawal."[6] The stage apprehended as this middle ground needs to be seen, not in traditional terms as mimetic of a "real" world beyond it or as expressive of the values of the "real" world that views it, but as generative – ludic, liminal, and irenic.

Wallace Stevens, in the poem "As at Theatre," dramatizes the imagination as playing upon what amounts in Stevens to a demurral, an acquiescence, in thinking upon the Protean nature of worlds appearing and disappearing, coming and going. Most clearly, and most essentially, the voice of the poem recognizes that "Another sunlight might make another world." The proposition echoes, of course, the Platonic opposition of worlds in Plato's metaphor of the fish, who live in an undersea that presumes its reality to be the substantial and the true (*Phaedo* 109b–e). If the fish could rise into Stevens's "another sunlight," they would discover the other dimension, where the sunlight is an immanent sign that this dimension is genuine. Stevens imagines the discovery of "the artifice of a new reality" as concerning both the human being and its world: the "candle of another being," the human subjectivity that is always withdrawing in its revelation; and the constituted universe, the "sphere … at the elbow of Copernicus."[7] The other world might wonderfully fulfil the hope for lasting reality, and might turn out to be "a universe without life's limp and lack." But the activity in play is, as the title of the poem establishes, an imagination of a stage world that could be a new reality: "The curtains, when pulled, might show another whole." Stevens would agree with Heidegger that the imagined other dimension is not already known but possible.[8] Whereas Heidegger's metaphors of the horizon and the activity of withdrawing make urgent the philosophical mode of endeavor, requiring meticulous care in negotiating the circle of interpretation, Stevens's concern is for the activity, the route taken, itself. The idea of a perfected universe might be the "philosopher's end" (perhaps recognizable as the certainty for which Husserl or Plato searches), but Stevens's concern is for the life lived and the quality of that life: "What difference would it make, / So long as the mind, for once, fulfilled itself?" Implicitly, for Stevens the event in the theatre is that event of the *theatrum mundi* as I have approached it. The fulfilment lies in the activity of discovering "another whole," in the event of contemplation or in the theatrical moment rather than in the outcome of knowing a final truth or finding only another disappointment. Thus Stevens's ostensible hedonism grows serious in showing a kinship with what happens in Gadamer's version of the theatrical event.

Gadamer's argument for the vital activity in the theatrical event traces its origins in ancient festival and locates its essence in the activity of play. Engagement in the theatre, the activity of experiencing the play, can be thought of as the activity seen in Husserl's "intentional" acts, and involving Husserl's *epoché*, the "bracketing" of "living convictions." It is not inconsistent with the aesthetic motive in Stevens's question about "what difference" it would make "so long as the mind … fulfilled itself," stressing the fulfilling nature of the event. For Gadamer, however, it is important to see that the "mind," that subject "fulfilling itself," is not an isolated subjectivity willing, thinking, and judging but a being fully engaged in and interdependent with the presented world of the play. The *epoché* becomes in Gadamer's conception of the theatrical event a giving of oneself to the event rather than the momentary giving up or suspending the influence of a prejudice, conviction, or belief. The audience engages in the way the player in a game participates, playing intently and with interest so fully concentrated in the game that we might reverse the thought of the participating agents – rather than the player playing the game, the game plays the player.[9] The dramatic event is, as with children playing games, not played for anyone –

and an audience – but rather in total absorption. This wholeness, the "meaningful whole for the audience," is an "openness toward the spectator [that is] part of the closedness of the play" (*Truth & Method* 109). Thus the "play remains a game – i.e., it has the structure of a game, which is that of a closed world ... but the play itself is the whole, comprising players and spectators." In the audience, who fulfils and is fulfilled in the play, "the game is raised ... to its ideality."

This raising to "ideality" is essential to the event, a point Gadamer makes clear in his depiction of the "transformation into structure" achieved in the "change, in which human play comes to its true consummation in being art" (*Truth & Method* 110). When the audience engages in the play, as being part of the structure, what appears is the "pure appearance (*Erscheinung*) of what they are playing," and the audience is part of the idea. The presentation's ideas involve conceptions of coherence, wholeness, and purpose; participation and *bildung*;[10] the work as presentation and its appearing among the manifold of appearances. Idea is not transcendent otherness but the objective factuality caught up in the play of thought and illusion – illusion now to be thought of as positive, as the dream is in "having dreams." Plato "had dreams"; Martin Luther King had "a dream." The idea of the *theatrum mundi* invokes a dimensionality that involves the audience both as observant participant and as idea of possible or imagined audience within the represented world of the play. The play not only manifests that structure of a play-audience relationship, but also plays upon it thematically, bringing it to consciousness in, for instance, masks, photographs, disguises, pageants, doubling of characters, audiences on stage, backstage, or beyond. The audience's involvement in the presenting is an apprehension of objectivities, a going to the things or "facts" themselves, *zu den sachen selbst*, as imagined and imaginative intentional objects, and they are not reducible as being either in the subject or out in the "objective" world. As an idea, the manifold of *theatrum mundi* is an essence, and it is of eventfulness in the play as a play of eventfulness. Gadamer's "transformation into structure" involves the gestalt of possible *erscheinung* in the idea of play. As metaphor, the essence is transformational and mediational, anticipational and unfinished. It retains the quality of always being in motion that is essential in Schiller's idea of play: Schiller insists that when "we begin to play" the play is beauty and not "merely in earnest."[11] It remains, as play-drive, always in motion and irreducible to what it mediates, the temporally conditioned sense-drive and the ambitious ideality of the form-drive. As idea, it is a particular manifestation of the idea of possible other worlds.

The *theatrum mundi* metaphor, then, is both a rhetorical device and a mimetic affirmation of Gadamer's view of the theater. The metaphor and the stage itself bear implications and invite applications of an idea of the world as a generative and revisable constitution of dimensions. Instances of dramatic literature in which the idea materializes range from Euripides's *Bacchae* through Calderon de la Barca's *La Vida es Sueño*, Ben Jonson's *Bartholomew Fair*, Shakespeare's *The Tempest*, to the consummate modern instance, Luigi Pirandello's *Six Characters in Search of an Author*.[12] Instances in contemporary popular literature include *La Cage aux Folles* and *Victor Victoria* and extend to more recognizably "serious" plays such as Bernard Pomerance's *The Elephant Man*. A major American playwright, David Mamet, invokes the metaphor, as does his British contemporary, Harold Pinter. Mamet's plays demonstrate the structure essential to Gadamer's conception of the theater, in that the mimesis often involves the negotiation between dimensions that engage us in ideas of *bildung*, coherence and purpose, and the experiencing of the event as *erscheinung*. The plays *Cryptogram, The Water Engine, The Shawl*, and *The Old Neighborhood* might be suggested as instances that open up the fluctuating relationships between worlds and reveal the mystery and uncertainty of their making connections. The audience, engaged in this *jeu*, becomes, as in Gadamer's reading, a participant fully engaged in and reflecting these ideas.

If the play serves as a narrative in the vein of stories in Socrates's meadow (in the "Myth of Er," *Republic* 10.614b–21d) and the romance of Nathaniel Hawthorne, then the stories or representations of this idea of mediation within the play's action become reflective of the idea as represented by the play in relationship to the audience. This statement as I make it here becomes, then, a story that imagines the participants in the event as assuming and playing roles. The emerging idea of a "real" audience is not, however, an assertion about psychological, scientific, or philosophical truth in our world but, as an idea, aims to achieve a mimesis of the play's mimesis. The images of the idea of the *theatrum mundi* in Mamet's *The Water Engine*, for instance, the reflexive idea of the interrelationships of reflective worlds, is an appearance of this essential interrelatedness. Mamet's prefatory statement proposes this playful doubling of realities in the original production, with some scenes "played on mike, as actors presenting a radio drama," many "played off mike as in a traditional, realistic play. The result was a third reality, a scenic truth, which dealt with radio not as an electronic convenience, but as an expression of our need to create and to communicate and to explain – much like a chainletter" (p. 5; "chainletter" is a character in the play). This "scenic truth" elevates radio to the idea of communication, and it reflects the play's other images of creating a world's reality through telling about it. The play's own factual being, of course, reflects this duality concerning language and a world, in that it was "first written as a radio play" (8) and then as a stage play. This doubling of realities is an essence in the idea of the *theatrum mundi*, and it remains for us a metamorphosed, perhaps ironic or paradoxical, echo of the Platonic duality of worlds. Throughout the play the alternate realities are played off against each other, not toward resolution but toward the sustained and unresolved posing of the question of reality. The emergence of other worlds (and other standards, values, modes of understanding) becomes a proposition by the other world or agent for it, that it might be or should be heard, that it requires attention. From Aristotle's treatment of tragedy we derive the understanding that the play reflects how our world is supported by the principle of "probability or necessity." Yet Aristotle might lead us to observe that the mystery or wonder of the other dimension shines forth in its appearing. Those who have read Greek tragedy as being "about" the absolute determinative power of the gods would reduce the stage to the status of a counter pointing to the "true" world of the gods beyond the play. The liminality of the dramatic event opens up the relationship between probability and wonder. It is this sense of wonder that is an insistent appearance in the play of the *theatrum mundi*; it is part of the essence in Mamet's *The Water Engine* and appears as a major design in the fabric of Mamet's plays.

In *The Cryptogram*, for instance, the boy John becomes an agent of negotiations between worlds, and the outcome of his dialogues with his mother Donny and the family friend Del is indeterminate. As an agent, John delivers a viewpoint on the world as a stage or dream, questioning the reality of both the world and the text of the world: "I was looking at my *book*. I thought 'Maybe there's nothing *in* my book.' It talked about the *buildings*. Maybe there's nothing *in* the buildings. And ... or on my *globe*.... Maybe there's nothing on the thing that it is of.... Or in *history*.... Or *thought*" (53). John's questioning reality in terms of where we exist allows the conclusion that "we are a dream" (54), as he declares, but also the proposition that there might be a hell and that "maybe we are there." John's mobility of thought is instantiated in his physical mobility, his traversing the staircase that leads from the living room, where Donny and Del are trying to fathom the mystery of Robert's abandonment of his wife Donny and son John, to his room, to which John is sent to sleep. His sleeping, more precisely his moments of transiting, or vacillating between sleep and waking, generates a sense of the impingement of other worlds. His room becomes a moonlit space like

Hawthorne's place of romance, a place where realities interpenetrate or confront one another. John in his room "heard voices" (75) that might have been those of the two people downstairs or of others outside. Concluding that "someone's troubled," John reports, he "went outside" where, looking back in his room, he "saw a candle … burning there" (76). The room had been faintly and mysteriously lit by the candle, establishing it as the territory between the darkness outside and the light of reality downstairs, the "up from" and the "down toward." The hard reality with which his mother must cope downstairs is only troubled by John's excursions toward a fuller perspective, and in an atmosphere of questions and uncertainty, John declares at the end of the play, as he is sent to bed, that he hears "voices. They're calling to me" (100). For John, the reality of his visitors affirms his view of reality, the questionable nature of the dimension that declares itself to be the real. John's tenuous involvement in this world of actuality leaves him vulnerable to the intrusions, to the moments of vacillation in the discovery of some other dimension.

Although similar representations of opposed and transitional worlds are central and overt in a play like *The Shawl*, in other plays the movement into another dimension might be metaphorical, paradigmatic, or hypothetical. The character in *The Shawl*, also named John, who is mediator to the spiritual dimension, performs negotiations between the worlds for others, as a professional mystic and/or as a con artist. The outcome of his machinations remains problematical, since his creation of illusions might be understood not in terms of probability but in the posing of the question of illusion itself. Although he characterizes himself as creating what only "looks like magic" (21), as deluding a client whose mind becomes "*freed* by 'magic,'" he ends up in a thoroughly ambiguous relationship with her. Under her guidance at the end of the play he seems to be surprised by her leading him to see more and by her affirming that his seeing of appearances is beyond his control. Approached in terms of the idea of probability, the questioning whether the evoked images are seen or imagined, whether the mystery can be explained as his discovering a true identity, whether she has become, as manipulator, a mystic or a con artist, are questions that all remain unresolved in his closing statement, "That is all I saw" (53). A major outcome of this play, as in *The Cryptogram*, is the irresolution concerning the consequences of bringing opposed worlds together, leaving the impingement as a site for considering possibilities, a meadow for telling tales.

In *The Old Neighborhood*, the movements between worlds are as important as in *The Cryptogram* and *The Shawl*, but the characters' attentions are on the moment and its relationship to the past. After separating from his wife, Bobby Gould has returned home to Chicago from California, a literal journey, but the three short plays under this title reveal the characters obsessed with the past, with memories and imaginative recreation of events and characters. The journeys in the play's action are, then, imaginative, metaphoric journeys into the past: first, Bob and his childhood friend Joey consider their middle-age losses and disappointments and how if they had been in a different world in the past they might be now more fulfilled. In the second play, Bob is led by his sister Jolly in the exploration of their painful childhood, and they together lament the residual pain and dissatisfaction of this world, memories borne by them from the past. Although Bob shares with both Joey and Jolly moments of happy memories mixed with painful ones, with Deeny in the third play the world evoked is a delight of tranquility and communion. The two characters are aware of their world's deficiencies and their own inadequacies, but they share a moment of harmony and love. The past evoked seems to nourish their spirits, as they explore their imaginations and memories for images of sustaining value. The idea of a garden becomes the paradigm of the happy place they have achieved, at least momentarily and metaphorically. They seem to have a delight that has been realized or at least constituted as the idea of a nurturing place in their cultural past – in the pastoral retreat, in the view of Paradise offered to Dante by his guide Beatrice. In *The Old Neighborhood*, then, the literal journey home that Bobby makes does not form a dramatic structure based in probability or necessity and ending in a congruous resolution. The first two episodes seem to represent a descent into the pain and disappointment of retrieving the personal and cultural past, and the third seems improbably to be a recapturing of the potential joy of an imaginative past. An audience sharing these journeys into a past world, perhaps recalling comparable ideas of the past in personal terms or as memories of such journeys as Dante's, is offered, finally, the glowing figures of the third play in a poetic resolution. In the appearance of the numinous world of Deeny's garden, the real world's imperative for the reassurance of probabilities and understanding on its own terms is suspended.

A sustained look at *The Water Engine* might reveal the coherence and persistence of these ideas in the way worlds play off against each other, the way metaphor reiterates the alternate views, the way character expresses itself and identities resonate, the way the idea of resolution and harmony appears in the condition of strife and vulnerability. The central metaphor in the play is, of course, the water engine itself. It encompasses the ideas of both the practical, the real, the actual world – of nature, science, progress, and power – and the world of dreams – hope, transcendence, possibility, and peace. The water engine was invented by Charles Lang, a dreamer who dares to realize the super-natural, the improbable if not impossible, in the natural. It will bring to the real world a quiet, seemingly disembodied power, "like a sailboat.... There are no more factories" (22); and in imagining their new life "in the country," Charles's sister Rita declares that "we'll have no machines" (33). When Charles tries to patent the engine in order to share with the world, in order to make a better world in collaborating in the dream of Progress, he is assaulted by the powers of greed, the Patent official Gross and his lawyer accomplice Oberman, who evidently are the agents responsible, finally, for killing Lang and his sister Rita. The idea of the water engine, its metaphorical power, is repudiated by the world of will and strife. Metaphor as a ground of understanding reveals through analogy rather than analysis, bearing in itself, then, the ideas of uniting, reconciling, connecting, harmonizing, as opposed to separating, distinguishing, isolating. The irenic illusion championed by Lang appears to be accompanied by naiveté and an unpreparedness to contend with the machinations of men of power in the real world. The water engine's illusion of providing power to transform the world embodies that unitive function of metaphor, and Lang speaks for the reconciliation of principles of human nature in the natural world. "If we will think correctly," we do not have to "distinguish between inorganic and organic.... All things come from hydrogen. They all come from the earth. As we do. We are made of molecules. We all are made of light. We are the world in this respect" (11). The thinking that reads human beings as being made of both "earth" and "light," that makes all things one in the natural world, accepts as natural the unseen and benign energy of the engine whose power is like that of a sailboat.

The essentially irenic world so imagined is not merely an antithetical world contrasted to the actual, but a site of transition and mediation between worlds. The idea of opposed worlds becomes a negotiation between those opposites and, like metaphor and the stage, generates a middle ground as a meadow, as a transformative moment in the moonlight, as a temporal event reconciling opposites. The fair, the site of dreams and affirmations, becomes a magnetic force, a center, like Ben Jonson's Bartholomew Fair, attracting dreamers and manipulators, pleasure-seekers and con artists, and it becomes the site of possible transformation and realization. The metaphoric doublings of worlds promotes extrapolation of these worlds-within-worlds, inherently realizing the idea expressed as the play-within-a-play. The fair that is the "Century of Progress Exposition" (11), as an ostensible realization of the Idea of Progress, lies within and reflects the Idea that is Chicago, about whom singers sing as the play begins, "Chicago, great and free, / Turning all the world to thee, / Illinois, Illinois."[13] Illinois, then, epitomizes the Idea of the "New World," the site of Progress, as opposed to "Old Europe" (54–56). As Murray, the editor of the newspaper, speaks for the "Principles which made this country … great" (54), the orator in "Bughouse Square" (an alternative spokesman within the world) contradicts. Murray's rhetoric demeans "senile Europe," which has lost its "forgotten vigor," and exalts the "Golden … West," its "prosperity to come"; The Speaker at Bughouse Square inverts the values, talking of the America that "never existed. It was all but a myth. A great dream of avarice" (55). Such passionate definitions of reality presume to deal with the "facts," with the world as it is. Lang's dream, however, probes to the heart of reality in imagining "molecules" and "light" to coexist in his human beings, in his scheme of truth; and such voices as Murray's and the Speaker's must make room for other voices. The voice of the "Chainletter" serves not merely as an omniscient narrator announcing the facts: "In September, 1934, a young man in Chicago, Illinois designed and built an engine which used distilled water as its only fuel" (14–15). It speaks as well for the mysterious principle, sounding like the Delphic Oracle. It repeatedly explains the necessity and technique of ensuring good fortune, predicting bad fortune for those who have broken the chain. It can, with disinterest, report in the end that "One man saw the plans for a machine which he was told would run on water as its only fuel" (60); and it can enumerate the successes of those who accepted, and the failures of those who rejected, the mysterious power of the improbabilities of causation in the chain-letter phenomenon. This version of reality reflects and perhaps complements the improbable idea of a machine that runs on water, the dream of the young man who thinks that such realized power will transform the actual world. The chainletter sounds like a godlike voice affirming more than this natural, scientific world's accepted version of reality.

Other voices that clamor within the fair and the world outside it, range in attitude and function between these possibilities of speaking literally or metaphorically, stating the facts or implying alternate truths, dealing with their own world or some other. They might, among the other voices, confirm their own world's reality and truth (or their version of it); but their voices might undergo a change as well, an enlargement or enrichment of meaning, in their repetitions. The Barker, who announces and provides guidance for the journeyers to the fair, has a transformative and elevating function as well, exalting the idea of our Science, as represented in the "Hall of Science" (47), to grand illusions. It is both "the greatest force for Good and Evil we possess" and "the Concrete Poetry of Humankind" (60). Thus this voice, like others, raises its world to the realm of the ideal. It also, through repetition, grows ominous like other ominous voices, warning that "this is our last tour tonight." This idea of ending is repeated in the killing of Lang and his sister, in the histrionic outcomes of maintaining or breaking chains, in Europe's loss and debility, in the way dreams are shattered, in the very way that the actual becomes transformed, both in exaltation and in decay. The Barker speaks about the future not only in terms of Progress, reaffirming the message of the Fair, but also in terms of loss: in "dilapidated office buildings … in torn and filthy manuscripts," in "the vestiges of this and other cultures." As narratives of the world convert accomplishments into the "concrete Poetry of Humankind," such speaking might also become the disappointed or fictionalising voices that maintain the past: "Technological and Ethic masterpieces decay into folktales." The Barker's voice becomes then, as others can, a voice from a different perspective, perhaps *sub specie aeternitatis*. The final announcement that "The Fair is closing" (61) is a literal statement about this fair, this night, and it echoes toward endings and closings, toward other worlds, worlds within and worlds beyond, and towards the world of the audience.

As journeyer to the site of the Fair, the Play, the meadow, the audience might hear these cacophonous voices as representing conflict, the strife of the world, the reality of destruction, loss, and confusion. It might, on the other hand, engage in sympathy in the apprehension of something more than the assertions about a known world of conflict and decay, those representations of threats and vulnerability to them in a menacing world. It is allowed to hear, in the play, not only the ominous but also the benign voices. The voice announcing that the "Fair is closing" gives directions for exiting and expresses the "hope that you have had a good day at the Fair" (52). Benevolent voices that are not heard in the play, coming from beyond, have been agents in the action, Lang evidently having been inspired and sustained by the voice of a woman. His conviction that we are made not only of molecules but also of light has come, he reports, from a woman: "We are all made of light, she said" (52).[14] In the world of menace, dominated by those like Gross and Oberman, that truth seems to be an overwhelming cause for dismay, but that truth is alleviated by the spokespeople for an irenic world.

The audience might become representative of the pessimism and grow suspicious even of innocent and benevolent actions, as when, after Rita is abducted, Lang speaks on the telephone with the neighbor from upstairs, Mrs. Varek. Doubting that the police who came are the real police and suspecting that Mrs. Varek is an accomplice, the audience is reassured when she reveals, after Lang hangs up, that she too is controlled by Oberman and Gross. For the audience, then, interpretation must range between what seems certain because it is present and visible, the objective and "factual," and the mysterious, the hinted, the metaphor-driven. In the pursuit of a hermeneutic toward cogent understanding, the preconstructed version of reality – perhaps naive, perhaps idealistic, perhaps dualistic or monistic, perhaps pessimistic – is grounded in its world's reading. The intrusion of another, as in the *erscheinung* of the mysterious or magical into the comfortably known realm of the scientific or practical, instigates a potential "change of mind" – retrenchment, modification, revolution, etc. In the construction or reconstruction of a world, "Technological and Ethic masterpieces decay into folktales" (60). On the other hand, "The Second Hundred Years of Progress" become "The concrete poetry of Humankind" (11). The consequent new, mediated, or reaffirmed truths will continue to draw from images of the past, a reading of events (this play in particular) recalling images of the tragic, the pastoral, the ideal, the terrifying. The audience, then, in this event of interpreting and seeing or making connections, might, as Lang imagines, "just think ourselves from one place to another" (52).[15]

That thinking ourselves "from one place to another" seems like a fanciful, provocative expression of how we negotiate the ideas of worlds that are the constructs, the consequences, of our constituting or "making" worlds. Reflection on and extrapolation of the uncertain impingements of worlds of stage and audience, the seesaw of the two worlds, might create in that medial space a collaboration that amounts to Gadamer's "transformation into the structure" of ideality.

*

As representations of reality, both the stage and the theater-dream metaphor affirm the purposiveness and the tentativeness of the challenging conception of the world as one among worlds. Theater's transformation onto structure, which is mimetic of phenomenological apprehension of essential experience, presents the illusion of momentary appearances and fosters a sense of the tremulous shifting of solid ground, the presumed solid, substantial reality we know. Such an involvement in the appearances of the stage involves the audience as well in a gathering into a momentary unity that is Gadamer's "transformation into structure" of ideality. Yielding to the play of the event, which is the temporal "site" analogous to the site of Husserl's intentionality, the audience as mediating agent would become a principle for making connections.

The problem of mimesis, then, retains that old duality so troubling to Plato; and it invokes both the need for ontological definition and the problem of language as mimesis that troubled Socrates in the *Cratylus*. Mimesis as engagement in that transformation into structure, however, as the audience's being raised into the play of idea that preserves an essence of Husserl's ideas of *epoché* and intentionality, becomes a negotiation of the gap between dimensions, their abutment or their impingement. It calls for circumspection, for cautiously redefining the idea of mimesis itself, which is not represented in the model of the illusory set off firmly against the real, a stage representing another dimension that is the absent reality to which it points, but in the event of generative structure.

Florida Atlantic University

NOTES

1 Speculations about the *theatrum mundi* metaphor might well benefit from the phenomenological attempt to notice essences in objects of thought or perception. Underlying the particular characterization of these impinging dimensions of theater and world is perhaps a more essential opposition of worlds – or dimensions – by which the theater itself comes to be thought of as a world interdependent with the "real" world. Directing us to "possible other world" theories, Mihai Spariosu sets up this opposition of worlds wherein a world of strife and will is opposed to the "ludic," the "irenic," and the "liminal." Hans-Georg Gadamer insists on the "importance of defining play as a process that takes place 'in between'" (*Truth & Method* 109).

2 Gadamer admonishes that "we must not overlook the mimetic character of Plato's dialogues. We are dealing here with a poetic presentation, which should never be measured against a one-sided criterion of logical consistency" (*Dialogue & Dialectic* 21).

3 Hillman identifies himself with the literary critics Owen Barfield and Norman Brown as being of the "mafia of the metaphor to protect plain men from literalism" (149).

4 Both the site and its texts, the stories generated in relationship to it, point back and forth to other dimensions. The story told by Er reflects, on the one hand, the story told by Socrates, the character in Plato's text, and on the other, the stories told by the travellers who meet in the meadow (*Republic* 10.614d–e).

5 In the Preface to *The House of Seven Gables*, Hawthorne describes the world of romance as the moonlit world where, in the faint and transformative light, daytime objects take on another reality and become things of thought.

6 In the representation in which an actual world or character is confronted by an illusory one and in which the illusion demands the question of its reality, as in ghostly manifestations, the appearance presents the activity of withdrawal; the secure locus of the real is being withdrawn as the illusory attracts, pulls thought toward its elusiveness, its withdrawal. In Heidegger's *Was heisst Denken* (translatable as "what is called thinking?," and as "what calls for thinking?"), as Heidegger turns around the question [*Basic Writings* 359–61]), the event of the withdrawal "touches" the human being "in the surely mysterious way of escaping him by its withdrawal" (350). The withdrawal is, in Heidegger's thought and in a play's representation of ideas, an event in a moment, a temporality. Heidegger's account of our holding to what holds us in our essential

being recognizes memory as our means of holding and giving "thought precisely because it remains what must be thought about" (345); but since "what is to be thought about turns away from man, has turned away long ago" (348), that "event of withdrawal could be what is most present in all our present, and so infinitely exceed the actuality of everything actual" (350). The play's presenting the withdrawal of the illusory from the actual is the essence of the play's being; in the doubling of the *theatrum mundi* metaphor, the play sharply calls attention to the question of mimesis and the need for discovery: not a divorcement from, a denial or falsification of, the actual, but the opening of play with it. In the imaginative play, the freedom gained is not a release from the actual but a liberation toward discovery and transformation of what must always be returned to as the actual.

[7] Stevens's mention of Copernicus seems to echo Ralph Waldo Emerson's thoughts about Kepler ("Experience" 208). Characterizing the world of experience as the subject's kingdom, Emerson sees the kitten chasing her tail, an action wherein she is "surrounded with hundreds of figures performing complex dramas, with tragic and comic issues." The kitten's self-absorption is essentially parallel to "Kepler and the sphere." Emerson's question, "what imports it whether it is Kepler" or the kitten? seems an anticipation of Stevens's use of the theater metaphor. Emerson has begun this essay with the pertinent question, "Where do we find ourselves?" and answers it with our being as if midway up/down a metaphorical stairway, as bemused as if in an interim world.

[8] Thomas Hines sees Stevens's poetry as developing from the Husserlian toward the Heideggerian. James Leonard and Christine Wharton continue the discussion of phenomenological implications in Stevens's thought. Johnson and Wharton comment that this poem reveals what is, for Stevens, an essential truth: that what is important for human beings is "an enjoyment of the art of belief itself" (143).

[9] An essence of play "is reflected in playing: all playing as a being-played.... The game masters the players (*Truth & Method* 106)." ... "The purpose of the game is not really solving the task, but ordering and shaping the movement of the game itself" (107).

[10] Gadamer's *bildung* retains Husserl's phenomenological ground in that, as *bildung* achieves a coherence that transcends the personal, it never loses the concrete or the object of seeing – "In bildung," he reminds us, "there is bild" (*Truth & Method* 11).

[11] In the 14th letter, Schiller begins his analysis of the "play-drive" (or "instinct"), coming to a firm assertion in the 16th that, as operative, it remains inevitably in motion, never settling down on the side of the "form-drive" or the "sense/matter-drive."

[12] Jackson I. Cope's *Theater and Dream* has become a classic excursion into the Renaissance generation and elaboration of this metaphor.

[13] Mamet's playing off against each other worlds of the natural, material, and scientifically verifiable, opposed to the mysterious, supernatural, or improbable, reflects the Fair's idealizing and spiritualising of science. Robert Rydell observes how the "guidebook issued for the Hall of Science" promoted visitors' viewing the "mural in the Great Hall" as representing the culmination of scientific progress imbued with spiritual fulfillment: "Fairgoers were ... invited into a scientific Eden to discover for themselves that the 'unity of the basic sciences' rested on the experimental method, or 'common sense,' and to realize that 'there is nothing mysterious in science'" (*World of Fairs* 101–102). The rhetoric of this guidebook is in harmony with the voices calling to the Fair in Mamet's play; the mystical is incorporated into the dream of science.

[14] The identity of this woman is not clear, but the "she" could refer to Charles's sister Rita. An extended narrative about this play's truths might pursue this thought. The exalted possible or

potential images of human beings include, in Mamet's plays, the often shadowy or withdrawing woman who, when looked at again and closely, begins to take on the power of spirit and guide, to represent a perspective as an alternative to that of a "man's world." Analogous figures of women in the past include the poet's muse, the spiritual guide like Socrates's Diotima, Petrarch's Laura, and, as I have mentioned above, Dante's Beatrice.

[15] I am imagining an audience's involvement as derived, to cite the source of much discussion, from Aristotle's talk about pity and fear and catharsis – considering questions of affect and involvement, both individual or psychological and communal. I keep in mind Gadamer's warning that such involvement in the drama does not promote the idea of the audience as engaged in "aesthetic differentiation," that is, standing apart and judging the play, its production and acting, etc. (*Truth & Method* 85-87).

WORKS CITED

Cope, Jackson I. *The Theater and the Dream: From Metaphor to Form in Renaissance Drama.* Baltimore: Johns Hopkins UP, 1973.

Emerson, Ralph Waldo. *Essays and Poems.* Ed. G. F. Maine. London: Collins, 1954.

Gadamer, Hans-Georg. *Dialogue and Dialectic: Eight Hermeneutical Studies on Plato.* Trans. P. Christopher Smith. New Haven: Yale UP, 1980.

– *The Relevance of the Beautiful and Other Essays.* Trans. Nicholas Walker. Ed. Robert Bernasconi. Cambridge: Cambridge UP, 1986.

– *Truth and Method.* Trans. Joel Weinsheimer and Donald G. Marshall, 2nd edition. New York: Crossroad, 1989.

Hawthorne, Nathaniel. *The Centenary Edition of the Works.* Ed. William Charvat et al. 11 vols. Columbus: Ohio State University Press, 1962.

Heidegger, Martin. *Being and Time.* Trans. John Macquarrie & Edward Robinson. New York: Harper, 1962.

Hines, Thomas. *The Later Poetry of Wallace Stevens: Phenomenological Parallels with Husserl and Heidegger.* Lewisburg, PA: Bucknell UP, 1976.

Hillman, James. *Re-Visioning Psychology.* New York: Harper, 1992.

Husserl, Edmund. *Ideas: General Introduction to Pure Phenomenology.* New York: Collier-Macmillan, 1962.

Leonard, James S., and Christine E. Wharton. *The Fluent Mundo: Wallace Stevens and the Structure of Reality.* Athens: U of Georgia P, 1988.

Mamet, David. *The Cryptogram.* New York: Vintage-Random House, 1995.

– *The Old Neighborhood: Three Plays: The Disappearance of the Jews. Jolly. Deeny.* New York: Vintage-Random House, 1998.

– *Two Plays*: The Shawl *and* Prairie du Chien. New York: Grove, 1985.

– *The Water Engine. Two Plays:* The Water Engine: An American Fable, *and* Mr. Happiness. New York: Samuel French, n. d.

Plato. *Collected Dialogues.* Ed. Edith Hamilton and Huntington Cairns. Bollingen Series 71. Princeton: Princeton UP, 1961.

Rydell Robert W. *World of Fairs: The Century-of-Progress Expositions.* Chicago: U of Chicago P, 1993.

Schiller, Friedrich. *On the Aesthetic Education of Man in a Series of Letters.* Trans. Reginald Snell. New York: Ungar, 1965.

Spariosu, Mihai. *The Wreath of Wild Olive: Play, Liminality, and the Study of Literature*. Albany: State U of New York P, 1997.

Stevens, Wallace. "As at a Theatre." *The Palm at the End of the Mind*. New York: Vintage-Random House, 1972.

ELLEN J. BURNS

AN EXPLORATION OF POST-AESTHETIC ANALYSIS: W. A. MOZART'S *DIE ZAUBERFLÖTE* BY INGMAR BERGMAN*

Best known among Roman Ingarden's phenomenological investigations are *The Literary Work of Art*[1] and *The Cognition of the Literary Work of Art*,[2] in which he describes his theories on the structure of and hierarchies in literary art. His hermeneutical theory, as presented in the latter book, includes a more fully developed notion of the aesthetic attitude. In previous investigations, he described the aesthetic attitude as comprising two phases.[3] In *The Cognition of the Literary Work of Art*, his notion of the aesthetic attitude developed into a triptych containing pre-aesthetic, aesthetic, and post-aesthetic phases.

While Ingarden describes the first two phases at some length, the post-aesthetic is the least developed, as Ingarden considers it to be the least significant: "The indirect means of our disposal for finding out anything more exact about .... [other concretizations] ... often fail almost completely for various reasons, although they are not completely without value."[4] We do learn that once individuals achieve closure – that is, concretize – the work, we can then compare their descriptions with descriptions by others. The comparisons that Ingarden examines span the whole of human behavior and emotions as well as moments of emotional reaction occurring within his/her behavior. They also provide the point with which I shall argue:

> From the reader's behavior ... we can draw conclusions about certain details of the concretization he [or she] has constituted. The reader's emotional reactions, in particular, can be instructive in this respect.[5]

If the post-aesthetic inquiry is limited according to Ingarden's view, its significance would be unnecessarily restricted. In this schema, we would concern ourselves only with the emotional response of the individual experiencing single works.

The two tasks here, consequently, are wide-ranging: to develop Ingarden's post-aesthetic phase and to report findings from its application in analyzing Ingmar Bergman's cinematic setting (1975) of *Die Zauberflöte* (1791) by W. A. Mozart and Emanuel Schickaneder.[6] The development of the theory proposed here has been guided, in part, by the work of G. W. F. Hegel and Nelson Goodman.

A.-T. Tymieniecka (ed), Analecta Husserliana LXXIII, 129–144.

Hegel's threefold description of the acts of consciousness – described in his *Phenomenology of Spirit* – parallels Ingarden's first two attitudes and provides a point of departure for developing the concept of the post-aesthetic attitude. Ingarden suggests that during the pre-aesthetic or introductory phase, "we establish the actuality of this work and its details."[7] During this experience, we "'become acquainted with' the work, 'get to know it somehow.'"[8] In Hegel, this parallels the earliest activity – describing the object as the thing in itself – where consciousness both acknowledges the distance between itself and the yet-to-be experienced work, and initiates the first attempts to bridge the distance. The aesthetic phase is a complex process, starting after consciousness has bridged the distance and begins its interaction with the object. Here, conscious reflection begins: in Hegel, the thing takes on its existence for us. The aesthetic phase continues as the object is constituted in consciousness.[9] Paralleling Hegel's notion of the thing in-itself-for-us, the structure within consciousness is completed, and the process of knowing and the now-known object merge.

The post-aesthetic phase, comprising reflective cognition of the aesthetic concretization in Ingardanian theory,[10] could be described, by way of an extrapolation of Hegel, as the thing in-itself as it exists for others. Once the structure of the work has been completed in a percipient's consciousness, reporting the structure – in a discussion, review, scholarly article, or other artistic work – provides other objects for experience. The post-aesthetic investigation, consequently, returns to the beginning of the aesthetic cycle as one experiences "other" works stemming from the original.

Post-aesthetic study, consequently, would survey those works (and not simply emotions), evoked from scholars and artists whose responses – *qua* works – present other objects available to consciousness. The shift to experiencing these other works, moreover, transforms an initial aesthetic inquiry into one of culture history. Besides experiencing these new works following the three aesthetic phases, concretizations not only complete the structure of the object at hand within consciousness, but also study the historical implications of these "other" works. What had occurred in a microcosm – an individual consciousness – continues in a macrocosm, as we trace the original in the history of artistic culture according to the "other works" emanating from it.

"Other works" can be described via two categories, hermeneutic and expressive. Hermeneutic responses generate such interpretative works as exegesis and criticism that we would anticipate in such printed – or online – forms as scholarly articles and books. The second category includes works of literature, music, visual arts, and cinema, as artists respond to the original with "other" expressive creations. These categories, as we will see, are not mutually exclusive, as an expressive response can encompass hermeneutical aspects. In this case, an "other" artistic work would not only serve as a medium of expression for an artist, but also as his/her interpretation.

Continuing our development of Ingarden's post-aesthetic phase, we will also turn to Nelson Goodman's notion of "worldmaking."

In *Ways of Worldmaking*, Goodman argues that "the arts must be taken no less seriously than the sciences as modes of discovery, creation, and enlargement of knowledge."[11] His tack is to compare the scientific orientation of knowing the world with those "worlds" to be known as they are generated within consciousness. Consequently, Goodman suggests that "if worlds are as much made as found, so also knowing is as much remaking as reporting. All the processes of worldmaking ... enter into knowing."[12] His views on worldmaking, consequently, contribute to an explanation of the "other work" phenomenon vis-à-vis *Die Zauberflöte*.

Although consciousness responds to works of art by "building" structures *qua* worlds, these worlds, according to Goodman, are not static entities. Worlds within consciousness evolve as such forces as memory and new experiences subtract, add, or offer new perspectives.[13] When artists produce works – or worlds to be structured within consciousness – a universe forms as works mirroring the original multiply.

Such is the case of *Die Zauberflöte*, where the original has generated a literal galaxy of worlds. Both categories – hermeneutic and expressive – of "other works" are replete with works arising from Mozart and Schikaneder's only collaboration. These other works, moreover, span multiple genera: they include literature,[14] music,[15] visual arts,[16] and film.

Filmed opera is not a recent phenomenon, as thousands of video recordings provide hundreds of operatic titles.[17] While the bulk of video productions merely store staged productions via the audio-visual medium, an increasing number transcend mere documentation, and themselves constitute an artistic subgenre. I refer to these trends as opera *on* film and opera *as* film. Opera *on* film includes those documentary productions wherein the film director's impact is of little or no consequence. Reviews of this subgenre are analogous to reviews of sound recordings; publicity focuses on performers and conductors, not the film director. The director, in fact, serves more as a "performer" of the work than as an *auteur*, as his/her camera, passively sitting in the audience, provides a virtual seat in the theater.

In operas *as* film, on the other hand, a director's role extends past that of a performer to that of an expressive re-creator. The director takes full advantage of cinematic techniques and often interpolates his/her vision in "other" works of their creation. Such well-known and established directors as Hans-Josef von Syberberg, Franco Zeffirelli, Joseph Losey, and Peter Brook have plied their cinematic skills and visions in producing operas as film.

The only director to choose the enigmatic *Die Zauberflöte*, Bergman set the *Singspiel* as a film for television, not the movie theater. Commissioned for Swedish Radio's golden jubilee, it functioned as a visual "family greeting card"[18] to the Swedish nation. Bergman's 1975 production, however, was not his first expressive response to the opera. In 1968, he set Act I, scene 15, where Tamino realizes his unfulfillment in the face of Enlightenment, as a toy theater production included in *The Hour of the Wolf.*[19] According to interviews and other sources, Bergman was struck by Tamino's stark alienation. This scene subsequently influenced Bergman's expressive and interpretative treatment in his opera as film six years later.[20]

Bergman combines the roles of performer and creator. As performer, he faithfully renders the world of Mozart's work. In these sections of the opera, he not only preserves and enhances the contradictions among the major themes and character relationships in the opera, but he also plies his cinematic skills to heighten its effect. As creator, he asserts his artistic license in other sections of the opera, thereby creating a world of his own devising.

Mozart's two-act opera develops cyclic, dialectical paths as the major characters negotiate a world combining darkness and light, truth and lies, magic and reality, magnanimity and malignity. The first act, situated in the realm of the Queen of the Night, brims with misrepresentation and malevolence as she viciously pursues the power of day to add to her dominion of night. Neither laws of the natural world nor human nature deter her. Deftly manipulating deceit and magic, she thwarts reason and reality. Bergman cinematically accentuates her magic and deceit in Act I, scene 3, as Tamino contemplates the portrait of Pamina given to him by the Queen. Bergman does not overlook the slander with which she introduces Sarastro as Pamina's kidnapper, however, as the Queen steals a glance to see if Tamino falls for her false pretense.[21] The conjunction of world and image is also significant as she glances after the word "schwach" ("weak"), since her lust for power drives the plot of the opera.

The portrait, symbolizing the Queen's promise of Pamina's hand in exchange for the power of day and the overthrow of Pamina's alleged kidnapper, ensnares Tamino's devotion. In "Dies Bildniβ ist bezaubernd schön" (I,3: 3), the enraptured Tamino sings to a portrait that quickens in Bergman's cinematic hands. Tamino's first meeting and fast acquaintance with Pamina are symbolized by the successive shots that take Tamino – and us – closer to the portrait and the image therein. At first glance, Pamina appears in a static full shot located some distance from the camera. Bergman continues with a traveling shot toward the portrait. Once arriving at a closeup shot of the portrait, the magic begins. Pamina comes to life, and, the more Tamino studies her image and realizes his love for her, the closer she appears, through another series of graduated close-up shots within the quickened portrait.[22]

The magical portrait, however, does not always mirror happiness. As seen in the Queen's recitative and aria "O zittre nicht, mein lieber Sohn," the portrait also sounds an alarm. Monostatos, the single evil entity in Sarastro's realm, haunts Pamina, who appears in the closest and most intimate shot (I,6: 4.87–95, preceding "auf ewig mein"). Consequently, Tamino begins his rescue mission, certain of his love for Pamina and his hatred for Sarostro. By the end of the first act, however, the Queen's magic cannot hold Tamino's allegiance. Once he arrives in Sarastro's realm, he is forced to reconsider the world as reflected by the Queen.

The chiasma of the opera, in musical numbers 14 and 15, begins as Pamina also realizes her mother the Queen's true nature through the text, coloratura, and visual effects in the Queen's revenge aria, "Der Hölle Rache kocht in meinem Herzen" (I,8: 14). Pamina, believing her mother to be her rescuer, assumes that the Queen's entrance signals her liberation. The paradoxical truth, however, is that the Queen takes her daughter, supposedly Sarastro's hostage, as her own hostage, storming the scene with a murderous command. Bergman visualizes the Queen's threat of breaking the bonds of familial nature as she threateningly encircles her terrorized daughter. Paralleling the aural coloratura of this most famous of arias, Bergman visually underscores her madness through several cinematic techniques. Through the editing techniques of eyeline matches, the sight of the malevolent Queen becomes irresistible. Pamina's terrified reactions to the sight of the Queen both initiate the series of eyeline matches and provoke our anticipation of seeing what she has seen. Bergman manipulates our anticipation, however, as he delays the completion of the matches.

Complementing the malicious characterization of the Queen with his *mise en scène*, Bergman mediates her "motherly embrace" with the knife intended for Sarastro's murder. It, along with Mozart's pyrotechnic coloratura, emphasizes the passages when the Queen threatens to break the mother-daughter bond: "so bist du meine Tochter nimmermehr" (II,8: 14.17–43), and "Verstoβer sei auf ewig, verlassen sei auf ewig, zertrümmert sei auf ewig, alle Bande der Natur" (mm. 53–63). The delayed eyeline match recurs as the Queen transforms into a ghastly persona with a pasty white face reflecting the coldness of the night's blue light. Physical confrontation and maternal threats complete the Queen's taking of her daughter as an emotional hostage in her war against Sarastro. While cinematically "performing" the scene revealing the Queen's true nature, Bergman also adds a dimension of re-creation.[23] While Mozart and Schikaneder called for Pamina to see the Queen's true nature in reality, Bergman has set it as a nightmare, since he intended his production for audiences that included entire families.

The chiasma of the opera continues as Sarastro immediately follows and contrasts the expression of the Queen's evil nature. Sarastro, in whose realm magic is conspicuously absent, comforts the shocked Pamina in number 15, "In diesen heil'gen Hallen." While the Queen's scene is steeped in a stark blue night, Sarastro's scene glows with warm colors. Holding Pamina in a fatherly embrace throughout the first verse of his aria,[24] Sarastro not only comforts Pamina, but establishes his true nature, which, like that of the Queen, is the antithesis of what Pamina had originally believed. Paralleling his adaptation of the Queen's scene, Bergman sets Sarastro's scene as a lullaby to calm Pamina after her nightmare.[25]

Despite their settings as nightmare and lullaby, Bergman's "performance" of the arias epitomizes the chiasmatic relationship between the two rulers. The Queen screeches for infernal revenge; Sarastro murmurs about forgiveness and redemption. The chasm is evident not only in the texts but also in the musical medium, as the Queen's highest note of the opera juxtaposes Sarastro's lowest note. Diverging distances are not only significant in the musical score or between the characters, but also between the scene and the audience. While the Queen is generally more distant from Pamina than is Sarastro, her scene proceeds as a number of close-ups. The close-ups intensify both the menace of the Queen's wrath and Pamina's horror-stricken reactions. In a word, the horror and terror of the scene are "in our face."

Distances in the next scene are distinctly different. Sarastro either embraces or is within arm's reach of Pamina. Bergman varies the distances between the audience and the scene, moreover. Close-ups, which had previously pulled us into the scene witnessing the Queen's malignity and Pamina's terror, now reflect Sarastro's and Pamina's calm. The variety in medium and close-up shots, moreover, provides the audience relief from the intense and constant close-ups of the previous scene.

While cinematic magic and reality differentiate the Queen from Sarastro in Tamino's and Pamina's eyes, Bergman – following Mozart's stage directions – manipulates light to describe the triangle of Tamino with the Queen and Sarastro. Tamino's encounter with the Queen begins in light (I,6: 4). The darkness seems only natural as the Queen arrives at nightfall. At the conclusion of her aria, however, light abruptly returns. The shifts in lighting work their magic on a befuddled Tamino, who wonders whether he is awake or asleep (I,7).

Tamino is equally confounded during his first confrontation in the reputedly evil realm of Sarastro. Unlike his encounter with the Queen, which sequences from light to dark and back to light, Tamino's encounter with the Speaker, Sarastro's representative, reverses the schema. The Speaker's lamp, illuminating a space brimming with books, pierces the darkness during his enigmatic conversation with Tamino. Just before extinguishing the symbolic lamp (I,15: 8.81–4, "Veilleicht find'ich ... hier"), the Speaker tells Tamino that he will not know the whole truth until he joins the realm. In addition to employing the engulfing darkness, Bergman heightens Tamino's despair through sound. An echo during the last lines of the scene – "O ew'ge Nacht! Wann wirst du schwinden?" – emphasizes the vast emptiness that haunts Tamino and intrigued Bergman (I,15: 8.141–3).

In the hands of Bergman *qua* performer, Enlightenment clichés are revitalized. He generates the world of Mozart's design as a dynamic field in which the eighteenth-century affirmations of light and reason transcend the ritualised concept of "Enlightenment." Those enchanted with Bergman's cinematic treatment of Mozart's work thus far might expect a similar treatment of other characters and narrative elements, but such an expectation would remain unfulfilled. As will be seen, Bergman interpolates his own vision of a world intended to greet the Swedish nation.

Bergman's "re-vision" shortens, rearranges, and inserts new material into the original. His cuts include three musical numbers[26] and text from several spoken scenes, resulting in an opera 30% shorter than the original.[27] The greatest degree of artistic license is evident in the architectonic changes in the final quarter of the opera. Bergman deletes musical number 16, the point at which Tamino's and Papageno's paths bifurcate. After that point in the original, the plot falls into two major segments addressing three subplots occurring in cyclical patterns. As each subplot is resolved, the opera achieves "stipulative" finales: Pamina and Tamino's triumph (II,29: 21.412),[28]

Papagena and Papageno's union (II,29: 21.744),[29] and the final finale, celebrating the Queen's demise and Pamina and Tamino's ascent to power.[30] The events in the world of Mozart's work are driven by the continuation of the trials.

As seen in the table below, Bergman alters and compacts the time line. In the world of his design, events are motivated by the couples' desire for union,[31] and the rearranged scenes simplify parallels between characters. In Mozart, for example, Pamina and Tamino's union precedes that of Papageno and Papagena. In Bergman, Papageno and Papagena are united first. He makes parallels more obvious, as seen in scene 17 where Tamino plays his flute during their second trial. Before Tamino's flute solo, Bergman inserts Papageno's second aria (II,19: 20), producing a temporally linear parallel between the characters. Bergman also "linearizes" Pamina's (scene 26) and Papageno's (scene 29) suicide attempts as they occur one after the other. In the world presented by Mozart, parallels occur subtly in cyclical patterns. Bergman makes these lines plainer as they appear in seesaw succession.

Bergman begins the second segment with two inserts: a *dreimalige accord*, three statements of a chord sounded by the low brass, followed by an invented scene in which a priest advises Pamina about Tamino and gives her the flute to deliver to him. He rescripts the prayer that follows (II,20: 18), originally addressing the gods on Tamino's behalf, as a hymn to brotherhood. Bergman, consequently, "linearizes" the plot by compartmentalizing the events of the plot rather than following their dialectical progression to resolution, and shifts the focus from Mozart's milieu to modern Sweden.[32]

Within the now linear superstructure of Bergman's opera as film, other changes resonate with its function as an inspirational address on brotherhood, art, and virtue to the Swedish nation. Taking advantage of the adages, or *Sprüche*, as written by Mozart and Schikaneder, Bergman highlights the pithy messages both by having performers step out of character to directly address the audience and by displaying the texts on placards.[33] Bergman also invents adages by rescripting existing lines. Before Tamino meets the Speaker, for example, he alludes to the virtues of leadership in an Enlightened society. Bergman substitutes, "Where art is protected and beauty may dwell, the people are happy, the master rules well"[34] for "Wo Tätigkeit thronet und Müßiggang weicht, erhält seine Herrschaft das Laster nicht leicht" (I,15: 8.51–6). Facing the camera, Tamino extols the value of art in society to the nation.

Bergman's artistic license extends to his characterization of Papageno, his relationship with Monostatos, and the characterization of Pamina. In addition

COMPARISON OF PLOT DEVELOPMENT
Act II, scenes 13–30

Key to abbreviations: QN = Queen of the Night, Pa = Pamina, Ppa = Papagena, Ppo = Papageno, To = Tamino

Mozart			Bergman*		
Act, scene	Musical no., mm.	Plot	Scene	Musical no., mm.	Plot
II, 13		Priests RE 2nd trial	II, 13		Priests RE 2nd trial
II, 14		Ppo fears the darkness in trial	II, 14		Ppo fears darkness in trial
II, 15		Ppa enters in disguise	II, 23	20	Ppo yearns for Ppa
II, 16	16	3 Boys' advice	II, 15/ II, 24		Ppa offers water; shows true appearance; disappears
II, 17		To plays flute, Pa enters	II, 25		Ppo runs after Ppa
II, 18	17	To seemingly rejects Pamina	II, 17		To plays flute; Pa enters
II, 19		Ppo brags about his silence	II, 18	17	To seemingly rejects Pa
II, 20	18	Priests' prayer to gods RE To	II, 26	21, 1–44	3 Boys anticipate new dawn
II, 21	19	2nd trial ends; Pa sees To before 3rd trial	II, 27	21, 45–189	Boys intervene in Pa's suicide attempt
II, 22		Ppo lost, calls for To	II, 29	21, 412–744	Ppo's suicide; united with Ppa
II, 23	20	Ppo fails test; yearns for Ppa	inserts accord		
II, 24		Ppa enters without disguise	inserts invented scene with Priest and Pa		
II, 25		Priest separates the couple	II, 20	18	Priests' prayer to brotherhood
II, 26	21, 1–44	3 Boys anticipate new dawn	II, 21		intro to 2nd trial used for 3rd
II, 27	21, 45–189	Boys intervene in Pa's suicide attempt	II, 28	21, 190–412	Pa and To pass final trial; celebration
II, 28	21, 190–412	Pa to To pass final trial; celebration	II, 30	21, 745–920	QN vanished, realm celebrates
II, 29	21, 412–744	Ppo's suicide attempt; united with Ppa			
II, 30	21, 745–920	QN vanquished, realm celebrates			

*Bergman combines the 15th and 24th scenes, adapts the 21st scene and deletes scenes 19 and 22.

to his primary quest of uniting with Papagena, a more subtle issue for Papageno is the dialectical tension between his human and bird-like qualities.[35] While he neither achieves Enlightenment nor transcends his avian persona, Papageno does achieve a humble degree of maturation, making him worthy of Papagena. Bergman obscures his development, however, by cutting much of the spoken dialogue that describes and characterizes the simplicity from which he modestly develops. Cut from the opera, for example, is his geographical ignorance. Besides knowing little, Papageno talks too much. Papageno, in fact, either speaks or sings more than any of the other major characters in the work.[36] Most of the 30% cut from Bergman's opera as film, in fact, consists of Papageno's logorrhoea.

Bergman's interference with the opposition between Papageno and Monostatos – ostensibly to make the opera more intelligible to children – is paradoxical. Mozart and Schikaneder called for Monostatos and Papageno – the most visually unique of the characters – to mirror each other in horror. Their mutual terror is clear in the original as they shrink away from each other, muttering, "Das ist der Teufel sicherlich!" (I,12: 6.53–71). Changing their mirrored – and therefore equal – actions, Bergman has Monostatos threaten and terrify a cowering Papageno.[37]

The most significant and ironic change in characterizations, however, is that of Pamina. Bergman, agreeing with recent feminist criticism, was convinced that the work was misogynist, as he saw in it "a disdain for women ... which is difficult to cope with.... To get around the disdain ... we have changed the text a bit...."[38] While his intent may have been antimisogynist, Ruth Coser argues that Bergman actually imposed a patriarchal solution that belittles Pamina's significance. By cutting the spoken material establishing the Queen as the widow of the Ruler of the Day,[39] Bergman follows the *faux* narrative,[40] casting Sarastro as the Queen's husband and father of Pamina.[41]

When Bergman "changed the text a bit" to lessen the perceived misogyny, he actually did the issue a disservice. The overall problem is not simply whether the solution should have been matriarchal or patriarchal, but the belittling of Pamina's existential plight. Inspired by Tamino's "existential moment" with the Speaker, Bergman, who had already incorporated the scene in his earlier *Hour of the Wolf*, emphasized Tamino's trials at the expense of Pamina's. Focusing his, and, by extension, our, existential eye on Tamino in his opera as film, he blights Pamina's existential plight.

Mozart's Pamina is no heroine "side kick" designed to applaud Tamino's heroics.[42] Her trials and development are as treacherous and fearsome as those of her male counterparts. Pamina's journey, in fact, is more classically existential, as conceived by Jean-Paul Sartre, than that of Tamino or

Papageno. Tamino and Papageno are conscious of their tests, and are repeatedly coached by Sarastro's representatives. Pamina, however, undergoes unannounced trials in everyday life and has little coaching. Awash in terrifying circumstances beyond her control, Pamina looks to herself for the resolution upon which to act. She neither gives up, nor looks to an external authority – the Priests, the Boys, the Ladies – for the answers that she eventually uncovers for herself.

Successfully negotiating the twists and turns in her trials, Pamina is initiated into the realm. She succeeds not simply because she is a woman, however, but because she is worthy: "Ein Weib, das Nacht und Tod nicht scheut, ist würdig, und wird eingeweiht."[43] For the audience of Bergman's vision, on the other hand, Pamina receives instruction from the Priest before the last trial.[44] In this "feministic" revision, moreover, Pamina becomes a mere pawn in a domestic dispute between Sarastro and the Queen, whom Bergman has cast as her father and mother. One could speculate that Pamina's simplicity accommodates Bergman's treatment of the work as a fairy tale. To protect the children of the audience, Bergman protects Pamina from the responsibilities of adulthood. Rather than relying on herself, she acts out the role of the dutiful daughter yielding to her father's wishes, and everyone lives happily ever after. A happy ending, presumably, for a happy nation.

While Bergman highlighted the obvious dramatic tensions between major characters in his role as performer, his changes as *auteur* distort the deeper, more subtle dialectical relationships and progressions in the opera. Consequently, the dialectical relationship between the influence of the Enlightenment on Hegel and that of Hegel on the Enlightenment (available in the opera'a depth as originally expressed by Mozart and Schikaneder), went to naught in Bergman's version of the opera story. On the other hand, Bergman was not only sensitive to the inspiring message that *Die Zauberflöte* shared with its audiences, but he also saw it as a medium to enlighten his own audience. The tension here, should we permit it, is one of ownership.

Bergman's expressive and hermeneutical response mirrors aspects of the opera's "Truth" and sets *Die Zauberflöte* as an icon in the making. As a performer, Bergman preserved the original nature of the opera; as an interpreter, he "published" his own vision of the work to address the nation. Bergman disseminates his message of hope, however, neither on printed pages of scholarly publications nor through lectures, but on celluloid.

Whose *Magic Flute* is it? Simply determining the "quality" of Bergman's work by judging its resemblance to the original assumes that Mozart's work is an absolute from which no deviation should be made. This world would suffer no change. In this view, the original becomes a canon, a "Truth," or an absolute. But Goodman reminds us that contemporary thought shifted *from* the assumption of a "unique truth and a world fixed and found *to* a diversity of right and even conflicting versions or worlds in the making."[45] Following Goodman's reasoning, the original would not exist in culture as an absolute or "Truth," but would transcend to the level of a cultural icon imbuing life into other worlds of experience. The *Flute* is ours.

Our post-aesthetic inquiry has proceeded through an aesthetic analysis of an "other" work stemming from Mozart's and Schikaneder's opera. In it, we have considered a synchronic moment in the history of artistic culture as the now opera-as-icon reflects, in varying degrees, the original. The logical continuation of this study requires opening the temporal lens to a diachronic study of the artistic, literary, and musical works summarized above. Continuing this investigation would entail a macrocosmic survey of works "about" *Die Zauberflöte*, as well as works about the works about *Die Zauberflöte*. The mirroring of the multiplicity of these works, consequently, generates a kaleidoscopic universe. Contained therein are works – performances, scores, and the like – canonically mirroring the original, as well as others stimulating the evolution of this cultural icon's universe.

As an icon, consequently, the title *Die Zauberflöte* would evoke not a single work but an evolving universe. That universe, according to phenomenological theory, is composed of consciousness-achieving structures ordered by the works considered. Upon achieving closure, according to Hegel, the boundary between consciousness and the object disappears. Not only would consciousness evolve as it encounters works evolving from *Die Zauberflöte*, but as the boundary disappears, consciousness becomes fully aware of itself as it achieves a final – albeit stipulative – structure. Transposing Goodman into a popularised mode of discourse, and concluding on a lighter note, opening ourselves to these worlds results not only in our "knowing large," but "living large" as well.

Schenectady, NY

NOTES

* I am happily indebted to Professors E. F. Kaelin and David Levin for their advice and assistance in the preparation of this essay.

1 *The Literary Work of Art: An Investigation on the Borderline of Ontology, Logic, and the Theory of Literature*, trans. George G. Grabowitz (Evanston, IL: Northwestern University Press, 1973).

[2] *The Cognition of the Literary Work of Art*, trans. Ruth Ann Crowley and Kenneth R. Olson (Evanston, IL: Northwestern University Press, 1973).

[3] These two phases roughly parallel the pre-aesthetic and aesthetic phases. The evolution of Ingarden's work on the aesthetic attitude is described by Bohdan Dziemidok in "Roman Ingarden's View on the Aesthetic Attitude," in *Roman Ingarden and Contemporary Polish Aesthetics*, ed. Pietr Graff and Staw Krzemién-Ojak (Warsaw: Polish Scientific Publishers, 1975), pp. 9–31.

[4] Ingarden, *The Cognition of the Literary Work of Art*, p. 412. Dziemidok's survey, moreover, does not mention the post-aesthetic attitude.

[5] Ibid.

[6] Ingmar Bergman's *The Magic Flute*, sung in Swedish with English subtitles, videodisc (Bel Canto Paramount Home Video, 1986).

[7] Ingarden, *Cognition of the Literary Work of Art*, p. 10.

[8] Ingarden, *op. cit.*, p. 6.

[9] Ingarden, *op. cit.*, xviii.

[10] Ibid.

[11] Nelson Goodman, *Ways of Worldmaking* (Indianapolis, IN: Hackett, 195), p. 102.

[12] Goodman, *op. cit.*, p. 22.

[13] Goodman provides a classification of the processes – composition, decomposition, ordering, deletion, supplementation and deformation – observed in the physical world as well as the arts (*op. cit.*, pp. 7–17).

[14] Literary works stemming from *Die Zauberflöte* include Paul Adler, *Die Zauberflöte* (Dresden, 1916), G. Lowes Dickinson, *The Magic Flute: A Fantasia* (London: George Allen & Unwin, 1920), and Alan Spence, *The Magic Flute* (Edinburgh: Cannongate, 1990).

[15] Musical works encompass vocal and instrumental works. This literature includes the sequel operas *Sarastro: Music-Drama in drei Aufzügen* (by Karl Eduard Geopfart, Leipzig, 1891), *Die Zauberflöte: Zweiter Teil* (by Martin Vogel, Bonn, 1990), and *Die Zauberflöte, Zweiter Teil: Das Labrinth, oder Der Kampf mit den Elementen* (by Peter von Winter, Bonn, 1798). Instrumental settings include transcriptions of the complete opera for the piano, as well as more than 200 transcriptions by such composers as Beethoven, Glinka, Liszt, Sarasate, and Sor. These adaptations have been scored for practically all orchestral instruments, as well as guitar and carillon. Transcriptions have also been published for such ensembles as brass quintets, wind sextets, orchestra, and concert band. For a compilation, see my "The Dialectical Structure of *Die Zauberflöte*: A Phenomenological Analysis" (Ph.D. dissertation, Florida State University, 1994), pp. 675–704.

[16] In addition to paintings of scenes in the opera found in Die Gesellschaft der Musikfreunde (Vienna), "re-visions" of the opera include *Randzeichnungen* by Max Slevogt (Edenkoben: Max Slavogt Gallerie, 1985), and a comic book illustrated by P. Craig Russell, *Wolfgang Amadeus Mozart's* The Magic Flute (Forestville, CA: Eclipse Books, 1990).

[17] For studies of the combination of opera and film, see Jeremy Tambling, *Opera, Ideology, and Film* (New York: St. Martin's Press, 1987) and Marcia J. Citron, *Opera on Screen* (New Haven: Yale University Press, in press).

[18] As reported by Peter Branscombe in *W. A. Mozart: Die Zauberflöte*, Cambridge Opera Handbooks (Cambridge: Cambridge University Press), p. 177, and Jeremy Tambling in *Opera, Ideology, and Film* (New York: St. Martin's Press, 1987), pp. 126–39.

[19] Bergman uses the Mozart-Schikaneder plot in a dialectical relationship with his protagonist, Johann. As the Tamino character, a "small abject character at the mercy of those who manipulate

him," travels from darkness to light, Johann travels an opposite path (Peter Cowie, *Ingmar Bergman: A Critical Biography* [New York: Scribner's Sons, 1982], p. 246).

[20] See Paisley Livingston, *Ingmar Bergman and the Rituals of Art* (Ithaca: Cornell University Press, 1982), pp. 233–35.

[21] Passages from the opera are identified by act, scene, musical number, and measure numbers. To correlate the discussion and film, the citation includes the first and last words of the text. For this passage, see Act I, scene 6, musical number 4, measures 67–68, occurring after the word "schwach" (I,6: 4.67–68, "schwach").

[22] The excerpts illustrating the magical portrait occur in I,3: 3.5 ("wie … gesehen"), 12–17 ("mein Herz … Regnung fühlt"), 24–26 ("soll … Liebe sein"), 34–37 ("O wenn ich … könnte"), and 40–44 ("ich würde … würde ich").

[23] Coser reports from an unpublished shooting script of the film that Bergman set the aria as a nightmare rather than reality to preserve the fairy tale nature of the work to which he was committed (Rose Lamb Corser, "The Principle of Patriarchy: *The Magic Flute,*" *Signs: Journal of Women in Culture and Society*, Vol. 4, no. 2 [Winter 1978]: 344).

[24] See II,12: 15.1–23 ("In diese … besser Land").

[25] This adaptation, however, creates a problem in continuity. If, as Bergman indicates, the Queen's revenge aria is a dream, then the knife that she gives Pamina does not exist. That it exists in the following aria is illogical.

[26] Numbers 11, 16, and 19.

[27] This is not new, however, as the libretto has undergone adaptations and cuts since its 1791 premiere. See my "The Dialectical Structure of W. A. Mozart's *Die Zauberflöte*," pp. 1–5 (note 15, *supra*).

[28] In C major.

[29] In G major.

[30] In Eb major.

[31] Cineaste Paisley Livingston reports that Bergman subscribed to the theory, held in many musicological accounts, that the libretto is flawed and that Mozart and Schikaneder radically changed the plot midway through its composition. Bergman reportedly incorporated changes to remove perceived inconsistencies (Livingston, *op. cit.*, pp. 237–42).

[32] These changes not only distort the cyclical narrative of the original, but they also destroy the important harmonic sequence that provides a sonorous foundation serving to unify the opera, the description of which is beyond the scope of this paper.

[33] *Sprüche* cover such topics as truth (I,7: 5.54–77), love (I,14: 7.18–34), and harmony (I,17: 8.327–50).

[34] The text provided here, of course, is an English translation of Bergman's Swedish text.

[35] The specific use of *Mensch* and *Mann* distinguishes between Papageno's male gender and his maturation as a man. His lack of development is clear as the priest addresses him as *Mensch* after failing the second trial (II, 23).

[36] A word count of the opera shows that Papageno dominates the soundscape as he speaks or sings 43.9% of the words in the work. He, however, is not frequently addressed (20.5%), and while he is his own favorite subject, as he refers to himself 68.8% of the time, he is rarely the topic of conversation (9%). A majority of productions obscure these tendencies by cutting the bulk of spoken material.

[37] The confusion is heightened, moreover, by the English subtitles, in which a terrified Papageno addresses Monostatos as "friend," not "fiend."

[38] Bergman as quoted by Coser, *op. cit.*, p. 338.

[39] See Act II, scene 8.

[40] The *faux* narrative in which Sarastro and the Queen are married occurs in some productions of the opera, as seen in the 1993 production by the Santa Fe Opera, as well as "other" works: *Opera, or the Undoing of Women*, by Catherine Clément, trans. Betsy Wing (Minneapolis: University of Minnesota Press, 1988); a comic book setting, *The Magic Flute*, in 3 volumes by P. Craig Russell and Patrick Mason (Forestville, CA: Eclipse Books, 1990); and *The Magic Flute: A Fantasia*, by G. Lowes Dickinson (London: George Allen & Unwin, 1920).

[41] Coser, *op. cit.*, p. 343.

[42] Judith Eckelmeyer initiates this line of thought in *The Cultural Context of Mozart's Magic Flute: Social, Aesthetic, Philosophical*, Vol. 1 (Lewiston, NY: E. Mellen Press, 1991), p. 23.

[43] See II, 28: 21.267–74.

[44] Between scenes 29 and 20 in Bergman's arrangement.

[45] Goodman, *op. cit.*, p. x, emphasis mine.

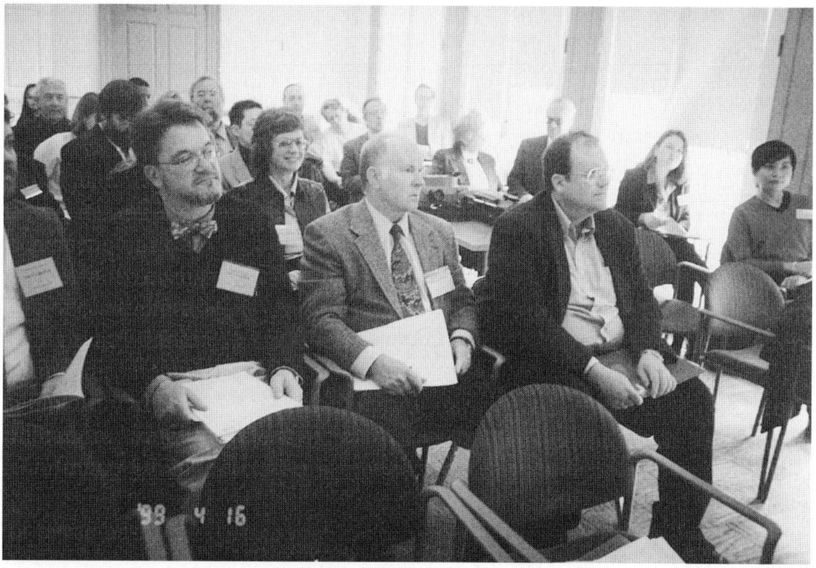

The audience during a lecture (in front, Gary Backhaus, Lawrence Kimmel, W. Melaney).

GARY BACKHAUS

THE FEEL OF THE FLESH: TOWARDS AN ONTOLOGY OF MUSIC

INTRODUCTION

The purpose of this article is to conduct a phenomenological inquiry into what musicians call "the musical feel." As a musician I "intuit" when a music performance "feels right" and when it does not; but upon thinking about this I realized that I cannot conceptualize what I mean when I make reference to the "feel" nor can I "point to" something specific that I mean. Is the feel of music, then, musical mysticism or metaphysical vacuity? As a musician I conceptualize the melodic modalities, harmonies, time signatures, rhythms, pitch, timbre, orchestration, etc., and as a former music educator I have critiqued performances for intonation problems, rushing, etc. But the feel of the music appears to be that "something, I know not what" that founds all of these components. So, this article presents a phenomenological investigation of "the feel" of music in order to uncover its significance.

The raising of this philosophical problem about music provides important insights concerning our fundamental relation with the world, including the relation with Others with whom we do not share conceptual meanings. The source for the notion of the universality of music is exhibited in the feel, for the lived-body participates in the feel of music without the need for understanding, that is, the feel involves the pre-conscious attunement of the lived-body. The source for the diversity of the vast literature of music is in the way it sounds to conscious acts of audition, for perceptual acts constitute content about which the understanding must learn. The understanding must make sense of the plethora of musical styles from the great variety of cultures and sub-cultures. Thus, there is unity-within-difference in the experience of music. This point is corroborated by Mikel Dufrenne, who states, "It is for the body that unity is given before diversity" (1973: 339). The great diversity of musical styles finds an original basis of unity in the feel of the music, for prior to conceptual distinctions, the lived-body has already surrendered to the feel of music. The lived-body attunes to (feels) the music across cultural diversity through its fundamental relation to a phenomenal field from which embodiment gains its qualitative meaning. The music of a people may sound strange, "Other-to-me," but my lived-body has already communed with their music by taking up forms of behavior in the body schema. The culture I may

A.-T. Tymieniecka (ed), Analecta Husserliana LXXIII, 145–170.

not understand, but the lived-body has already opened to the "foreign" music (a primordial *mit-sein*) through an ontological source that claims, one and all, what Maurice Merleau-Ponty calls "Flesh." So, the feel of the music reveals to us something significant of our fundamental relation to the world-originator, Flesh, which is the primordial source for the fundamental meanings of all music.

A phenomenology of the feel of music attempts to apprehend the foundational structures in the experience of music, which begins through an examination of the intentional structure of the lived-body. The experience of music unfolds through the *a priori* meanings that emerge out of the morphological structure of the lived-body and its patterns of behavior. The placial qualities of the lived-body consist of a formed relation to the phenomenal field of music. Cognates and neologisms of the word "place" are employed rather than "space," because the spatiality addressed is inherently qualitative. The feel of music involves somaesthetic, visceral, and kinaesthetic components that are qualitatively placial. Phenomenological psychologist Erwin W. Straus promotes the fundamental doctrine that "all perceptual experience is controlled by modes of the spatial as basic forms. We respond to distinctions in the pathic sphere" ["pathic sphere" is Straus' word for "the feel"] (1966: 19). But the lived-body's placiality emerges concomitantly with the placiality of a primordial environing field whereby what is meant by "the feel" becomes a component of the ontology of the Flesh, which is the non-dualistic foundation that provides the structural ground for the meaning-genesis in the experience of music. So, the dehiscence of Being gives rise to the "theatre" of life on the "stage" of the world. The feel of music is traced back to the fundamental structures of Flesh, which are prior to the conceptual distinctions of subjectivity and objectivity.

CONCEPTUAL MEANINGS VERSUS THE EXPRESSIVE MEANING OF INTONEMENT

Since verbalization is a much more common experience than musicianship, yet they share the quality of sound, a situation involving the spoken word will serve as the descriptive starting point for this investigation. By watching my own children engage in the familiar taunting game of name-calling, I have recalled a particular "philosophical puzzle" that occurred to me as a child engaged in the same activity. This may seem to be a quirky point of departure; nevertheless, the solution to this problem squares with the ontology of music that I want to espouse. As a child I once wondered about the difference between calling someone a "moronic idiot" and an "idiotic moron." I delighted in switching the order of the words and practiced the first phrase and then the second over and over, but I was unable to articulate the difference that might assuage my puzzlement. When I performed the two phrases in statements, e.g. "you're an idiotic moron," and, "you're a moronic idiot," I felt attuned to the difference, but what exactly constituted the difference that I felt, I could not say.

I looked up the meanings of the words in order to work out the conceptual distinction. I found out that a moron is someone with an IQ between fifty and seventy and an idiot is someone with an IQ not above twenty-five. Since I understood the grammatical function of adjective and noun I concluded that an "idiotic moron" was an underachieving moron and a "moronic idiot" was an overachieving idiot. In the game context of name-calling these appellations are merely instruments for baiting and insult, so the ascription serves a non-literal function. Distinguishing the definitions of the words does not provide the solution to the problem, because I had employed the two phrases without knowing the objective meanings and I had already *felt* a difference, even though I could not uncover the root of the difference. Moreover, "the names" were just as effective without the receiver of my taunts knowing their specific meanings either. What mattered was that the words functioned as "names" in the context of name-calling. Nevertheless, this generic function does not imply that the words themselves substitute for one another, for each one had its own special effect.

This strategy of distinguishing the conceptualized meanings can be analogously applied to the experience of music. Suppose we want to account for the difference between the ascending and the descending minor seventh. By defining what is meant by the words "ascending minor seventh" and "descending minor seventh" and by singing out the names of the notes, e.g. "a" then "g 1," the experienced difference is not captured. This fact is readily obvious, for even without this theoretical knowledge one can be attuned to the felt difference, which is especially apparent with untrained people who are quite musical. But what exactly constitutes the experiential difference is our problematic, and the music pedagogue passes over this felt difference. Music theory is a highly conceptualized affair, a mathematical nomological science whose purpose is to train people "to know the structure of music." However, as Dufrenne so succinctly states, "Objects do not exist primarily for my thought but for my body" (1973: 337). The neglect of the feel in music pedagogy is based upon a fundamental lacuna in a psychology that founds educational practices and theoretical taxonomies. "Psychologists have talked almost exclusively about sensation but never about sensing. Of experience as a whole, it has been always the *gnostic* and never the *pathic* moment that has been remarked and considered" (Straus 1966: 11). Felt experience is not considered to contain content worthy of pedagogical attention. Felt experience is viewed perhaps as a primitive aspect of behavior, that which is shared with brutes, and which is transcended by human consciousness and thus deemed irrelevant.

All too frequently, music teachers and aestheticians make claims that a true appreciation of music requires an understanding of its theoretical components. Upon the implications of this intellectualist basis, it would be possible to take a deaf person and train him/her in music theory so that he/she could label everything on a musical score, and this would constitute a rich music appreciation. In opposition to the intellectualization of the appreciation of music, the teenager who retorts that "it has a good beat and you can dance to it, so I'll give it a ten," is closer to the essential nature of music. The music student, whose learned intellectualism has her labeling the theoretical components, has not gained any ground in music appreciation. This theoretical abstraction is analogous to looking up the meaning of the words "idiot" and "moron," in order to become attuned to the "musical feel" of the words, which has been shown as an otiose exercise.

The heuristic of the name-calling example teaches us that the conscious acts of discriminating and naming do not bring us any closer at all to grasping the felt difference in sound, words, and music. Since the feel of the music is not the content of music theory, as our example concerning intervals implies, one may be misled into thinking that the so-called "feel" for which we are searching is a merely "subjective" phenomenon. But the feel is not an appraisal that answers the question, "How do you feel about x?" Nor is it the emotional responses that music elicits, although emotion is a necessary moment. The feel involves the fundamental meaning-structures of sensibility.

What puzzled me as a child had more to do with the real act of name-calling, not the significance of the ideal concept. What I had in mind fulfilled the function of name-calling, but the difference that I felt had to do with the specific gestures enacted by the body, which had to be found in the actual performance of the specific names called. When my cousin and I had enough of "playing nice," we would circle the dining room table opposite each other. The name-calling game that commenced (this was staged by us as a form of entertainment) also included bodily movements and facial gestures. Although these affectations were enhancements to the game, they were neither sufficient nor necessary to fulfill it. This follows from the fact that name-calling can still be accomplished in the dark. What constitutes the necessary condition for the game is what I will call "gestures of intonement." Even though vocalization is necessary, this form of association called "name-calling" need not necessarily involve a direct appellation. If one child says something and the second child responds in a certain tone, "ladi-dadi-dadi-da," this vaguely implies some form of disapprobation towards the first child. Substituted for a name is the implication that the other child is a kind of entity, a name only tacitly alluded to, who would deserve such a response, and the mocking nonsense syllables indicate this. This gesture of intonement, which includes no words and thus no conceptual meanings, is sufficient to express its gestural intention, which is a meaning that involves an attitudinal posture, the appropriately felt fulfillment of intention. Of course, mute gestures can also be a form of disapprobation, and "just looking at" another child can be so effective as to reduce that child to tears. It is obvious to putative experience that rests and periods of silence in music are quite effective as well. Name-calling is one component in a whole performance ritual of rivalry amongst children. The gestures of intonement in name-calling are only one aspect that includes other bodily gestures in a total meaning context of interactions. But, by substituting "gestures of intonement" for "sound perceptions," attention is drawn to embodiment, which brings us to focus on the lived-body versus the acts of consciousness.

GESTURES OF INTONEMENT AS FORMS OF BEHAVIOR

The problematic concerns the sound performance through which the differences in the intonement are felt, and this involves specific organs and gestures of the lived-body. What is felt does not issue from a conceptual signification intended through the act of calling a name; it is the specific gesture of a word or a phrase that is *felt through the lived-body*. The lived-body experience is what I could not identify but felt as a child as I repeated the two phrases. Merleau-Ponty states, "Before becoming the indication of a concept it [the word] is first of all an event which grips my body" (1962: 235). It is the enactment of the word that determines whether it has the appropriate feel for what is to be expressed. One experiences "a whole rearrangement of the bodily schema.... Words have a physiognomy because we adopt towards them..., a certain form of behavior which makes its complete appearance the moment each word is given" (ibid. 235–36). For children involved in name-calling, the word need not reach conceptualization. Children remain more attuned to the physiognomy of words without the conceptual distance that ruptures the primordial attunement into the subject/object distinction (meaning intention/object intended) borne in the signifying acts of intentional consciousness. "The child is in accord with the world, comprehending the gestures of the language of others as soon as he is capable of certain modes of behavior, long before repetition has been able to establish stable associations and fix them in him" (Dufrenne 1973: 335). The distinction which I had been in search of between "moronic idiot" and "idiotic moron" has to do with the particular qualities in the gestures of intonement, that is, qualities that manifest in embodiment through the verbal gestures. I mean by "verbal gesture" not merely the sound as a disembodied abstraction, but the sound as produced and experienced through the lived-body. The conceptual meaning of the words is thus a latent manifestation made possible through the embodied expression. So, the puzzle that faced me as a child was a Bergsonian-like problematic: what I had felt in intuition I could not capture in conception. As Straus points out, "The pathic is a characteristic feature of primordial experience; it is for this reason that it is so difficult to understand conceptually, being the immediate present, sensually vivid, still preconceptual communication we have with appearances" (1966: 12).

If only we could feel the nursery rhyme as adults the way we felt it as very young children. Even if we get beyond our intellectualization to the attunement of our body, the adult body schema does not feel the rhyme as the child does. I can remember to a degree the feeling in the gestures of intoning, "Help murder police! The baby fell in the grease! The baby died, my mother cried. Help murder police!" But I can no longer live that intonement as I once did. Yet, a child intoning the rhyme allows an adult to experience the gestural feeling once again through being attuned to and thus "in-formed" by the intoning gestures of the child. This is an important point, for if this "in-forming" were not possible, one would not be able to attune oneself to the specific qualities of musical instruments through a bodily schema, a form of behavior. I cannot enact the sounds of a clarinet, because obviously I am not one, but I can be attuned to the sound issuing from its embodiment and thus feel the sound as an "in-forming" of my embodiment, as I do through the child who enacts the nursery rhyme. So, one does not have to enact or be able to enact the gesture of intonement oneself in order for the lived-body to feel it, that is, to live through a form of behavior.

With instrumental musicians, there are bodily movements and gestures that are executed in the act of playing the instrument. But these gestures, although a necessary component, are not what I want to call "gestures of intonement" – just as "making faces" is necessary in the game of "name-calling." Intonement is the response of the instrument's voice, which issues from its peculiar embodiment as it sounds a passage of music. These intoned "gestures" issuing from the "physiognomy" of the musical instrument establish the qualities of embodiment, the enacted forms of behavior, that play through the body schema of the listener. Of course, the qualities of the musician's gestures are a necessary component for the quality of the tones elicited from the instrument. As Dufrenne aptly describes, "Felicity belongs in the body of the artist if not in his soul.... The phrase 'thinking with one's hands' applies to all artists, but especially to the composer improvising at the piano, or the painter at the easel" (1973: 340). It is insightful to note that when listeners are attuned to the work of an accomplished instrumental musician, they frequently retort, "That guy (or gal) can make that thing talk!" But the response in the body schema for the listener will be quite different according to whether a mandolin or a saxophone is doing the "talking," and the manner in which it can "speak" the musical passage. These differences are due to the structure of the instrument producing the sound and to the way in which sound is produced upon the instrument, e.g. plucking, striking.

LIVED-BODY INTENTIONALITY AND ATTUNEMENT

The felt difference is a preconscious genesis of meaning, but the schematic expressivity of the lived-body accompanies all mental acts. Qualities of vocalization, sound, and music are not merely audile perceptions made possible through sensory contents; the qualitative genesis of sensation is to be traced to the schematization of the lived-body. "With the notion of the bodily schema we find that not only is the unity of the body described in a new way, but also, through this, the unity of the senses and the object" (Merleau-Ponty 1962: 235). If the senses are treated as separate functions of perception, the investigation remains on the level of the subject/object dichotomy, which means that "the feel" will remain in obscurity. And without tracing an auditory phenomenon back to its feel, audition remains an abstraction. In the subject/object dichotomy, sensation is either a state or quality of consciousness or consciousness of a state or quality. But the feel requires discussion of qualities inherent to embodiment, not merely to presentational qualities of perceptual evidence in consciousness. "Each of the alleged qualities – red, blue, colour, sound – is inserted into a certain form of behavior" (ibid. 208–9). These felt qualities are prepersonal meanings in the anonymous intentionality of the lived-body. Merleau-Ponty states, "If I wanted to render precisely the perceptual experience, I ought to say that *one* perceives in me, and not that I perceive.... I am no more aware of being the true subject of my sensation than of my birth or my death" (ibid. 215). Felt experience is meaningful prior to the meaning-bestowing acts of consciousness. Moreover, the lived-body is not a nascent subjectivity. That which is felt is immediately meaningful, for it is not passively acquired sense data that await the organization of consciousness.

The conceptualization of music is not its expression, which must be manifested through the meanings of embodiment. Some musicians appear quite ignorant about what they are doing, inarticulate, or even stupid; in fact, some exhibit all three of these characteristics. But this proves that musicality is not the child of consciousness, but an attunement of embodiment in its essential relation to music. The feel issues as a meaningful form of behavior played upon the body schema, which is not a conscious activity. It is in the felt experience that the musician and those attuned to music must dwell. Attunement to music, in this sense, does not mean that we must live in pure intuition and "lose" consciousness. But it does mean that consciousness must be subordinated to sensibility so that it can let it be, that is, consciousness becomes attuned to the attunement that has already been managed by the lived-body.

THE A PRIORI STRUCTURE OF PLACIALITY AND THE BODY SCHEMA

When we speak or sing, we are the instruments that produce the sound, which means that our physiognomy necessarily must conform to certain patterns of behavior to emit the intended sounds. It is the synergy between the placial behavior formed through the musculature and the placiality in the perception of the sound that makes verbalization an insightful starting point for the investigation of the feel. Placiality is the expressive foundation for the gestures of intonement. When one is uttering "idiotic moron," the sound starts at the front of the mouth and the tongue rises to the roof. The next syllable, "i," sounding as "e," brings the tongue away from the roof and the following "o" opens the mouth up. The mouth then closes with the lips slightly open and the "tic" must be performed abruptly by the tongue, which begins its motion behind the teeth and ends towards the back of the mouth. "Mor" begins with the lips closed and the sound moves softly toward the throat. "On" requires the mouth to open and to smoothly close with the tongue against the roof.

The placial qualities of the sound-producing musculature resonate as a behavioral formation throughout the body schema just as sound resonates throughout the body of a musical instrument. Saying a word entails a gestural terrain, and each word's "geography" forms a body schematization. According to Merleau-Ponty, the original genesis of language consisted of gestures of intonement that expressed the attunement of the lived-body with the qualitative meanings of the world. These gestures of intonement became a means of communication and gradually were sedimented as a language. Language carries ideal meanings that were once concrete gestures that have become faint vestiges in the halo of consciousness. The discussion of the feel of the words appears trivial, for we have lost our fundamental relation to it and it is difficult to reawaken to the attunement to the physiognomy of words. However, conscious attunement to the feel of language reopens the fundamental expressivity of the gesture of intonement.

Those who are interested in the physical conditions for the production of vocal sound such as speech pathologists would describe the formations and movements that I just described in a more precise scientific way. Our interest does not reduce to a formational analysis of the physical apparatus that constitutes the components for sound production; rather, it lies in the significance of the placial qualities that are involved in the activity of performing these sounds. Merleau-Ponty states, "All senses are spatial if they are to give us access to some form or other of being, if, that is, they are senses at all" (ibid. 217). With musical instruments, there are technical motions that the player must perform. But with vocalization, we have the experience of being the instrument. The positions of the mouth, lips, tongue, teeth, tensions in the vocal cords, etc., are in one sense only the conditions for the possibility of the particular vocalization. But in another sense the changing formations of the musculature emit meaningful placial qualities that are intimately connected with the placiality of the sounds, which is a relation that cannot be disconnected in vocalization. When a word is heard or said, its physiognomy resonates through the body schema based on the placiality in the gesture of its enactment. So, a particular placiality of embodiment "fits" the word. Moreover, one has the experience of the placial qualities of these muscular formations even when the words are "said silently" or when one hears them spoken by someone else. We experience ghost gestures through our body as if we ourselves were saying the words out loud. Words speak to us through their inherent physiognomy, that is felt.

> The types of micromovements I am calling "ghost gestures" are not necessarily overt or visible movements,... but as it were the ghost of a gesture – a kind of inner "quasi-gesture," a schematic

inner vector or tendency toward movement that can persist in the body even when the large-scale gesture that the ghost gesture schematically implies is not actually being performed (Behnke 1997: 188).

In the specific sense, ghost gestures are sensations that remain in the organic kinaesthesia even after one has completed certain tasks, such as the lingering sensation of the movement of the boat after boating. But ghost gestures can occur, if I attune myself, when I remember the sensations of boating or sensations of the ghost gestures after boating. So, ghost gestures allow for the fact that the placial qualities of words are felt regardless of whether or not we are in the act of speaking.

The placial qualities of the sounds are felt along with auditory imagination. Don Ihde states, "The auditory presence of language is an almost total constant of experience" (1983: 32). Inner speech is an explicit form of auditory imagination. Auditory imagination is "broader than *inner speech* since one can imaginatively "hear" music, another person's voice, etc." (ibid.). Ihde further claims that linguistic thinking is an on-going accompaniment of experience. Yet, this should be qualified, for the auditory imagination of musicians is much more filled with music, and thus much more non-verbal than the inner experience of non-musicians. Whether auditory experience is productive, perceptual, or imaginative, and whether it is filled with the content of music or the sounds of the life-world, the feel is the postural attitude of the bodily schema. Attunement synergizes the experienced placial qualities of the musculature formations with the placial qualities of the sound. So, what the musicians call the feel of the music is a species of the placiality that is manifest in all auditory phenomena, which all of us experience in its various modalities. When one "hears a melody in one's head," there is a feeling in the throat as if one were actually singing the melody. "Sensation as it is brought to use by experience is no longer some inert substance or abstract moment.... [It is] a particular manner of being in space and, in a sense, of making space" (Merleau-Ponty 1962: 221). The point that is established here is that to talk of the placiality of auditory phenomena is not a metaphor, which is the view of many theoreticians who remain on the level of consciousness; placial qualities are the very being of the schema of sensation. And, the placial qualities of sensations are the primordial field by which we engage the world.

It is not the sound qua sound that captures the meaning of the feel, which erroneously seems to follow when the teleology of musicianship is the consideration. In the end it is a particular quality of the sound that the musicians are after, yet they are only after the quality of sound that feels right. If it were merely the sound (which should now be understood as an intellectualist abstraction), we would be satisfied with a linguistic description of the sound of the words, and a description of sound as merely a temporal phenomenon, in order to describe the musical feel. As we have developed, the founding structure of the differences in the qualities of sounds is not based on the conscious acts of audition that constitute a heard object. The lived-body does not merely supply the sense organs for conscious acts. Sensations can only be qualitatively differentiated through their inherent placiality, as felt meanings that have their "perceptual" origin in the lived-body schematization.

The body physiology does not merely causally respond to objective stimuli transmitted to the ear, as a reductive empirical standpoint would explain audile perception. It is not that some external object produces sound waves that impinge upon our organs of audition that extend through pathways to the brain which then result in certain responses in our nervous system. The meaningful discernment of sounds by "the ear" is made possible by placial qualities formed by the lived-body schema. But the differences in the meaning of the placial qualities playing upon the body schema emerge in relation to the structural properties of the placial qualities of the place-world. "Our own body is in the world as the heart is in the organism" (ibid. 203). "The sensible has not only a motor and vital significance, but is nothing other than a certain way of being in the world suggested to us from some point in space, and seized and acted upon by our body, provided that it is capable of doing so, so that sensation is literally a form of communion" (ibid. 262). This point must be insisted upon, unless we are to slip back into the subject-object dichotomy that serves the epistemological distortions of the transcendental subjectivity of intellectualism and the causal explanations of reductive empiricism.

PLACIALITY AS THE PRIMORDIAL FIELD

The qualitative placiality of the lived-body only gains its meanings through a reciprocal relation with the qualitative placiality of the lived-world. The primordial phenomenon *is* a placial field that is qualitatively differentiated. Using vision as a way to illustrate this covering over of the primordial field in the epistemological rupture of conceptualization, Merleau-Ponty states, "Instead of living the vision,… I break the link between my vision and the world, between myself and my vision, in order to catch and describe it. When I have taken up this attitude, at the same time the world is atomized into sensible qualities, the natural unity … is broken up, and I reach the stage of being unaware … [of the primordial nature] of a visual field" (ibid. 227). So, the placial qualities in the synergy of audition and embodiment cannot be formed through stimulus and response, as causal analyses would explain perception. Rather, placial qualities are inherent to a field that takes the body out of itself and into the environing world. And the world is available for meaningful attunement through the lived-body. Sounds are worldly entities; sounds are not constructions or representations either of consciousness or of the lived-body. But the sounds are not objective either, for the objective account breaks apart the primordial field, and thus objectivity is the creation of conceptual abstraction. The placial qualities that are felt draw the lived-body out of itself into an autochtonous field, i.e., a spontaneously manifesting phenomenal structure that is neither subjective nor objective. As Dufrenne states, "We are in the world – by forming a subject-object totality in which the subject and the object are not yet distinguishable" (1973: 339). Thus, the intentionality of the lived-body is de-centered, for the forms of behavior cannot be partitioned (as in stimulus-response) from the environing field. The de-centered embodiment is an opening to a field of which it is neither author nor spectator. This means that the fundamental experience of music needs to be described from this de-centered structure, for the phenomenal field is prior to the subjective/objective bifurcation.

THE DYADS OF IMPLACEMENT AS QUALITIES OF THE FEEL

The placial differences in the meanings of the gestures of intonement are founded upon the placiality inherent to being-in-the-world. Implacement is the collusion of the lived-body and place (Casey 1993). Our description of this collusion commences with three dyads of implacement, above/below (up/down), ahead/behind (front/back), and right/left. These are *a priori* structures that are made possible through the asymmetry of the lived-body. Yet they are not subjective, for these structures of the primordial field are the ambiguous relation of the world and the structure of embodiment. As heterogeneous asymmetries, they form three somatic axes that extend through the body in directions that are differentiated placial qualities. These meanings are not merely metaphors; they are grounded upon the fundamental relationship of embodiment with its world. Starfish do not have the dyads of ahead/behind or right/left, only of up/down. So, if an entity had the self-consciousness of a human being but the embodiment of a starfish, it would both conceptualize and feel the world quite differently. It is from the nature of our embodiment that we are capable of apprehending the meanings of these placial qualities, which then can be reductively quantified in abstractive mathematical conceptualizations of space. These somatic placialities are "rationally purified" in the homogeneous system of Cartesian coordinates. Descartes' mathematical system is an abstraction and its alleged real objectivity is fallacious. Likewise, the mathesis of music which had already begun with the Pythagoreans and which is perfected in the work of Joseph Schillinger (1978), is a quantitative abstraction which does not causally account for the qualitative meaning of music, as the materialist claims. The feel of music consists of the qualitative morphology of placial meanings, which is not reducible to mathematical formulas of exact essences. Placial qualities are the source for qualitative conceptual significance. The "up" gains its qualitative meaning due to the upright posture, which must resist gravity, and the place of our sense organs on our bodies. It also gains its meaning from the open sky above and the bodily motility of climbing. The lived-body in collusion with place is responsible for providing the fundamental source for spiritual meanings, the language of values, and the felt meanings in the placiality of music as well. Significance arises in the collusion of the lived-body and place, which serves as the *a priori* meaning-structure already setting parameters for consciousness and the interpretations and constructions of culture.

In speaking the two phrases "idiotic moron" and "moronic idiot," the difference in sound is founded upon the difference in the placial qualities of their intonement. "Idiotic" starts in the front of the mouth and moves back towards the center. In the middle of this word the gesture moves up and expands and then moves down in a rounded cavity. In the formation of my own mouth, the first syllable starts slightly to the right of center and the second and third syllables rely on the left side of the tongue. Although I can perform the first syllable on either side of my mouth, I can only perform the last syllable, "tic," on the left. I cannot even image that sound on the right side of the tongue; it is as if that feeling is rejected. This has to do with the preference we establish in our embodiment for the left or the right, which is the basis for the subtlest articulation of place qualities, for it has to do with handedness and the closest zone to the body, the zone of manipulation. I think the phrase, and I feel the ghost gestures through my vocal apparatus, and "tic" is ghost gestured on the left side of my tongue, which is slightly askew and sits against the upper teeth on the left. I am incapable of forming the "mirror" sensible image on the right side of my tongue. This idiosyncrasy of mine is analogous to that of a musical instrument. The top of a cello may be carved in such a way that certain pitches resonate better by some portion of its vibrating surface than others. So, my body admits of a style and the instrument offers a peculiarity of tonal response through the "behavior" of its resonating body.

But what is gained from analyzing the two phrases is the placial nature in the gestures of intonement. The placiality of up and down, for example, cannot be captured in concept, but must be experienced. In language the feel is subordinated to the grasping of concepts, but with the medium of music, the feel is intensified. We are aware of the feel of music at least through the motivation to move our bodies or we are aware of the feel indirectly through the emotion that accompanies it. The lived-body feels the significance in the placial quality of sounds, words, and music and interprets it through forms of behavior that organize the body schema. Sometimes the placiality of a passage of music can lead to very specific imagery. In Bartok's *Concerto for Orchestra*, there is a section in response to which I feel as if my body were being tossed in the air. In Stravinsky's *Firebird*, there is a cymbal crash that makes me feel as if I had jumped into a pool of warm ginger ale every time I hear it. But the feel of music need not lead to such specific imagery in order to involve placial qualities. In fact, to arrive at a description of the placial qualities of the feel, I ask myself why the music brings about certain specific images. These felt qualities can change rapidly or the music may set up a specific "groove" such as that of the sixties British rock group, The Kinks, whose basic musical feel had been characterized as a "wall of sound."

Language as a vehicle for concepts motivates the disregard for the attunement of the lived-body and instead motivates one to focus on ideal meanings. A similar intellectualization of music for its own sake hardly seems like a meaningful endeavor, although it can be done. The intellectualization of music should serve only to enhance a greater appreciation for the lived-body's attunement to music, just as reason for a religious person is secondary to faith and is only used to gain a better understanding of what has already been gained through religiosity (*Gefuhl*). But what happens is that a "legitimate" discussion of music is the form and analysis of music theory, which in its scientific endeavor abstracts from experience.

A musical instrument articulates sounds through the specific nature of its construction, and the lived-body of the listener attunes itself to that embodiment. The sounds produced involve patterns of vibrations through the instrument, which are experienced as placial qualities. It is not that we metaphorically call quicker vibrating frequencies "higher" or "above" lower vibrating frequencies. Rather, the lived-body is attuned to the body of the musical instrument, which, if it were subject to an intentional structure of its own embodiment, would feel placial qualities too. Instead of feeling the collusion of body and place, it resonates with it. But its resonance sings, for example, the feeling of being pushed from behind or pulled along, or initiates its own movement as it climbs higher or leaps down from above. A musical instrument resonates with the placial feel of embodied-being-in-the-world through the medium of music.

There are other dyads of implacement, such as here/there and near/far. When the lived-body moves from "here" to "there," making "there" now "here," that movement can be one of ease or difficulty. The performance of the gestures of intonement thematically displays the tensional arc between "here and "there." Great tension is experienced in the performance of tongue twisters. The word "idiot" is far easier to complete (here-to-there) than the word "idiotic," and this is not due to the addition of a syllable. The word "moron," gives the feeling of moving further out in the far sphere and drawing nearer to the far, whereas "idiot" seems to remain within the near sphere but quickly achieves a little distance at the end, if the "t" is pronounced. In music, melodic contour can unfold with ease or through great difficulty.

Notational indications in music do not indicate the placial qualities of the feel. For example, *pianissimo* does not indicate the feel of the passage. Certain passages performed at that volume level sound as if they were far in the distance. At other times a *pianissimo* passage feels intimately near. A *sforzando* can sound like a distant explosion or it can feel like a sudden jolt from the center of one's own body. What is interesting is that the feel, which is the true experience of music, is not indicated by musical notation, which should serve to indicate the intellectualist bias inherent to it. Musical notation is on the level of literacy, which directs acts of consciousness to produce sound qualities that are discernible to the conscious acts of listening. But the two sharps indicating the key of B minor do not indicate the feel of that key. The *accelerando* does not indicate whether we are running into the distance far away from our original place or whether we are fast approaching our destination. Such is accomplished in the feel of the music, which does not occur through a legislation of consciousness. If a pain is felt in the right foot, consciousness cannot dictate that the pain be felt in the left hip. But, this *a priori* does not mean that a composer does not attempt to create a feel. Of course the feel is exactly the intention, but the composer does not accomplish this in the meanings of the musical notation. The feel is lived in the ghost gestures and transmuted into notational conventions. In fact, the spatial configurations of musical notation, if anything, thoroughly obfuscate the feel of the music.

The spatial distance of the octave in musical notation looks much greater than the interval of one whole step, but the feel of the intervals may be just the opposite. Unfortunately, the scope of this article does not allow the space to describe a piece of music in terms of the dyads of implacement, but readers are encouraged to become consciously aware of what the lived-body has already accomplished in "understanding" a piece of music prior to a notational convention. Music transcription distances consciousness from its meaningful source, that is, conception is constructed according to notational convention rather than an attunement to its source.

TIMBRE AS BODILY ATTITUDES

To investigate the placial qualities of timbre is of the utmost importance, for a link is established with the appraisal function in the forms of behavior of the attuned lived-body. Timbre is defined as tonal color, the tone quality of a particular sound. It is thought by some that timbre reduces to the series of harmonic tones. But this fails to take into consideration the physical structure of the source and the manner of its sound production. The bow rubbed across the cello strings or the air pulsating through the metal tube of a flute produces a material quality that cannot be reduced to harmonic quantification. For even with colors, the hue, shade, tint, and tingeing produce unique qualities that cannot be separated from the texture of their specific material presentations. The timbres or colors of sound qualitatively organize the physiognomy of the body schema. One just has to call to mind the screeching of fingernails across a slate surface to be assured of this. It is the case that tonal colors and visual colors are named according to perceptually differentiated features, but this is only possible because the body has already adopted an "attitude" toward them. Merleau-Ponty discusses the significance of experimental findings in which brain damage shows the effects of sensory excitations on muscular tonicity. Since the excitations are not integrated into a comprehensive situation, that which usually remains unseen in the physiognomy of the body schema becomes overt behavior of the inflicted individual. For example, "the gesture of raising the arm, which can be taken as an indicator of motor disturbance, is differently modified in its sweep and its direction according to whether the visual field is red, yellow, blue or green" (Merleau-Ponty 1962: 209). A general distinction of the placial qualities of motor physiognomy concerns adduction and abduction. "The significance of adduction is that the organism turns towards the stimulus and is attracted by the world ... of abduction, that it turns away from the stimulus and withdraws towards its centre" (ibid.). Abduction and adduction as spatial qualities of sensation set up a certain rhythm of existence. "I am brought into relation with an external being, whether it be in order to open myself to it or to shut myself off from it" (ibid. 213–214). But the settling of this component of the felt response is already at work in hearing a single musical note, which "is merely the final pattern of a certain tension felt throughout the body" (ibid. 211). As the body adopts different attitudes to sweet and sour flavors and then again to the sickeningly sweet and the extremely sour, so does the timbre of music affect bodily attitudes. One need only compare the unaccompanied viola to the oud of classical Persian music to experience the difference of attitude that the body takes toward the timbre of these instruments. But once in that attitude, just as with the complex flavors of well prepared dishes, the flavor varies around a theme.

The appraisal of sound, then, is also a form of behavior. Music provides an array of tonal colors to which the body schema takes up an attitude of appraisal. Tonal colors are not absolute, because they relate to one another in a complex whole and are influenced by their placiality. The lived-body feels differently about a bright color that emerges from above or below, or one that pervades the horizon, or one that occurs in minute splashes across the immediate zone of operation (the near sphere).

Implacement also involves forms of dwelling and pre-positional (prepositional: alongside, between, through) preparations to forms of dwelling. Abduction, taken as a more global appraisal, is the body schema that corresponds to hestial forms of dwelling, dwelling-as-residing; and adduction corresponds to hermetic forms of dwelling, dwelling-as-wandering. In music, hestial dwelling can be revealed in the establishment of the home tonality, and hermetic dwelling through modulation. These forms establish the interrelation of appraisal and implacement, for forms of dwelling prescribe an appraisal, e.g. returning home and setting out on a long trip, and require different global forms of appraisal. Such appraisal of placial meanings occurs in music. But to experience this, awareness of the lived-body is essential, or the conscious judgement that the music has returned to the home key is a lifeless analytic matter of fact.

SYMPATHETIC RELATION AS AN AUTOCHTONOUS FIELD

Sensation involves sympathetic relation, communion, or surrender. At the level of surrender the field shows its primordial nature, which manifests as an autochtonous gestalt. This spontaneous self-organizing field is neither the product of subjectivity nor of the objective world, for it is ontologically prior to both. The phenomenal world is the primordial world and both subjective and objective orientations theoretically partition the phenomenal field. Yet, surrender does not yield a coincidence as if the musical instrument and the lived-body converged. Merleau-Ponty describes how the two poles of the field reverse roles. "There is an objective sound which reverberates outside me in the instrument, an atmospheric sound which is *between* the object and my body, a sound which vibrates in me 'as if I had become the flute or clock'; and finally a last stage in which the acoustic element disappears and becomes the highly precise experience of a change permeating my whole body" (ibid. 227). The primordial field is the ground from which the lived-body and the other phenomenal entity are established in a commonality by which their roles can be reversed. This identity-within-difference is already established through the lived-body itself, which is exemplified in the reversible roles of the two hands alternating between being touched and touching. At no point do both hands coincide and accomplish both tasks at once. Rather one touches the other, and short of coinciding, changes its role and is touched by the other. The meaning of being de-centered can now be made more explicit. The Other allows me "to see" myself from the vantage of its being. The reversibility is that the Other is seen by me and the Other sees itself through me, and I see the Other and I see myself through the Other. Through my lived-body the sound "hears itself," that is, it is taken up by a vantage it cannot live. Through my body schema, I am in-formed by the sound-phenomenon that I cannot be, but yet feel myself through that sound that now establishes itself through me.

When I hear the timbre of your voice, my body takes on the attitude to receive it, that is, surrender to it, by forming a behavior. But in full presence of you my body also surrenders to the contour of your body and to the contours of your gestures. And if both of us are attuned to one another, our bodies enact a dialectical field between us. The interaction "between" is analogous to a musical duet. But what is of import here is that the placial qualities of the body schema are a synergy, a sensorial commune. I hear your voice: I see the contour of your body and your gestures. But my body schema attunes to the complete presentation from which the placial qualities of the feel are formed. As my cousin and I circled the dining room table calling each other names, the sensory experience included the visual and kinaesthetic as well as the aural. Not only did my body schema surrender to the sounds, but also to the contour of his body and his gestures. Similarly, the visual experience of hearing music live enters into the total experience. The cellists Janos Starker, Lynn Harrell, and Yo Yo Ma each bring a different visual experience to the listener, which enters into the sensory feel. The surrender of the body to the experience will be modified by the difference in their visual presence. Our bodies become attuned to the sound of the embodied instrument as the sounds play through our bodies in a placiality that communes with the shape and materiality of the instrument. But visually, we also see the instrumentalist, and our sensory experience takes up their style and communes with the contour of their body and their gestures as they play the music. How we attune ourselves to the feel of the music will be influenced by the visual appearance of the musicians and the entire environment. So, the synaesthesia of the senses cannot be ignored, even when our interest is in the feel of the music. Yet, the feel of the music can be attended to in a way that attunes the lived-body to it while other components of sensory experience fade into the sensory horizon.

THE SYNAESTHESIA IN SOUND PERCEPTION

Merleau-Ponty states, "Synaesthetic perception is the rule, and we are unaware of it only because scientific knowledge shifts the centre of gravity of experience, so that we have unlearned how to see, hear, and generally speaking, feel, in order to deduce, for our bodily organization and the world as the physicist conceives it, what we are to see, hear and feel [i.e. touch]" (ibid. 229). But before consciousness has a chance to create theoretical preconceptions about its experience, it is quite an ordinary experience to say that "one sees the hardness and brittleness of glass, and when, with a tinkling sound, it breaks, this sound is conveyed to the eyes. One sees the springiness of steel, the ductility of red-hot steel, the hardness of a plane blade, the softness of shavings" (ibid. 229). Although we use specific adjectives that we associate with specific senses, when we are attuned to our actual experience, the adjectives are transferable to each of the senses. "When I say I see a sound, I mean that I echo the vibration of the sound with my whole sensory being, and particularly with that sector of myself which is susceptible to colours" (ibid. 234). Under the influence of certain drugs, such as mescalin, the synaesthesia becomes transparent to consciousness. "The influence of mescalin, by weakening the attitude of impartiality and surrendering the subject to his vitality, should therefore favour forms of synaesthetic experience.... The sound of the flute gives a bluish-green colour, the tick of a metronome, in darkness, is translated as grey patches" (ibid. 228). Colors are always colors of things in the horizon of place. The placial qualities of music can evoke coloration of its placial field through synaesthesia.

Music also evokes textures associated with the sense of touch. I hear the metallic texture of the banjo, the waxiness of the classical guitar. Perhaps the body schema forms behaviors of taste that are evoked by music. The saxophone has an earthy flavor. Music can be served cool or hot. The vibraphone sounds like a cold drink. But none of these descriptions need to be "accurate" in order to indicate something to us. The consequences of synaesthesia for our inquiry are that the differences that concern the feel of music are not to be thought of as completely distinct from other forms of behavior. Not only are all the senses placial, but their characteristics are also transitive. This legitimates a whole wealth of descriptive language that can be employed in describing the feel of the music. Some sounds do feel dull and others sharp, or some do feel murky and others vibrantly clear. But, there is the danger of slipping into poetic meandering. What is it about wax that is different from wood or metal, so that the classical guitar sounds waxy? The description must disclose phenomenological evidence. These "morphological" characteristics suit the feel of music, whereas the exact essences that are set forth in the mathematization of music contain nothing of the feel.

THE KINAESTHETIC COMPONENT OF THE FEEL

There are temporal aspects that are necessary to address in order to show the placial qualities that are more emphasized in music than in other phenomena. In performing the gesture "idiotic moron," there is a sequence of placial qualities, which involves temporality. The word "idiotic" requires a more rapid sequence than the word "moron." The performance of the "r" in "moron" feels inordinately sluggish; the musculature seems to thicken while it is in the process of enacting the "r." With the word "moronic," the sluggish fluidity of the first two syllables finds its pause on the syllable, "ic," which is a sound whose "telos" is just short of the choke point in the throat. Yet, the "ic" can be played slowly, if one's accent makes a dipthong of the vowel, or else it speeds by and feels like it is allowed to be swallowed. The word "idiot" involves a quick punctuated rhythm that can end with a sound that we emit from our lips when we are disgusted. Consciousness for the most part is only minimally aware of the feel of words and the aspect of their kinaesthetic flow. In ordinary communication, if I become tongue-tied, then I become aware of the rhythmic flow. Otherwise, only with special uses of language, poetry, lyrics, or slogans do I live in the rhythm of words.

What has been uncovered concerning the placial qualities of implacement in the gestures of intonement will not elicit any surprise from poets, and yet they are the ones who are attuned to the music of words. However, poets are not engaged in the phenomenological description of the attunement. In the analytic attitude that concentrates on the music of poetry, the metered phrases of the long and short syllables, the rise and the fall of the voice and the sounds of the words are the elements of discussion. But the essential root of the music of the words, the connection of music to the nature of embodiment, is experienced but not captured in the analysis. A phenomenological description of the feel consists of a description of the placial qualities that are immediately sensed through the lived-body's kinaesthesis. Sensations "present themselves with a motor physiognomy, and are enveloped in living significance. It has long been known that sensations have a 'motor accompaniment', that stimuli set in motion 'incipient movements' which are associated with the sensation of the quality and create a halo around it, and that the 'perceptual side' and the 'motor side' of behaviour are in communication with each other" (ibid. 209–10). Merely the audile perception of a consciousness would not be able to experience the feel of temporality even though judgments about the before and after and the "swell" of a meaning context can be articulated. "Meaning is not primarily something that I think with detachment but something that concerns and determines me, resonating in me and moving me. The pure signification that I contemplate without adhering to it will arise from this more primitive signification, which convinces me because *it sets me in motion* [my italics]" (Dufrenne 1973: 335–36).

The analysis of the music of poetry is done for the sake of consciousness, which takes a distance from its experience in order to theorize it. But a description of the immediate experience of the feel of poetry or music would have to be like the running-off of placial qualities. The problem however is that one cannot articulate in words what occurs much faster than the tempo of our verbal capacities. Of course, short passages can be described in this way, because one can retain the experience in memory so that the running-off can be re-experienced until the aspects of the feel are articulated in language. But description of longer complex musical works is possible through the help of memory and reenactments of the musical experience. When someone says, "You're an idiotic moron," the word-sounds are connected to the running-off of gestures that are a sequence of placial movements. Kinaesthesis feels this mapping process in a temporal depth. So the phases of the gesture are not an atomized sequence of individual syllables or tones; a meaning-context is built up that establishes a whole, an envelopment of feeling. But the danger is to treat the running-off from the perspective of internal time consciousness rather than as an expressive behavior of the lived-body.

The experience of music motivates us to move our bodies in an outward expression, as a dancer or conductor. "A rhythmic series of tones compels us to move in a way that is characteristically different from everyday walking, running, and jumping.... Without any additional activity on our part, we are captivated and carried away by the musical rhythm. The link..., is a completely direct one" (Strauss 1966:11). What is to be described is the solicitation of the feeling of movement through the specific character of the rhythmical map. But the next danger is to direct attention to kinaesthesia as if it occurred "in" the body, rather than as one pole of an autochtonous phenomenal field, which is primordial.

As the one receiving the taunt, "you're a moronic idiot," if I am an attuned listener, I feel the words. I "feel" the taunt from the way it is played by your body. My body feeling the words plays those sounds through its kinaesthetic attunement to the sound as it plays upon you. But this does not mean that I can only be attuned to the audition of another lived-body. I can be attuned to any body. I do not merely hear, I kinaesthetically feel, the difference between musical passages as they set up patterns of movement in the body schema. The lived-body maps the appropriate kinaesthesis in its attunement to the temporal patterning of the musical structure. Music is an "architectural feat" that the lived-body experiences as a process of building. The mapping is a dynamic process that deepens the phenomenal field as a "geographic" event of contextual meaning. So, music pervades an environment in a kinaesthetic mapping that temporally unfolds.

THE ENVIRONING FIELD OF MUSIC

The feel is a relational field that manifests through two coeval poles. Implacement, the collusion of lived-body and place, means that the feel permeates an environment and is modified by the placial quality of the temporal process. The musical source and the listener are bound with the phenomenal field as reciprocal conditions. The placiality of sound is different from the placiality of sight. However, it is not a spatial metaphor to say that sounds are higher or lower, though it would be if all that we experienced involved the conscious awareness of audile perception. As has been established through the synaesthesia of sensation, the sounds are felt as higher or lower due to the structure of our embodiment, the lived-body as an instrument for the participation in sound. Care must be taken to properly distinguish characteristics of the phenomenal field of sound from the field of sight. When something is looked at, over there, it takes its place in the horizoned environment. I am here but it is there, next to that, in front of something else, etc. From my "here," a domed visual field is formed. I see a cello over there; it is an object in placial relation to other objects in the visual field. It too has a perspective, for if I were there, it would be my "here," and the visual field would extend from that "absolute" position. Each "here" is the source for a visual field, but it is not itself the field.

When we hear a sound we frequently can point to its source; we visually see and we hear where it is coming from. The cello is over there, the place from which the sound originates. Yet, the sound envelops us; it surrounds us. The sound also envelops its source. "Color clings (phenomenally) to the object while the tone produced by an object separates itself from it" (ibid. 8). The sound is not there, nor is it here. Rather, the sound is the field. The visual cello is not a field, but the sound which originates from it is just that, a field. Strauss states, "Colors appear opposite to us, over there, confined to one position; they demarcate and differentiate space into partial spaces, appearing side by side and behind one another. Tones, on the contrary, approach us, come to us, and surround us, drift on; they fill space, shaping themselves in temporal sequences" (ibid. 7). The separation is the key to the distinction concerning the visual and auditory fields. "Thus the sound is experienced as the effect of an action because it is of the essence of sound to separate itself from the sound source. While color and form … constitute the object, sound, both as tone and noise, merely points to the object and only indicates it" (ibid.). Music is an ongoing activity that promotes a highly kinaesthetic experience of the enveloping medium.

The cellist plays a melodic passage that starts from the open low C string and ends at the octave of the G string. The melodic passage soars upward in temporally even intervals, but the melodic contour skips steps during its climb, which gives the feeling of gaining more ground at various stretches along the way. The body of the instrument voices these placial qualities from the standpoint of its peculiar structural make-up. The feel of the musical passage plays through its body. The cello "finds out" what it feels like to elicit this melodic content as I find out what it feels like when I hear the melody that fills the environing field. But not only does the cello resonate the feel, its body is the source of a sound field. The environing world feels the sound as well, that is, resonates to it. The sound resonates through the door, which means that the door feels the sound according to its structural properties. The door plays the melody through itself, even though it did not originate it. What makes the feeling of the lived-body different than that of the door is that its structure involves intentionality; the feel is inherently meaningful for the lived-body. This means that the lived-body is capable of interpreting the feeling of the music from a vantage that is opened to the structure of world audition.

THE LIVED-BODY AND THE EMOTIONAL EXPERIENCE OF MUSIC

Music allows the body to feel the structures of implacement. The body feels the up, the down, the far, the near, the over-there-alongside. The cello voices itself through the lived-body; the lived-body becomes that voice in its kinaesthesia. The cello plays sounds like the leaves that are played by the wind as they shimmer in the breeze. The kinaesthesia that I feel through my body is attuned to the sound-mapping by a receptivity and postural attitude of its body-schema. There is a synergy that the field expresses. As a lived-body that feels kinaesthetic meanings, I am one pole and the cello music is the other, but one is not the subject and the other is not the object of audile perception. This structure is important in terms of the emotion of music, for the emotions elicited are not "merely" subjective. I do not find a piece of music festive, and you find it filled with tragic loss. The audile field already prepares the lived-body to experience forms of behavior for which consciousness is moved to make emotional responses. The key is to remember that the intentionality of the lived-body is prepersonal. If we were asked to move our bodies the way, for example, Vaughan Williams' *In the Fen Country* makes us feel (emotionally), you would not find someone doing "the twist." The lived-body attunes to the "physiognomy" of the music through forms of behavior that play the music through the body schema. The music, then, becomes meaningful through a field in which it participates along with the lived-body. These gestures of intonement, felt through the body schematization, are interpreted pathically by consciousness.

However, the conceptual meanings of words (as they are meant in context) play a part in the emotion. But this means that we must realize that the feel of the music of the words is to be distinguished from the passions that are elicited by the meaning of the words. Yet both tone and conceptual meanings are sources. Even the conscious judgement that a musician is good contributes to the emotions. The feel of the music is not to be confused with the emotional response, but of course, there is an intimate connection. The emotional response is what the feel feels like, that is, the running-off-building-up of placial qualities aesthetically calls forth the lived-body to form behavioral gestures that are consciously interpreted through emotional feeling.

THE FEEL OF THE FLESH

The Flesh is the primordial ontological ground. It is neither a subjective nor an objective foundation; it is neither a transcendental subjectivity nor a material mechanism. But the Flesh is also not a simple monism. The Flesh is a phenomenal field that, due to a fundamental dehiscence, allows for the emergence of the subject/object dichotomy. The feel of music has allowed for the exploration of audile perception from its source in the intentionality of the lived-body. But the intentionality of the lived-body is not a subjective constitution; it arises as a component of the autochtonous field. The lived-body allows for the phenomenal world, in this case the world of sound, to be experienced, which means that a phenomenal body is experienced by another phenomenal body. The primordial relationship is one of reversibility and not the coincidence of the two poles. Coincidence would constitute a simple monism. Reversibility is the character of an autochtonous field, a field that is not one of cause and effect, but one in which meaning is participatory. The lived-body neither causes nor is the effect of the feel of music. We must say along with Merleau-Ponty that the music feels itself through me.

The lived-body is attuned to the world. The body schema enacts the meanings of sensation as forms of behavior that exhibit *a priori* structures. These meanings are essentially placial and temporally run-off in an encompassing meaning-context. Music intensifies the motor physiognomy by which the placiality of the body schema enacts motor forms of behavior.

The preconscious, pre-personal lived meanings open us to the world. The lived-body's attunement to music is already a surrendering to an artifice of an Other. The universality of music is a communing in Flesh. We can respond to music of cultures far removed from our own, for the intensification of motor physiognomy allows for an immediate attunement to which consciousness cannot help but respond. From the feel of the Flesh, world harmony emerges.

Morgan State University

REFERENCES

Behnke, Elizabeth A. 1997. "Ghost Gestures: Phenomenological Investigations of Bodily Micromovements and Their Intercorporeal Implications," pp. 181–201, in *Human Studies*, Vol. 20 No. 2, edited by Anthony J. Steinbock. Dordrecht: Kluwer Academic Publishers.

Casey, Edward S. 1993. *Getting Back into Place: Toward a Renewed Understanding of the Place-World*. Bloomington: Indiana University Press.

Dufrenne, Mikel. 1973. *The Phenomenology of Aesthetic Experience*, translated by Edward S. Casey. Evanston: Northwestern University Press.

Ihde, Don. 1983. *Sense and Significance*. Atlantic Highlands: Humanities Press.

Merleau-Ponty, Maurice. 1962. *Phenomenology of Perception*, translated by Colin Smith. London: Routledge & Kegan Paul.

Schillinger, Joseph. 1978. *The Schillinger System of Musical Composition,* Volume I: Books I–VII. New York: Da Capo Press.

——. 1978. *The Schillinger System of Musical Composition,* Volume II: Books VIII–XII. New York: Da Capo Press.

Straus, Erwin W. 1966. *Phenomenological Psychology Selected Papers*, translated in part by Erling Eng. New York: Basic Books, Inc., Publishers.

ETHAN JASON LEIB

FOREMAN FOR EVERY MAN: *PEARLS FOR PIGS*

In *Lava*, Richard Foreman utters, as if talking to himself, "I understand you're looking for an excuse for a lot of fuzzy thinking." Foreman tries to avert all criticism of his thinking by showing us that he knows that he can be interpreted as believing in contradiction only so that he does not need to resolve the world of paradox in which he lives and writes. But we must ask whether Foreman's acknowledgment of his "fuzziness" really absolves us of scrutinizing him? And, of course, it does not. Here, we will rigorously trace Foreman's thinking in spite of his desire that we refrain from thinking too hard about his "message"; in spite of his wish that we occasionally abandon "mental rigor" for the sake of allowing ourselves to experience inspiration. He describes what he wants from us with the following analogy in his *Unbalancing Acts*: As a child's top spins, "the top makes a humming noise, and the individual pictures [on the top] can no longer be distinguished. All that's seen is a blur. But the child becomes fascinated by that energetic blur." Foreman begs us to become fascinated with the energy of the hum and to stop concentrating on the individual pictures, or in this case doxa, on the top. Yet, as Foreman must recognize, if the pictures are there to begin with, the energetic blur is, in important respects, not revealing truth. Foreman sometimes forgets this. This is why we have slowed down the metaphorical top in this essay, just as Foreman slows down that top in his new production *Pearls for Pigs*, which is all but a handbook for understanding his work. But this slowing down is only temporary, as Foreman struggles to re-spin the top at the end of the production to reestablish the mystery he so desperately tries to engender.

In *Pearls for Pigs*, the Maestro, an obvious representation of Foreman himself, speaks thus:

> I say things that don't make sense, but now you understand my motives, so it makes perfect sense ... God damn it! That's why I wanted to keep things to myself because now that I explained myself, our relationship has changed. But I liked it the way it was when I was just your problematic object of examination. Now that I explained myself, look ... at me! Suddenly cast as a wise man holding the balance of things on the tip of my tongue while you sit beholden, waiting for me to perform my particular therapy. No! I liked it better when I had you off balance, trying to figure me out.

Foreman expresses discontent at how much *Pearls for Pigs* must be a "pandering to the masses," so to speak – the title of an earlier Foreman play.

A.-T. Tymieniecka (ed), Analecta Husserliana LXXIII, 171–186.

In *Pearls*, Foreman slows the top down to such an extraordinarily slow pace that the individual images which he is trying to impress resonate too clearly. This belies the mystery that he feels is crucial to his theater. Yet, it behooves us to use this opportunity to understand Foreman by observing the gyrations of this play against the background of earlier productions. In *The Other American Drama*, Marc Robinson writes that Foreman's "*What Did He See?* is full of longing for return, for a time when all things still wore their mysteries."[1] That return is precisely what *Pearls* reiterates but with the frustration that, indeed, the mystery has been unraveled.

Foreman's "pandering" and explicitness in *Pearls* notwithstanding, he describes his experience in *Reverberation Machines* as continually one of "hummm, that's not quite right" as he backs away for new angles of approach. The blur to which Foreman appeals is a "hummm" which is a constant state of unknowing, a constant flux, a struggle for new perspectives. Yet, what Foreman tells us he hopes for in *Reverberation Machines* is that his art does not "'speak' of this … but rather EMBODIES this. To speak of such things is invariably to drag the new back into the language of the old … and the old language is always (we are habituated to it) more 'powerful,' more 'seductive,' to our mental mechanisms, than the new insight." What makes Foreman unique is that his theories are embodied in his work of art, and not merely written in prosaic or poetic form. When Robinson writes that Foreman produces anti-theater pieces or works which somehow end up in the theater, he is not quite right, to employ a Foremanism: To experience Foreman, one must be in the theater. The Maestro asks Pierrot, his clown companion, to read his book to him out loud. He exclaims: "If I hear it spoken by a human voice, I might comprehend it better than just by reading it alone." It is crucial that Foreman be read to us, acted for us. As literary criticism, suggesting that all is a matter of perspective and that fragments ought to take precedence over metanarrative, Foreman's insights are ones we have heard before (from any post-modern theorist since Nietzsche). As poetry, Foreman's writing is far inferior to that of most French symbolists with similar ideas (Rimbaud, Valéry, or Mallarmé, for example). As analysis, Lacan and Freud far surpass Foreman's rather basic psychological penetrations. And as philosophy, any French post-structuralist (Derrida, Lyotard, and others) could have written treatises which look a lot like *Lava* or *Pearls for Pigs* and, indeed, they have. Yet, to make such treatises work on stage as drama is Foreman's particular contribution to the post-modern corpus. Foreman has taken the post-modern and made a post-modern *drama* of it. Though he may want to think of himself in the philosophical-psychological historical tradition, Foreman is far more important and unique in theater than he ever could be in academia.

WHAT FOREMAN MEANS

A Foreman play, as Foreman himself reminds us in *Reverberation Machines*, "always makes sense. Sense *can't* be avoided. If it first seems to be non-sense, wait: roots will reveal themselves." There is no such thing as before, beyond, or outside of meaning for Foreman as Robinson might have it. By making us feel comfortable with a certain usage of a word, Foreman can lure us into embracing a specific referent for a specific word in a specific context. And then he destroys that comfort by making us question our habitual embracing of a specific meaning. Only once we see what is at stake in each word we choose can we even have a chance at saying what we mean, at authenticity, perhaps even at control.

But the picture is not so univocal. Foreman believes that part of his theater is about being "a reverberation machine" which shows us how "everything is secretly present in everything else." And if this is true, then we can never have control over what we mean, precisely because the world is a reverberation machine which unifies all meanings by showing them as interrelated, thus divesting them of the intentional signification of the speaker. Ultimately, Foreman suggests that the specific context in which a word appears is random and, therefore, careful consideration of word choice is only an illusory form of control. But, in this vein, Robinson overstates the case regarding Foreman's randomness. It is not that Foreman "won't add." Rather, Foreman's addition is not linear; his ideas agglomerate one atop the other, never reducible to a formula, because context is always changing as time is always changing. But, maybe "random" is a more apt description if we think about a strictly mathematical model: The random number generator in mathematics does have a few seeds and an algorithm, an equation. The numbers which the algorithm produce simply *appear* random. At once we can perceive a randomness as well as see its pseudo-random quality, as Foreman reveals to us an algorithm of sorts, a logic.

Indeed, Foreman's theater is saturated with rules. For Foreman, rules allow for the possibility of any meaning whatsoever. Without a logic, sense does not need be avoided because it does not even exist as a potentiality. And it is a logic of modality which Foreman cultivates. He prefers a logic which juxtaposes opposites because, for example, as he writes in *Unbalancing Acts*, "feelings of hatred do not arise in a situation in which love is not possible."

Foreman's use of irony is more total than Ibsen's or Shakespeare's (as is the experience of his theater in general),[2] for his irony is not directed at a person, but toward every word uttered. Proposition **X** can only enter as something meaningful against the background of not-**X**. To create possibility always means to confront the opposing drive beneath any drive which makes the original (is there any original after the reduction?) possible.

Foreman displays all possibilities of meaning on a field by changing the subject all the time. But, since the subject is the field, according to Foreman, we can never really change the subject. As we learn in *Reverberation Machines*, "The subject of the work itself [is] the continual change-of-subject, interruption, re-beginning [which] reflects the true shape and texture of conscious experience, which recognized and reflected in the work of art puts us in the very 'place' where being-human becomes a free and creative way-of-being." It is precisely that psychic freedom which Foreman wants to engender so that we might speak (at least) more playfully, which would translate into a heightened degree of authenticity.

And yet, with all the psychic freedom in the world, Foreman writes, "Whatever I say necessarily misses the mark. Because I, myself, don't choose the terms in which I explain what I'm feeling about myself or the world or anything imaginable, since of necessity I use a language forced on me. Invented by other people." Yet Foreman does not mean to reserve this problem with communication only to his own situation. Rather, this describes the human condition.

SLIPPERY TWISTING

Foreman is always twisting, contorting, and interfering with the narrative of his plays to make paradox a revelatory experience. Slipperiness, as a state of being which continually undermines what is written into the text, exists so that the gears of imagination will not lock up, so that we never remain stuck in unimaginative interpretation which inhibits psychic freedom. Ironically, while recognizing the limits of psychic freedom, Foreman uses contradiction as a means to accomplish the freedom, which itself cannot not escape the hold of contradiction.

Foreman thinks of his work in *Unbalancing Acts*, "as an attempt to crack open the prisonlike shell of the particular reality we have convinced ourselves imprisons us. Backing up, slamming against the wall and striking an odd position is an attempt physically to crack that shell." And this break occurs through a "ricochet between levels of meaning," to employ Foreman's own analogy. The vaudevillian Maestro in *Pearls* oddly throws himself against a vertical mattress repeatedly, a physicalizing of the psychic process of ricochet, of reverberating meanings. In this way, Foreman, in the words of Schlomo Leviticus in *I've Got the Shakes*, wants to "Hold it! Right there" in mystery. This imperative occurs after Leviticus proclaims, in proper Foreman fashion, "That's not quite right!" To hold all further investigation with the qualification of "That's not quite right" is precisely the type of mystery that Foreman wants to preserve. In *Pearls* we hear the same locution as we see a similar circus-like set.

The poles which appear in the middle of the sets in *Pearls* and *What Did He See?* are the gears which turn the world in mystery, and we need slipperiness as a lubricant so that they do not get stuck. Everyone on stage gyrates, turning around in his place and turning the cranks on the ends of the poles. The poles also undermine the convention of setting. In *What Did He See?*, we learn from Foreman that "had there been no bizarre pole in that living room, and the actor had instead spun wildly around the post of a normal banister, it would not have referred to [a] wider level of meaning. It was only when the actor ran to manipulate the pole that he seemed to be entering another level of the set." The poles, like the strings separating the stage from the auditorium (an idea he steals from Brecht) are tools for evoking alienation. Yet, the most powerful alienation effect is the apparent randomness of the script, the narrative (or lack thereof) itself, which comes to make sense. By showing us that what we think is nonsense makes perfect sense, Foreman creates a most powerful alienation, far more powerful than nonsense and contradictions alone could be.

Foreman's use of music is also a unique incarnation of the ideas he seeks to reify and embody in his theater. Foreman writes in *Unbalancing Acts*: "In Hollywood movies, they will almost always use music to reinforce the emotional atmosphere, intensifying the most obvious thrust of the scene. I, on the other hand, [am] interested in introducing music that displaces the emotional quality of the scene." In *Pearls for Pigs*, at the moments of greatest poignancy, loud blaring circus music is played. The drama is meant to be taken seriously and, at the same time, to be dismissed as laughable melodrama (and, indeed, every character on the stage laughs at Maestro in his most overblown moments). We cannot forget, however, that they bow to him as well.

Light, too, is manipulated for inversion's sake in Foreman's world, as it both elucidates and blinds us when too bright. In *Pearls*, however, the blinding light is often directed at the actors, not aimed outwardly to the audience, as is usually the case. (This is Foreman's technique for engendering self-consciousness in the audience.) Though *Pearls for Pigs* is a "pandering to the masses," it is not as much a manipulation of the audience as it is an occasion for extreme lucidity for Foreman himself vis-à-vis his own self-deception about what he is really accomplishing. When the Maestro is asked to do something with all the characters on the stage, and he says no, the lights blind everyone present on-stage as the Maestro walks off and the action stops (until the Maestro returns). This moment is a testimony to Foreman's degree of self-knowledge that he is doing nothing with his characters on stage other than staging his own autobiographical concerns.

Robinson acknowledges that for some moments Foreman's characters "seem knowable – difficult and mercurial, but knowable." Because we need something that we can hold on to so that something is at stake for us, we are given an opportunity to become acquainted with Foreman's characters at the outset. We are given the makings of a typology so that we can comfortably fit the characters into some habitual categorization, only so that Foreman can defy our expectations. The Pierrot character is dressed in a clown suit and begs mockery. Yet, he is revealed to be the Maestro's "most important witness," which both heightens the clown's place in the drama and lowers our expectations of the Maestro. The audience must connect with Foreman's creations (mirrors?), but without excessive attachment. We ought not to come out like the magicians at the end of *Pearls*, dressed in the clothing of the Maestro, the character with whom we might identify too closely. After all, the magicians are mocked as those who misunderstand the Maestro's "message."

Foreman's reverberation machine produces a double bind, a mixed message. Foreman writes in *Unbalancing Acts*, "The child is given contradictory signals; that's the double bind. Bateson theorized that such crossed messages are at the root of schizophrenia. While I may not employ this strategy to drive people clinically crazy, I do find it useful to employ externalized double binds because the frustration they create demagnetizes the spectator from normal avenues of conceptualizations." To subvert habit is to find *impulse* for Foreman. And when habit is rejected, the possibility for more inventive behavior is opened. This is consonant with Foreman's stage direction of "invented behavior" which recurs throughout his later plays.

CATEGORY THREE, OR DON'T SAY "AH!" YET

"What I'm after," writes Foreman, "is what in *Lava* I call 'category three,' which is a third possibility between logic and randomness. It's somewhere between narrative development and pure chance." Between the two realms of meaning and contradiction, a third possibility emerges as the content of Foreman's plays. In *Pearls for Pigs*, the characters on the stage consider the ramifications of the expression "Ah!" as it relates to discovery, to comprehension at a given moment. *Pearls* suggests that "Ah!" is the impulse of understanding, but it is precisely the moment of self-deception because the "Ah!" must always come *after* the actual impulse. Indeed, the "Ah!" kills the impulse. And it is not so that audience members may say "Ah!" that Foreman explores the impulse. Rather, it is the impulse itself which Foreman tries to cultivate, all while he goes to great lengths to interrupt us before we might explode with the fateful "Ah!"

Foreman is "proposing that there are new ways to navigate" which he is trying to explore with his plays. He hints at this mode of navigation in *Reverberation Machines*, using the analogy of "quantum jumps," suggesting that there exist "parallel rivers of discourse," and indicating that you can "jump back and forth between those rivers." The energy of that jumping, that shifting, is what drives thought. And his plays are "not ABOUT THOUGHT, but ABOUT WHAT DRIVES thought," a formulation devised in *Reverberation Machines*.

Foreman's later articulation from *Unbalancing Acts* is that he attempts to "demagnetize impulse from the objects it becomes attached to. We rarely allow ourselves the psychic detachment from habit that would allow us to perceive the impulse as it rises inside us, unconnected to the objects we desire. But it's impulse that's primary, not the object we've been trained to fix it upon. It is the impulse that is your deep truth, not the object that seems to call it forth." And when we experience that truth, we have the most hope for creative and inventive behavior. For this reason, Foreman sees himself as working in the Steinian tradition (in which one has a response to a situation, and immediately goes back to the original situation and tries to respond again – and begins again).

By aiming at the cultivation of personal impulse, Foreman desires a particular type of self-knowledge. He writes that "the impulse is the explorer-consciousness, staking out new claims, finding a way to be self-generative, not simply reactive." The Maestro proclaims that he is the most intelligent being precisely because he knows the most about his own life. Yet, to gain self-knowledge is to gain knowledge about humanity. The personal becomes universal as our wristwatches unify us to all human time. Like the personal time of our wristwatches which always refers to some more objective clock, Foreman explores personal impulse because it must give testimony to the general human experience of impulse. Indeed, in *Pearls*, the characters are fixated on the relationship between the clock on the wall and their wristwatches. Just as, for Foreman, making art is intimately personal though imbued with the hope of transcending himself in the universalization which art makes possible, so too, viewing art ought to be able to accomplish the same in the best possible world. But often this movement of the audience must be cultivated through its unconscious.

THE UNCONSCIOUS ACT OF ART

To focus attention on the impulse, Foreman subverts the usual way we may anticipate the impulse's becoming manifest. Foreman, however, does not like the drama of a character speaking "Let's go," while the stage directions suggest that no one moves. He prefers to make it physically impossible for the movement to occur because then we are not led to ask, "Why don't Gogo and Didi leave? The stage exit is right there!" Yet, Foreman acknowledges a similarity with Beckett insofar as he claims in *Unbalancing Acts* that

> despite all our great plans and monumental projects, what are of real weight are our obsessive fidgets. Everyone working in this tradition shows that the incidental stuff left out of the goal-oriented narrative of your life is actually the crucial, potent, soulmaking material. And that's also the insight of psychoanalysis; how the seemingly unimportant slips of the tongue, dream images, meaningless details you think tangential to your real concerns, are in fact the royal road to that unconscious that determines your true fate.

Thus, Foreman believes in false starts. Like the many "takes" which constitute *Pearls for Pigs*, false starts are like film takes which are edited out before printing. But instead, leaving them on the cutting room floor, Foreman stages failed beginnings because he thinks they contain truths that he may be repressing. Because Foreman is convinced of the Lacanian concept that "truth emerges from the mistake," he is always eager to stage his unconscious workings-out of the play which are manifest in early versions.

Fundamentally, Foreman sees himself as lazy. But he *wants* to be lazy because laziness is conducive to rumination and day-dreaming. Foreman thinks that truth can only show itself when one is receptive to truth. And one is only receptive to truth in situations wherein one is not self-consciously constructing a narrative, but allowing the narrative to emanate from within, allowing it to "RADIATE," albeit "intermittently," to employ Foreman's humble description. Like the situation of the free-associating analysand in analysis, the theater-goer watches Foreman's unconscious at work and often feels like Foreman is spewing nonsense. And we are expected to ask again and again with the Doctor in *Pearls for Pigs*, "What drives you?" At some point in the associations, the analyst, in this case the audience, can make sense out of the nonsense. But the audience does not really *make* anything. The sense is always there and at some beautiful moment, the sense is discovered. The coherence, Foreman hopes, reveals itself.

Yet, Foreman also wants to induce a dreamlike state in the viewer. In *Pearls*, when the Maestro delivers a long philosophic monologue, all the people around him fall asleep, and eventually, so does the Maestro himself. But the sleep does not only point to a boredom with or mockery of philosophy and the whole philosophic enterprise. Sleep also functions positively. Foreman wants to put people to sleep, but also to awaken them to their own meandering dreams. As Strindberg uses the structure and form of a dream in his dream plays to allow for playful imagination, Foreman uses his own dreaminess "to fight entropy." By showing the "disorder" of the dream, we are forced to confront the orderedness which the unconscious strand embodies.

Just as we have difficulty remembering a dream, the details of Foreman's plays are often difficult to recall. Foreman's theater is just like a dream which you experience with intensity but, try as you might, fail to recall; that is, unless you wake and write, which is exactly Foreman's technique. But trying to trace one's mental experience of watching a Foreman play belies the forgetfulness Foreman seeks to engender. He writes in *Reverberation Machines*, that "people who work in time are making things for memory, are not clear about here and now." Foreman aims at a forgetting which allows one to focus on now, on the impulse.

Though he draws a lot from its reservoir, Foreman's relationship to psychoanalysis is not univocal. As embodied in the Doctor in *Pearls*, psychoanalysis wins a Pyrrhic victory: It succeeds at forcing the Maestro to confront himself, but it deciphers mystery in a way that misses the truth of the matter. Foreman employs the basic psychoanalytic insight that dreams are revelatory, but refuses to systematize alongside Freud because he thinks psychoanalysis is too reductionist and causes man to be imprisoned. For this reason, the Doctor is the target of much criticism and mockery (he holds a golf club throughout and strokes at decapitated heads), even though he does succeed in teaching the Maestro something about himself through his analysis.

Foreman's plays of the early '90s are an examination of the return to childhood. This technique of therapy (of a return to the narcissism of the child) has its roots in the psychoanalytic tradition, but *Pearls for Pigs* is a movement forward, after denial, to an owning up. To Robinson, all of

Foreman's work "looks like a long sequence of denials – an ongoing assertion of what he won't let his theater become." Yet, although *Pearls* contains elements of this type of denial, Foreman acknowledges, in a bright light, that his theater is, after all, not wholly new and inventive. The backdrop of the set is a stage mounted atop the main stage of *Pearls*, with a prominent curtain (a recurring theme in *Pearls*) draped in front of it. The curtain remains closed for most of the performance, even though there is much conversation about what is behind the curtain. When the curtain is lifted and a blackboard with writing appears, we can easily read the "overtones" to *Pearls*. Foreman acknowledges that he is in the tradition of the theater of fragments (Stein), the theater of catastrophe (melodrama), the theater of lost hope (Beckett), as well as the theater of sleep (Strindberg). Later in the show, a circular platform on the blackboard upon which these descriptions are written is spun to embody the way Foreman takes from each of these types of theater and then spins (and contorts) them together to make them unrecognizable. The synthesis of these types of theater is what makes his theater unique. *But to synthesize is not to move beyond the tradition*, and Foreman knows this. We know that Foreman always struggles with synthesis, as he tells us in *What Did He See?* through Will, who is obsessed with Hegel, the philosopher of synthesis. We are also told at the end of *Pandering to the Masses* that the meaning of that play can be found in Hegel's *Phenomenology of Spirit. Pearls* is a step forward from the plays of the early '90s because, in the words of Lacan, there is a return of the repressed which is part of the cure.[3] Foreman returns to what he really means to say and begins to appreciate his place within the tradition instead of attempting to break free. He fights vigorously to overcome himself through his art. But through being true to his unconscious in *Pearls*, he better understands that which he has been repressing, the truth of his derivativeness, the truth of his anxiety of influence.

But the anxiety of influence is more than simply the artist's perpetual concern with outdoing and transcending his predecessors. Harold Bloom's *The Anxiety of Influence* also characterizes the strong artist as one who "begins by rebelling more strongly against the consciousness of death's necessity than all other men and women do." The artist, and Foreman is no exception, struggles to mark out his own territory and is always raging against time, against death. And yet, with Freud, Foreman appreciates that the artist also has a death drive. Foreman believes that "to produce, especially in the realm of art, is to turn the live thing (the impulse) into the dead thing (the work of art)." Elsewhere, he writes that the "hidden subject of *all* theater [is] death because the actions of a play, memorized and repeated in front of an audience rather than issuing from real spontaneous life, [are] dead actions." And to reflect that belief, the sets of his later plays are strewn with skulls, from paintings on poles, to rotting heads lying on lintels. In *Pearls for Pigs*, the image of the guillotine along with images of blood and stabbings recur time and again as the Maestro willingly walks into the hands of his executioners (Columbine, his lover/muse, Pierrot, his magicians, and us). Yet, the heads in *Pearls* are freshly cut off; the flesh has not been worn away by time. Death is somehow not as old in this play. The act of murder is fresh. The Maestro is killing his ideas by bringing them to life in the play, and in giving away his mystery, he feels that he may as well be dead. Death here is felt as real imminence, not simply as melodrama. The death instinct is what drives Foreman to write and "perform," and because death is such a real possibility in this play, the threat forces Foreman to be more explicit than he usually is. It forces him to pander to the masses, to give away his pearls to the swine.

ART AS POLITICAL/ART AS TRUTH

"Bad, boring art," writes Foreman, "tries to convince (usually in the realm of feeling)." Foreman, by contrast, always tries to avoid being moralistic, which, in some way, allies him with the standard post-modern doxa of perspectivism. He writes that "art should begin the process of freeing men by … laying bare the fact that it is a web of relations that exist, only." Thus, unexpectedly, an obvious political liberatory program lies underneath Foreman's theater. He may not want people to leave the theater and join a revolutionary group as Brecht might have (Brecht was probably more of an aesthete than a politically motivated artist, but I digress), but he does have some political program. And Foreman reveals as much theatrically and most clearly in his sets. In *Unbalancing Acts*, Foreman writes that "theater is a public meeting; it is a ritual. In order to suggest the ritualistic, ceremonial nature of the theater during the performance, the set, whatever the ostensible locale, should also refer to a place where you might come to meet with a lot of people: a hotel lobby, a lecture hall, a schoolroom, a doctor's office, or a courtroom." Appropriately, four out of the five suggestions are places of learning and judgment. Could there be a set more didactic than a lecture hall, more infantalizing than a schoolroom, more terrifyingly vulnerable than a doctor's office, or more judgmental than a courtroom? And indeed, the setting of *Pearls for Pigs* evokes many of the aforementioned associations. From the blackboards to the omnipresent Doctor, and the books to the study hall tables, the Maestro's didacticism is never understated.

Foreman knows that he is an artist and that artists lie by nature. But, as he reminds us in *Unbalancing Acts*, "as Picasso said, art is a lie that tells the truth.... Art is a perspective; all perspectives are lies about the total truth; so art is a lie that, if strategically chosen, wakes people up." Foreman simply tries to focus upon telling the right lies to effect the particular human emancipation that he seeks. He does believe, as the Maestro tells us in *Pearls*, that "the important work of the theater is to be done ... elsewhere," in life. The lies of the theater are supposed to serve life. No different from Plato who endorsed the noble lie for the sake of order and the betterment of the aristocrats, Foreman aims to discover which lies are noble and which are base and to keep those who have the potential to stay awake (an elite of some kind) out of dogmatic slumber.

But the work of art is not just for others. Foreman believes that the "work of art is always a picture of one's ideal world, a postulated utopia." As the dummy in *Pearls* testifies, the work of art is always a representation of the artist himself in one form or another; it is a glorified portrait of the artist. In the very last sequence of *Pearls*, a dummy gallantly rides across the stage on a faux-horse, baffling everyone on it, including the Maestro himself (not to speak of those witnessing the Maestro, including the audience). The dummy, the Maestro's ideal ego which he actually wears at certain junctures in the play, turns out, in the end, to be something which has emerged from the Maestro, though he can barely recognize it. The work of art stands as a mirror to the Maestro's life, which is why he needs no mirror. He cannot recognize his reflection anyway. The Maestro asks himself why he goes on, why he continues to be in the play, and his response is that he loves his dummy, his own utopia, his work of art. When the Maestro asks us what kind of mirror each one of us is, he suggests that we ought to compare our lives and our ideals. He asks us how well we incarnate our ideals in *our lives, our own works of art*.[4] *Pearls* begs us to measure our actual lives against what we want to see in our own reflection and reminds us that, if we cannot undergo a rigorous evaluation, the artist, Foreman, will do it for us.

A PORTRAIT OF THE ARTIST

A "MIND ATTACK," as it is articulated in *Pearls*, is the only proper reaction when one comes face to face with the reality that all is a matter of perspective and that one can never know when others, as actors in the drama of one's life, are improvising and saying what they feel, or articulating pre-programmed formulations. Madeline X, in *I've Got the Shakes*, exclaims, "If I went crazy, all my problems would be solved," much the way Cloy says in Beckett's *Endgame* that for a moment he had thought he was in his right mind, and then he became lucid all over again. The Maestro in *Pearls* experiences the same sort of insanity and proclaims that he will sacrifice his own sanity so that the members of the audience themselves need not suffer a mind attack. This places the artist, in Foreman's world, as a figure in our culture who is so in touch with reality that he can and must suffer for us. And we, as onlookers (witnesses) to that suffering, live vicariously through the work of art, the work of the artist.

The Maestro suffers a mind attack so that we will not, *even though taking him seriously means that we must*. And if we do not have the apparatus to emerge from a mind attack (which is the consequence of the type of nihilism Foreman embraces at times), we must ask whether it fair for Foreman to willingly and intentionally launch the mental machinations which cause a mind attack. The assumption Foreman must make is that only those equipped to emerge therefrom can understand the play and take it seriously in the first place. Unlike the Platonic philosopher who needs to be dragged back into the cave after he gains his own particular wisdom and undergoes his own particular therapy of climbing the mountain, the poet-artist has the need and desire to come down the mountain, reenter society, and speak to the community. In *What Did He See?*, this imagery of coming down from the mountain and proselytizing to the masses emerges most vividly. The character Lili recalls "the story of the well-known Buddhist monk ... who spent maybe a hundred years alone on a mountaintop to get calm and peaceful, and then, the first time back, he goes into the marketplace and gets bumped around." Much like Nietzsche's madman (did he go insane from a mind attack?) in *The Gay Science*, or Zarathustra in his *Thus Spake Zarathustra*, Foreman the artist feels the need to speak from an over-abundance of wisdom, yet meets adversity when the community to whom he speaks cannot absorb or understand his message. Will's advice to the artist is that he "should never have come down from his mountaintop." But, of course, the artist can never take his own advice because he must speak in spite of himself; it is hardly his choice. Sometimes he even speaks just for the sake of expression, not minding if he fails at communication. Is Foreman guilty of this egregious form of self-indulgence? With Pierrot in *Pearls*, we must concur that "it is too early to know."

Will remarks, "I made a serious commitment to a spiritually motivated being I never met – you know how many people laughed behind my back? So I made it part of my act." In *Pearls*, the magicians bow to and laugh at the

Maestro all in one scene, showing us that even though Foreman knows he will have to be laughed at and mocked in the marketplace, there is hope for the artist's words to be revered. Of *What Did He See?*, Foreman writes that the play was about "how to deal with the realization that the goals you work toward aren't worth a candle, while nonetheless understanding that giving up is morally unacceptable." Even though the artist struggles to be understood, knowing well that he never can be fully clear, he cannot give up. What is crucial is how he manipulates the audience in his effort not to be laughed out of the marketplace completely.

THE MAESTRO OF MANIPULATION

Foreman's first principle is that, as in a dream, we must "drift to coherence." This enhances the effects of joy and delight in finding the coherence. If the coherence is too obvious, no one can think that he has access to a secret. And discovering a secret induces the excitement of being let into a secret society of understanding. For the audience to feel "creative pleasure" is one aim of Foreman's theater. He wants his viewers to feel that they have received special knowledge or wisdom from the play to which no one else in the audience is privy. If an audience member believes that everyone understands some particular moment of the play, there is no effective creative pleasure. As *Pearls for Pigs* is exoterically directed to "Dr. Fishman," each member of the audience is supposed to feel that he is Dr. Fishman. Though theatergoing is a group activity, the experience is meant to be personal and individual, because if people are made to feel special by a particular discourse, they are less likely to laugh it out of town.

Yet, in the pandering which is *Pearls*, there is a recognition by Foreman that everyone in the room ought to be able to experience creative pleasure. It is overt, if that is the right word (or is that not quite right either?). In *Pearls*, Foreman has decided to become more accessible than he usually tends to be. We know that this bothers him tremendously, as the Maestro prefers to be a "problematic object of examination." Yet, we need not thank him for having allowed "our relationship [to] change," even though he has the Maestro ask the audience to utter a silent "Thank you" for his efforts. Because Foreman knows that "the truth of art is in the audience's, the individual's, awakened perceptions," he knows he must undergo the pandering which is so painful. And by showing us the Maestro's pain in being understood, he dramatizes his own frustration and makes the audience members feel sympathetic in the process. In this way, he manipulates the audience members to repress their moments of understanding and can, ultimately, preserve the feeling of creative pleasure because it comes only rarely, between the moments of repression. *Pearls* does not lose too much by being accessible, because it is manipulatively accessible.

Foreman is never lazy with regard to the construction of the space on his sets. He tells us in *Unbalancing Acts* that he "assume[s] that the spectator is watching the entire stage at all moments of the play, so [he] tr[ies] to make a stage picture in which every inch of the stage dynamically participates in the moment-by-moment composition of the piece." The stage allows the audience members to make choices and find meaning in whatever they choose to fixate upon. Though Foreman believes that he is not autocratic by allowing the audience's mind and eye to drift, he fixes everything on stage so that no matter where one glances, one confronts not just meaning, but Foreman's own meaning. He pretends that there is an element of non-control in his directing, and that directing is located in the dialectic between control and non-control, but one is hard-pressed to believe this after attending any production, including *Pearls*. He wants a proper relation to ambiguity, but the audience becomes so amazed that Foreman has layered his production so meticulously, that it becomes distracted from the fact that there may be ten more meanings added to the ten that Foreman has already provided. Though Foreman wants the audience to be manipulated into confronting possibility as an abstract concept, it often becomes too difficult to avoid simply (or not so simply) confronting the particular possibilities Foreman suggests. This manipulation tends to subvert psychic freedom even though Foreman does achieve a self-consciousness of the viewer which causes him to fixate on the layered quality of most images and words. He tries, through self-consciousness, to force the viewer away from accepting traditional meaning and to learn new strains of discourse even if they seem strained and far-fetched at times. Once we break the old moorings of speech by becoming self-conscious about our cognitive faculties, we might reclaim authenticity by absorbing some of the light from the play's radiance.

For Foreman, the power of self-consciousness lies in its ability to disappear. Indeed, only by drawing in the members of the audience, can he alienate them into self-consciousness. His work is a kind of tornado which pulls everything toward its spin without letting anything actually succeed at getting to the center. He forces the audience to lose its self-consciousness in the elaborate spectacle, which is precisely what he asks us to question in self-consciousness. By telling us through the Maestro that he has "run out of stories" and that "all that is left is a hole," and by giving us the makings of a traditional narrative in *Pearls* nonetheless, he makes us oscillate in and out of self-consciousness. And only in this oscillation does self-consciousness have any power to effect change, to have affective ramifications, to emancipate. It is the speed of the oscillations, in and out of self-consciousness, which gives the experience of watching a Foreman play its potential to succeed in its liberatory program.

It is up to each individual, however, to decide ultimately whether his personal wristwatch is synchronous with Foreman's own.

NOTES

[1] All citations of Marc Robinson in this essay refer to his chapter about Foreman in his *The Other American Drama* (Cambridge: Cambridge University Press, 1994).

[2] Reminiscent of Wagner, Foreman strives for his theater to be a total experience, affecting as many *senses* as possible.

[3] See Lacan, *The Seminar of Jacques Lacan: Book I: Freud's Papers on Technique*, Chapter XV, "The Nucleus of Repression" (New York: Norton & Company, 1988) 187–199.

[4] This idea mirrors Alexander Nehamas's interpretation of Nietzsche in his *Life as Literature* (Cambridge: Harvard University Press, 1985).

SECTION IV

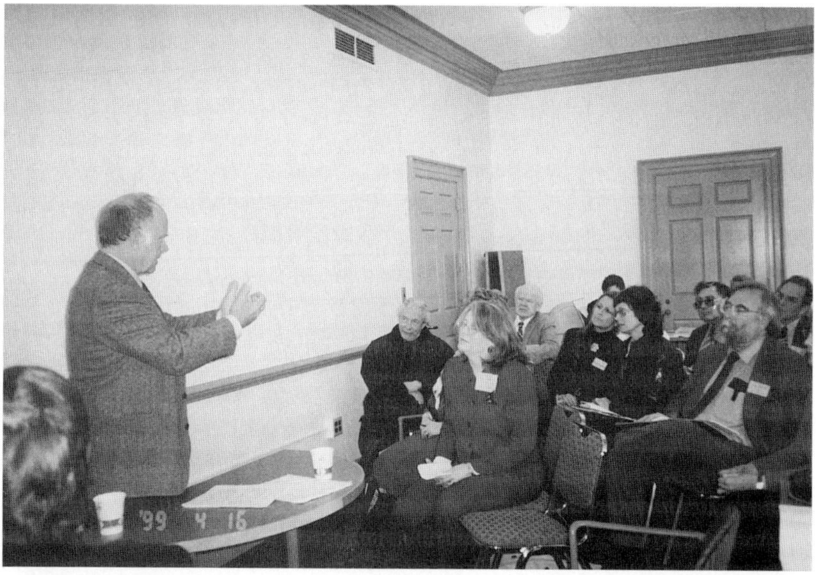

Lawrence Kimmel lecturing.

LAWRENCE KIMMEL

RECONCILIATION AND HARMONY: THE PHILOSOPHICAL ART OF TRAGIC DRAMA

PREFATORY REMARK

In the performance of art, one can begin at the beginning, but in a discussion of art one must begin somewhere in the middle. Here, it is with the conviction that art in whatever form, though it may surprise the senses and quicken the spirit, disturb our thinking or revoke a thoughtless ease, still, its full expression restores a sense of presence and wholeness to our being. That is, every art form has a point of closure in a harmony of the spirit. Even tragic drama, which brings the darkness of human character into a glare of recognition and acceptance, must find harmonic resolution, or fail as art. In this paper I will focus on the *idea* of classical tragic drama, the sense in which it discovers and reflects the convergence of possibility and inevitability in human life and world. It will be my claim that a primary aim of tragic art, paradoxically, is to reconcile the human being to his nature and to the world of his making. This mode of reconciliation fits a larger pattern and task for the whole of the fine arts: to disclose and create a harmony of Man and World.

I

The poet's expression "All the world's a stage…," wherein we its actors have our roles to play, does more than suggest a reason for the universal access and enduring response to Shakespeare's drama. The claim strikes a common chord of agreement that does not depend on poetic license – it seems less a metaphor than a simple fact. In Shakespeare's work, this world/stage metaphor creates a space in time for the comic and tragic strife and stridings of kings and knaves, wise men and fools. Our lives and language seldom live up to the dark splendor of the poet's tragic heroes and villains, but we recognize with a certainty of heritage the poor player who frets and struts his hour upon the stage. Shakespeare's extended metaphor, stretched by the regularity of its use, loses its figurative force to become an obvious truth: Yes, we *are* like that, and all our yesterdays but light the way to dusty death. Metaphors of world and stage, life and drama are staples of the world of

A.-T. Tymieniecka (ed), Analecta Husserliana LXXIII, 189–200.

literature within the world of the arts, and further, within the world of our shared ordinary lives.

The boundaries of the many worlds we inhabit are not enclosed concentric circles, but circles that overlap in their different dramatic spaces. But there is an insistent notion – both popular and philosophical – that a more comprehensive account needs to be given for the boundaries and substance of the *real* world. Is the boundary of this other, single, final world the totality of shared experience? Or the totality of *possible* experience, whatever that might be? A sensible way to address these questions is to acknowledge that "the real world" is also a metaphor, which leaves its meaning indeterminate and open to a range of viable interpretations. That means, in turn, that the boundaries of any alleged "real" world are the variable limits of human imagination. Wittgenstein, in his curiously mystical *tractatus* on philosophical logic, began with the proposition that "The world is the totality, not of things, but of facts." The world of his reference is the scientific world in which "The world is everything that is the case" – that is, the totality of true propositions. But the world in which we live and dream, rejoice and despair – the *lebenswelt* – has no such limits. If there are boundaries to this life-field, they are temporary and moveable bounds discovered in the creative and sharable domain of the arts.

There are both a real and a fictive sense in which the drama of our lives is played out on the world's stage, or on the many stages of the world. Nor are the boundaries set between dramatic play and real life. To think of oneself as having a life, of living in and sharing a world, to think of this life as coherent, as constituting a story, already requires an aesthetic frame of metaphor. Life and language, labor and literature are all domains and activities that mutually create and sustain. Forms of art and genres of literature mark perspectives through which the staged drama of human being can be conceived.

II

In what sense do the arts constitute and effect a harmony – of perception, of sense, of imagination? Harmony is an internal value to the arts; it is also an affective goal of the arts, to bring Man and World, life and thought, into accord. However natural the original impulse to art – the contemporary child at play in the yard, the ancient Lascaux hunters in the shadows of their caves – the fine arts now clearly enlist freedom as an essential value, a release from the routine functions of our lives. The appeal of beauty in the experience of art calls for a suspension of the ordinary business of living, a release from the imperatives of survival, the projects of progress. We accept an invitation and fall instead, if only for a moment, into the glory of the sunrise as if we ourselves had somehow accomplished it, or into a fleeting fantasy that it was done specially for us. How could the gods not continue their concern through such acts, having first given us the gift to perceive the world in this way? Our minds do not struggle to make such an overarching picture; the sense of harmony comes of its own. The rooster must have a similar sense in his response each morning to the rebirth of the day.

The natural wonder of the world requires no willing suspension of disbelief. Beauty, in the unadorned context of life and world, is more like an impulse of belief, of spontaneous surprise and wonder. We take delight in the things we see and hear: a color becomes an image, a sound a melodic line, several sounds a chord, and the whole resolves into a harmony in the freedom of time, in the distance of unhurried reflection. The development of perspective in the language of perception extends to metaphor, which becomes a line of poetry, which reconnects with image and sound, vision and melody, the harmony of Man and World. It is not difficult in this way to complete the circle and provide a lyrical appeal and expression for a natural harmony. The trouble begins when we try to substitute, instead, a detailed explanation. Then we require analytic arguments, and the effort itself seems to betray the initial impulse that gave life to the idea. Even so, let us make a try at putting the lyrical idea into some kind of cognitive frame, to put the vision of the arts into the language and perspective of philosophy.

When I was a child I recall hearing an adult conversation in which it was remarked or reported that the expression "cellar door" was the most beautiful sound in the English language. I also recall thinking, "Who could decide such a thing, and how would they do it?" (I did not question whether or not this was true; that did not seem important at the time.) The singular thing about this remark for me was, and remains, that of being keyed to a question about the beauty of words, how they sound, how the sound fits with the thing, the possible, secret thing. In this case, part of the beauty of the words "cellar door" must have included what the thing conceals, what strange or forbidden things are hidden there to be discovered. There was, I think, a sense of harmony born in the experience of thinking about this remark, in thinking about the beauty of words, words as sounds, ideas and images, music and meaning, words and world, life and language. I am sure imagination was engaged long before this time, but I think of the self-conscious awareness of constructing a world of beauty through the languages of art as dating from that moment.

III

Human identity, if there be such a thing "writ-large," is not a result or product, but a continuing process in which the cultural activity of the arts plays a primary role. It is common now to voice political reservations about any project that offers more than a provisional description of human identity for a limited use, a use that admits to the inevitable bias of its own time and place. Even so, a view persists that the fine arts, free of politics, constitute a language different in kind but no less universal than mathematics. This view of expressive and receptive capacities common to human beings lends weight to the idea that beauty may be as much a constituting value as truth, with neither being reducible to the exercise of power. These two fundamental concerns of beauty and truth, taken together, provide a foundation for a universal culture of the human that should arguably also include the values of the good and the sacred, as well as the useful. Having said this, we should remain aware of Veblen's scathing and specific inversion of any claim to the universality of these values in his closing chapter of *The Theory of the Leisure Class*. *Only* aware, however: I deliberately leave Marxist concerns aside in what follows. To make that turn would take us down an alleyway of discussion that would not easily offer a return to what I want to discuss in tragic drama.

Great works of art are not limited to an expression of the beautiful. Such works are concerned as well with truth, though not in the validating sense of science, with the good, though not in the justifying sense of ethics, and with the sacred, though not in the doctrinal sense of religion. One could perhaps even argue that the fine arts – sometimes called the useless arts – are also concerned with the useful, though not in the instrumental sense of utility.

If culture is an individuated process, it can also be considered, abstractly at least, a universal project – that of civilization. Although civilization seems clearly to be a product of creative labor, the role of the arts has remained ambiguous and subject to criticism. Both Plato and Freud seem to argue, for example, that the civilizing of the instincts, the taming of passions that disrupt the security and growth of community, can only come at the expense of artistic freedom. Politics and the arts appear at root to be opposing impulses. Plato's ancient anxiety about the "divine madness" of the artist legislated censorship, and Freud's modern analytic of the infantile indulgence of the artist counsels sublimation.

There are however, compatible alternatives to political censorship and therapeutic sublimation. Politics and art align when one considers the ideal of human community to be *freedom under law*, and the parallel requirement of art to be the free play of *disciplined* imagination. There is an impulse to order in the freedom of both kinds of cultural activity, even if the conception and role of freedom differs. A harmony of Man and World does not require, indeed precludes, unison. Art is surely the most vital resource against either a one-dimensional world or one-dimensional man. There are inevitable differences and conflicts that constitute human nature and the human condition of plurality. In a healthy community no less than a great work of art, however, such differences must be reconciled or otherwise resolved into a common key of accord. Such is the task of creative activity in the parallel worlds of politics and the arts. We will consider one ancient and continuing resource in the resolution of discord in the service of cultural harmony: classical tragic drama. But first a codicil may be required for use in the musical metaphor of harmony in this essay.

I realize, of course, that for much of twentieth century music, it is problematic to speak of harmony or accord. The replacement of modal keys with tone rows shifts the idea of the coherence of a work to *cadence* and *progression*, rather than harmonic resolve. If there is no key, there is no root for a resolution. Since there is nothing to get back *to*, creative tension is left unresolved within the medium itself. There are similar problems in contemporary painting. Although I bend the metaphor of harmony to a classical frame throughout most of this essay, I hope my remarks on creative tension within classical drama will suggest ways in which, by extension, resolution is still an essential interest of the arts. Consider only the convergence or resonance of the pitch and rhythm of contemporary music to our fragmented experience and our sense of the complexities in our lives and worlds. There are both cultural symptom and effect in Schoenberg's abandonment of traditional tonality to structure a new world of musical expression. In Bartok's strained and haunting quartets, or in Berg's musical portrait of Woyzeck as a contemporary Anyman, musical mode fits cultural mood in which we are reconciled to the radically open texture of our changing lives. If this is not a classical harmony of man and world, it is ground for a continued search for it.

IV

Tragic drama may seem like an unlikely place to look for harmony between Man and World, for we think of tragedy as precisely that point of rupture between the "overweening passion" of the individual will and an indifferent or hostile universe. Tragic drama would thus appear as a distinguishing art form in which the human situation is one of relational conflict: the individual is at odds with or contending against the universe, the gods, and others. The tragic world further threatens to dissolve into a disharmonious striving with and within the self. In classical tragic drama, the individual, as hero, pits herself as an exception to the relentless logic of the inevitable. In this vision, the human spirit is bound by discord. The heroic effort, as we spectators know from the outset and as the hero comes to recognize and acknowledge, is hopeless in its very conception. Fate is a province of Man independent of will. Destiny cannot be fashioned by an individual, however great in spirit. For Man, Fate is to be suffered; it cannot be commanded. The drama of tragic literature is not that of popular sports or political life. The excitement of the contest is not who will win, but how great the aspiration and noble the comportment of defeat. Yet there is something healing in the experience of tragic drama. It seems to have two defining moments for the spectator, perhaps moments shared with the whole of the fine arts, only here in an articulated intensity: one of provocation or disruption, and the other of recognition and reconciliation.

Mythic literature, which precedes and informs classical tragic drama, is full of stories in which passions of pride and ambition are fed by the fancies of human imagination. Prometheus stole fire from the gods and brought it to earth and mankind, but his eternal suffering as a consequence served as no warning lesson to the hero. Rather, the daring adventure and scope of the Promethean offense appealed to the imagination of this most arrogant of creatures, Man. The act of Prometheus is a tragic gift of *possibility* – an apparent rift in the fabric of inevitability that cloaks mundane existence. In this archetypal act of willful imagination, the tragic vision of life and human identity is cast. The lesson would seem to be that whatever act separates and thus defines the individual will, thereby violates the natural harmony of life itself.

Tragic drama claims a unique and fundamental place in the arts primarily because it focuses on a seeming disjuncture of life and meaning; it brings into question whether there is or what may be the meaning – the depths of sense and intelligibility – of human life. We most often are moved by and remember images of the rupture between life merely as time, and life as meaning. If life is reduced to an empty, petty pace of endless tomorrows, then, indeed, it is without meaning. A tragic vision through twenty centuries, from Aeschylus to Shakespeare, affirms this truth: the agonies of the *Orestia* are still echoed in *Macbeth*, where life passes its brief hour upon a stage full of sound and fury. But if life were only sound and fury, there would be no drama, no tragic vision at all, only a darkened immersion into a random cosmic flux. Tragic drama produces a clear perception of human life, however dark, which involves judgment as well as description. In the aesthetic distance of its expression, a fundamental passion of human existence is disclosed in the individual's temptation to exempt herself from the binding order of social normality, to disdain the easy identity of political community.

Dramatic art is a primary resource in identifying the uniqueness of human existence however human identity requires reflection, whether the defining activity is labor or laughter. Fine art in general may constitute a singular mode of human expression, but it is also a reflective process in which a refractive image is formed out of the mirror of nature. *Poiesis*, the genius of art, transforms human identity so that it is as much creatively formed as reflectively found. Artistic perception is already critical reflection, from and through the depth and distance of the mirrors' *taine*.

A great work of art is both open and complete. It is open to interpretation and a diversity of possible understandings, and complete as a meaningful expression of human existence. A work of tragic drama finds its coherence at the heart of conflict, in the gap between the possibilities and inevitabilities of action and events. A kind of sublime harmony is achieved in recognition and acceptance, in the resolution of tension between an unlimited human aspiration and the prevailing limits of life and world. A human being, however heroic in will, is only defined against the greater power and larger fabric of fate or destiny. Whether the overweening passion is pride, ambition, or jealousy, whether the motive is love or vengeance, a tragic flaw not only embodies the spiritual limits of Man, but also yields recognition of the common character of human aspiration. Several elements define movement in tragic drama. A tragic vision takes shape in the initial violation of the sacred or natural that leaves the hero isolated and alienated from the ordinary resources of human community and divine benevolence. A transcendent and countervailing force then resists this breach of natural order and restores a balance to the world. These two movements figure in the dramatic quest for human identity. The drama of conflict is resolved through acknowledgement of, and reconciliation to, the limiting conditions of human existence.

V

It is a familiar idea that fine art is a primary means through which the drama of human life finds a moment of coherence and closure; that we are reconciled to nature, to the world, to ourselves, through great works of art.

But in the process, art must both break the norm of expectation, and restore the balance of perception. Art breaches sensibility, then seals the breach. Great art reminds us of the more that we are not, then reconciles us to the who that we are, realizing the singularity of this transcendence. Music soothes the savage breast, but it also provokes passion in the civil breast. In the visual arts, the most extreme passions are brought into a still moment of a work, vital and alive, generating a response that transcends a given moment or event. In Michelangelo's *Pieta* and Picasso's *Guernica*, both sorrow and horror find a space of beauty or sublimity in which elemental passions become a condition of human existence: a lesion is opened in the spirit and then healed. Keats' great *Ode on a Grecian Urn* is a poetic image of crystallized motion, passion fixed by art for eternity on an urn designed to hold the ashes of a once living being. We are moved by the poem in that moment of yearning and youth that it expresses – forever still in the beauty of its movement. A reader's mind resonant with the vision of the poem, in harmony with its expression, inhabits a soul in grace.

The metaphor of harmony operates at many different levels in different ways in different arts, of course. But there are some common threads. First of all, a work of art is in harmony with itself. This aesthetic imperative is common both to composition and performance. Such a work is also in accord with an aesthetics of experience, in harmony with the complexity of its subject, whether civil war or crucifixion, and with the diversity of its audience, whether secular or religious, perpetrator or victim. Great works of art contribute to the larger harmony between Man and World – they bring the human mind and spirit, individually and collectively, into accord with the whole universe of nature. There may be no one aspect of the fine arts that accomplishes this. Capacities and effects differ with the art of music, or sculpture, or literature, and within the genres of each art.

Drama, whether ancient or modern, Sophocles, Shakespeare, or Stoppard, portrays human life as conflict. Tragic drama further uncovers an essential experiential aspect of the human condition in the isolating depths of an individual tangled in the web of her own striving. Whether the offending passion of the heroic figure is that of pride, ambition, or vengeance, we as spectators are brought into the common suffrage and suffering of its excess.

In his *Poetics*, Aristotle discusses two forms of reconciliation characteristic of ancient classical tragedy that relate to the harmony of Man and World. The first is the dramatic closure within the drama itself, a realization on the part of the hero – or the chorus – acknowledging a greater, transcendent power of destiny in the form of divine justice or simply divine interference. Although Aristotle's remarks are made in the context of classical Greek drama, the critical idea of dramatic closure has a continuing history in tragic literature. Shakespeare exemplifies the same point in as many ways as his tragedies require, in the various persons of callow lover, riven prince, jealous Moor, ambitious usurper, or crazed old man. The tragic irony of Shakespeare's drama is often a wisdom put into the mouths of fools, sycophants, and villains. Think of Polonius' counsel to his son Laertes, "This above all: to thine own self be true...," or Iago's ironic baiting of Brutus, "The fault... lies not in our stars, but in ourselves, that we are underlings." The classical theme of heroic self-realization is given a contemporary twist of parody in the familiar voice of T. S. Eliot's *Prufrock*, who confesses the pathos of everyman's fall from the grace and greatness of tragedy: "No, I am not Prince Hamlet, nor was meant to be...." Prufrock is a character who casts himself as a self-conscious caricature, a bystanding spectator of his own inadequacies, who hears the snicker of the eternal footman holding his coat, as he himself stands anxiously back and can only observe the moment of his greatness flicker.

Aristotle cited a second aspect of tragic drama relevant to reestablishing a harmony of Man and World in the therapeutic experience of the audience through the characteristic emotional breach and reconciliation of catharsis. The empathic terror and pity generated in the viewer in the context of the drama is absolved through its effect: it is, after all, only a play. Even so, and even here, Art does more than imitate; it alleviates and legislates life.

VI

There is an important reciprocity of metaphors in life and literature. Theatre draws from the drama of life, and life draws from the drama of the stage. Shakespeare often makes this connection explicit. Whether he is right that we men and women are players in a drama with seven acts, it is no prodigious feat of imagination to see our common world as a stage for the very real drama of human existence that we lesser figures act out, with and for each other. Macbeth's despairing lament "Tomorrow, and tomorrow, and tomorrow..." cadences the possible fate of whoever contests the inevitable. Macbeth's synoptic vision of the human condition portrays the fragile uncertainty of human action from the perspective of eternity. From such a perspective, the sweep of life itself, Macbeth's and yours and mine, is but "a poor player, that frets and struts an hour upon the stage, and then is heard no more." In such a vision, the whole of existence does indeed appear as the aforesaid meaningless tale. This need not be for us, as it is for Macbeth, the final word. It is, however, a word within the world of literature that reaches to the metaphoric depths of human self-understanding.

For the audience, of course, there is not simply sound and fury, but the poet's coherent vision of it. Cacophonous confusion resolves itself into an aesthetic flow of sense. Beyond the snarled consciousness of the heroic character is a lesson drawn from the creative fabric of human imagination. We are enabled by the poet to see the world through the eyes of Macbeth, but we also see the world of Macbeth through the eyes of the poet. The audience, as participant-spectator, has a transcendent view denied the dramatic character. Whether or not the tragic hero is reconciled to her fate, we as viewers are enabled by the art of the drama to be actor *and* chorus, participant *and* spectator. Our spiritual reconciliation is a response to dramatic art, not retributive fate.

The harmony effected by art is one of feeling and thought, of dramatic experience and cognitive understanding. We experience with the hero the horror of, and reconciliation with, his fate; but we enjoy as viewers the harmony of this existential construction and resolution. The aspiration of the tragic hero, the story of her actions, the drama of her courage in the face of existence, is a portrayal of the human being as a creature caught between angel and ape. Shakespeare famously put this point of tragic vision in Hamlet's lyrical expression: "What a piece of work is a man… in apprehension how like a god… this quintessence of dust."

Tragic drama is thus an investigation and passionate expression of limits – the heights of human aspiration and action, but also the depths of failure in human character. Achilles is the precursor in Greek epic to Oedipus in Greek tragic drama: the mortal flaw of Achilles' heel becomes the namesake clubfoot of Oedipus. In both, archetypal feet of clay betray the man who would be king. In mythic literature it is Prometheus who serves as the prototypical character of tragic culture: a figure heroic in aspiration and action, and also victim to the wages of that heroism. The heroism of the tragic figure is incomplete without the courage and nobility required of her fate.

In the tragic vision of great literature, the raw paradox of the human condition is both conceptually expressed and artistically resolved. The lesson of tragic drama contains a germ of paradox in which the high is brought low and in the low, exalted – not as a lessor god, but fully as a human being. The singular importance of the human comedy in tragic literature is that reconciliation is effected not through divine grace, but human perseverance. It is only in the vision of failed aspiration brought to action and turned awry that the depth of human greatness is wrought. A defining character of human being is discovered not only in the recognition that we are not gods, but also in the resolution that we shall not merely be animals. The harmony of Man and world in tragic drama is a sustained tension resolved only in the perception and acknowledgement of the viewer. Nietzsche expressed this dramatic tension as a requirement that tragic art worship at the shrine of both Dionysos and Apollo. Light and darkness must both be preserved, or the drama fails in a dismal half-light where shadow figures blur our vision.

How does the freedom entailed in the experience of art connect to the theme of reconciliation and harmony of Man and World? The moment of great art is thrilling, exhilarating, a response to the creative potential of difference, a realization of possibility, of freedom itself. It is only in imagination, through the articulated resources of art, that Man can become what he conceives himself possibly to be, and not merely subject to the crushing or banal imperatives of the natural world. Man is a creature that not only thinks, but laughs and dreams, and through creative works of art fully realizes – makes real – the whole of what he is. The art of tragic drama consists in the resolution of a primal conflict between the will of man and the dominion of god. The dramatic portrayal of a will reconciled to its own limits fully achieves, if only for rare moments, the full sense and meaning of what it is to be a human being.

Art cannot claim the whole of such experiences: falling in love, the wonder of birth, an act of friendship, simple humanity, all these can claim a transcendental moment of fulfillment. There is no point or profit in reducing all such moments to the category of the aesthetic. The production and experience of art, however, make such moments the conscious focus of their activity. From a cosmic perspective, life itself is an absolute value. From a human point of view, however, this is not so. We are all aware that not every life is worth living, and that there are worse things in life than dying. The plot of action in tragic drama begins with conflict and ends with death. If tragic drama can be said to have a thesis, it is that neither conflict nor death can be avoided or denied, and that whatever harmony Man achieves in the world must come through acknowledgement of, and reconciliation to, these indelible conditions of human existence.

Perhaps not surprisingly, this means that the goal of tragic drama is philosophical – its affective occasion is the pursuit of wisdom. Recall that following Socrates, the second and third reputations for wisdom among the ancients were two tragic dramatists. Sophocles, in seeming distinction from the rational professions of the philosophers of his time, insisted that true wisdom comes only through suffering. A case may be made, however, that there exists common ground with both Plato and Aristotle. These philosophers and the tragic poets all agreed that in order for us to genuinely know something, truth must become part of our very life and character.

A tragic vision of human existence may not be essential to wisdom, for there is wisdom in joyful celebration of life as well as practical prudence in action. But the great lesson of tragic drama is that the harmony of man and world is grounded in the fundamental conditions under which his life is given, as a creature bound to earth and time. This harmony depends on more than political rationality, a balance of justice through the exercise of reason. A more profound wisdom than negotiation, moderation, or compromise in action or policy is one that fully acknowledges and reconciles man to the ontological foundation of his being toward death. It is to this level of wisdom that tragic drama aspires, and the harmony it yields through reconciliation is a highly wrought complex of passion and reason that, fully realized, constitutes a vision of human existence that is deep and enduring.

Trinity University

INGRID SCHEIBLER

ART AS FESTIVAL: TRANSCENDING THE SELF THROUGH THE WORK OF ART

INTRODUCTION

The original conception of this paper was to put Martin Heidegger and Hans-Georg Gadamer's perspectives on the work of art into the context of the theme of the "festival".[1] Both figures conceive the encounter with a work of art through a critique of the subjectivization of traditional aesthetics. Further, as an alternative to a subjective basis for aesthetics, both Heidegger and Gadamer conceive the encounter with a work of art as a transcendence of subjectivity – that is to say that the encounter with the artwork has the character of an *event*. The work of art is like a festival in two ways. First, in the festival, the focus is not directed at the individuals, but at their participation in something (an event) which transcends their subjective standpoint. Second, it is, of course, central to the festival celebration that it is *shared*; participation is a sharing in an event, and thus involves community. The celebration is of something a certain group holds in common, and what is being honored, in turn, can be said to hold them in common.

Given these initial points of analogy, I first sought to use the theme of the festival as a way of examining some of the philosophical issues in Heidegger's and Gadamer's approaches to the work of art. Specifically, I wanted to look at the way in which both figures conceive the work of art as an encounter which, like the festival, involves a transcendence of subjectivity in an encounter with an event – in this case, the artwork – which the individual does not direct, but rather in which he/she participates. Yet, while this is an important part of the project which follows, I found that putting the theme of festival into play also provides a useful *critical* lever, especially in the way that the festival itself raises important issues of community, which can be used to problematize this issue in both Heidegger's and Gadamer's views on the artwork. More specifically, reflecting on the festival celebration raises the questions (i) of the nature of this community, and (ii) of the relation of the community created in the festival event to the community of the everyday, and thus also, in the context of this paper, of the relation of the aesthetic and the political.

In what follows, then, I would like both to put Heidegger and Gadamer into dialogue with each other concerning their views on the work of art, and

A.-T. Tymieniecka (ed), Analecta Husserliana LXXIII, 201–228.

also to begin to place them both in a dialogue with the issues of community raised by consideration of the festival.[2]

The first step, then, is to look at how an encounter with a work of art enables the individual to transcend the self in a way that is transformative. Doing this, for both Heidegger and Gadamer, does not mean that we start from the position of subjectivity. Rather, each asks how it is that the work of art can function as a site where the standpoint of subjectivity is surpassed, "*aufgehoben*", in an encounter with a greater context? To answer the question at hand requires a focus on the mode of being of the artwork itself. This is, for Heidegger and Gadamer, because the work of art (and our encounter with it) has the character of an event, a type of presentation in which we take part. Indeed, in this essay I am going to follow a lead of Gadamer's, that the work of art has the structure – specifically for him, a *temporal* structure – of a festival celebration.

In looking first at Heidegger, and then Gadamer, both approaches to the work of art demonstrate a clear concern with the first characteristic of the festival – the experience of self-transcendence. The second aspect of festival celebration – that it is a *shared* event – is also a prominent internal moment of both accounts of the work of art. But, although it is not the primary purpose of this paper to undertake a full-scale comparative analysis of Heidegger's and Gadamer's approaches to the work of art, I will emphasize the different ways in which the experience of self-transcendence and the transformation that takes place can be shared; the different ways in which an encounter with the work of art is related to the community and to the everyday.

The claim that the work of art can be both subjective and universal, has its reference point in traditional aesthetics in Kant's *Critique of Judgement*.[3] More recently, phenomenological and hermeneutical approaches to aesthetics – especially the work of Heidegger and Gadamer – posit a critique of traditional aesthetics, and what is seen as its subjective basis. For Heidegger, Nietzsche (himself a strong critic of Kant's aesthetics) is a central target in the critique of subjectivized aesthetics; for Gadamer, it is Kant himself.[4]

A phenomenological approach to aesthetics provides an alternative to this subjectivization. By starting its thinking about art from within the artwork itself – from an ontology of the work of art – Heidegger and Gadamer offer insights into the way that, at the heart of our encounter with art, there lies a transcendence of the self.[5] For Heidegger, this self-transcendence in an event takes place through a recognition of our ineluctable relation to what he called the event of Being, and which is manifested in the work of art as a site which grounds a people's history. For Gadamer, it is an experience of plenitude and truth that is related to our self-understanding.

I now want to sketch in broad contours Heidegger's efforts to inaugurate this shift in a critique of traditional aesthetics, and his own effort to think of the "origin" of the artwork as a way to move from a subjective aesthetic experience to one which recognizes our indebtedness to Being. I will focus initially on Heidegger's analysis of "the problem" and then turn to his view – in the essay, "The Origin of the Work of Art" – of an alternative. Next, I will look at how Gadamer, in *Truth and Method*, uses the concept of "play" and a reading of art as a "festival" celebration, with its own distinctive temporality which we enact, to explore the nature of the work of art. In conclusion, I will raise a question about the connection between the aesthetic and the political.

HEIDEGGER'S ACCOUNT OF THE WORK OF ART

Heidegger's essay "The Origin of the Work of Art" (1936) posits a new relation between the work of art and truth.[6] This notion of art is distinctively "Heideggerian" because it is thought against the perspective of either a subjective or an objective extreme: that is, it is thought neither from the standpoint of the artist as creator or genius, nor is it thought simply from a subjective standpoint of the feelings or experiences produced in the viewer/recipient. Further, for Heidegger, as we shall see, the work of art is also not merely an "object".[7] It is not evaluated only in terms of its ability "as a beautiful object" to produce these subjective states.

The question of the dominance of subjective aesthetic experience is raised in the Epilogue to Heidegger's *OWA* essay in terms of the destiny of art. Here, Heidegger says that in "aesthetics" we are to look at the way man *experiences* art to give information about art: "Everything is an experience. Yet perhaps experience is the element in which art dies".[8] Heidegger adds that this death is a slow one, taking several centuries. In the Epilogue, Heidegger invokes Hegel, who says, "Art is and remains for us, on the side of its highest vocation, something past".[9] If aesthetic experience now rules, then art is indeed something of the past, and Hegel is correct. But it is not yet a foregone conclusion that the rule of the subjectivization of art through experience is determinant:

> The truth of Hegel's judgment has not yet been decided; for behind this verdict stands Western thought since the Greeks, which thought corresponds to a truth of beings that has already happened. Decision upon the judgment will be made, when it is made, from and about this truth of what is.[10]

To determine whether Hegel's claim is true, we first need to consider the nature of art, and for Heidegger, this means asking the question about its "origin" [*Ursprung*].

In addition to Heidegger's critique of the dominance of aesthetic experience, a second – and more general – contextualizing point is that the *OWA* essay, and Heidegger's views on the significance of art (written just after this essay),[11] should be seen in relation to his more general project of thinking and retrieving the question of Being. In the *OWA* essay, Heidegger makes the work of art one instance, or site, where the enigmatic event of Being – the distinctive event of truth, a revealing and concealing – takes place. Here, the relation between Being and human beings, a peculiar type of interdependence, is manifested and can be brought to awareness after a long tradition in the West of oblivion to this question of Being. It is this interdependence, a "co-relation", thematized for us in the encounter with the work of art, that I have termed "self-transcendence".

Heidegger's reflections on the work of art, and the tradition of aesthetics in general, then, move beyond what he considers their "metaphysical" determinations; since the beginning of the nineteenth century (according to him) the metaphysical determination takes the form of a "subjectivization" of aesthetics. That is to say, his views on modern aesthetics must be understood as a part of his more general account of the modern period, which is part of the longer trajectory of a forgetfulness of Being and the dominance of a metaphysics of presence. The modern period is dominated by the rule of the *cogito*, the modern subject, and what can be called the dominance of "modern subjectivism". What I have been describing in terms of subjectivization of aesthetics is, then, the manifestation of this subjectivism in the sphere of art.

There is a chapter on the history of Aesthetics, "Six Basic Developments in Aesthetics", in Heidegger's *Nietzsche*, v. 1, *The Will to Power as Art*, where Heidegger characterizes modern aesthetics as an expression of modern subjectivism. The account of the modern period Heidegger gives is the third of five basic developments in the history of knowledge about art, the origin and formation of *aesthetics*. It is a development that does not flow directly from *art*, or meditation on it; rather it is something that involves our entire history: the beginning of the modern age. Further, it is this modern development that makes meditation on the beautiful slip into a preoccupation with man's state of feelings.[12] To quote at some length:

> Man and his unconstrained knowledge of himself, as of his position among beings, become the arena where the decision falls as to how beings are to be experienced, defined, shaped. Falling back upon the state and condition of man, upon the way man stands before himself and before

things, implies that now the very way man freely takes a position toward things, the way he finds and feels them to be, in short, his "taste", becomes the court of judicature over beings. In metaphysics that becomes manifest in the way in which certitude of all Being and all truth is grounded in the self-consciousness of the individual ego; *ego cogito ergo sum*,... I myself, and my states, are the primary and genuine beings. Everything else that may be said to be is measured against the standard of this quite certain being. My having various states... participates essentially in defining how I find the things themselves and everything I encounter to be.[13]

It is here that the subjectivization of the aesthetic begins: with modern aesthetics, meditation on the beautiful is tied to man's state of feeling. For Heidegger, it is Nietzsche who is responsible for the subjectivization of aesthetics.[14] Heidegger notes that what occurs alongside this formation of modern aesthetics is the decline of great art, "great in the designated sense".[15] This doesn't mean that quality is declining in a real sense, but that art gives up its *essence*; it loses its immediate relation to the basic task of representing the absolute, to establishing the absolute definitively for historical man.

Having seen how Heidegger views modern aesthetics as part of the way that the *cogito* (the subjective states of the ego) becomes the locus of authority for the truth of reality in the modern period, and so obscures the question of Being he is concerned to retrieve, I would next like to turn to Heidegger's own alternative account of the work of art.

As a way into this account, I would like to look at another chapter in Heidegger's volume, *Nietzsche: The Will to Power as Art*. In Chapter 15, "Kant's Doctrine of the Beautiful: Its Misinterpretation by Schopenhauer and Nietzsche", Heidegger challenges Nietzsche's reading of Kant.[16] Nietzsche's approach to art gives a central significance to the producer and artist, and this dictates Nietzsche's own criticism of Kant's aesthetics. Heidegger quotes Nietzsche in the *Will to Power*,[17] where Nietzsche takes aim at Kant, saying that:

Our aesthetics heretofore has been a woman's aesthetics, inasmuch as only the recipients of art have formulated their experiences of "what is beautiful". In all philosophy to date, the artist is missing....[18]

So, philosophy of art for Nietzsche means "aesthetics", but "masculine aesthetics" and not a spectator's or "woman's" (recipient's) aesthetics. It's the perspective of the artist/creator which provides the standard of what is beautiful.

Heidegger, however, sees this as a misinterpretation of what Kant's requirement of "disinterestedness" on the part of the spectator means.[19] In his discussion of Kant, Heidegger criticizes Nietzsche's reading of Kant's aesthetics not only as that of the "spectator" – the woman/recipient and not the active creator – but further, as one in which the spectator is disinterested and, hence, for Nietzsche, disengaged. For Nietzsche, Kant's aesthetics suggests an attitude of *indifference*. In contrast to this (supposed) state of indifference, Nietzsche contends that the aesthetic state is one of rapture [*Rausch*].[20]

In contrast, Heidegger finds Kant to have discovered and made a call for an encounter with a work of art in which we suspend all construction of the object. Kantian disinterestedness, rather than being a relation of disengagement or indifference, is instead a most committed form of engagement with the "object", and even to refer to the artwork as "object" is to miss the achievement Heidegger sees in Kant. What Heidegger views so favorably in Kant is important for our context precisely because it is a moment in the history of the West, of modernity and aesthetics characterized by subjectivism, where subjectivization is resisted; that is to say, in Kant, Heidegger locates a discovery that he himself has made. In Kant's approach, Heidegger says:

> in order to find something beautiful, we must let what encounters us, purely as it is in itself, come before us in its own stature and worth. We may not take it into account in advance with a view to something else, our goals and intentions, our possible enjoyment and advantage. Comportment toward the beautiful as such, says Kant, is *unconstrained favoring*. We must release what encounters us as such to its way to be; we must allow and grant it what belongs to it and what it brings to us.[21]

For Heidegger, Kant's awareness of a pure encounter with things[22] is a "magnificent discovery and approbation" of aesthetic behavior.[23] The crucial point for Heidegger is that it is wrong to interpret disinterest as meaning that all essential relation to the object is suspended, that one is disengaged. For this does not provide an alternative to the basic subject/object structure of modern aesthetic experience. Rather, Heidegger claims, in this suspension of interest, "*the essential relation* to the object itself comes into play. The misinterpretation fails to see that now for the first time the object comes to the fore as pure object and that such coming forward into appearance is the beautiful. The word 'beautiful' means appearing in the radiance of such coming to the fore".[24]

This should sound familiar to readers of Heidegger's account in the *OWA* essay, where the artwork's mode of being is seen as "coming forth out of itself" [Greek: *phusis*].[25] This is an essential moment, one in which the work is no longer experienced out of the ground of the spectator's subjective feelings and experience. Rather, for Heidegger, the work of art is itself an event, a happening of truth. Further, it is an event which is foundational, a founding event, a ground.

In the *OWA* essay, Heidegger approaches the question of art's origin from out of the mode of being of the work itself. He defines art as a "setting into work of truth",[26] where truth is defined as a revealing disclosure [*aletheia*]. Art's mode of being has two elements – the sheltering agent Heidegger calls "earth" and the disclosive "world". These two elements are in a relation of conflict [*polemos*], yet this conflict is not violent. Their relation is more of an interdependence: earth and world work with, yet against, each other in a striving [*Streit*]. In this striving, each lets the other be what it is: "In setting up a world, the work sets forth the earth".[27] For Heidegger, this striving, the conflict and opposition of clearing and concealing, is where truth establishes [*sich einrichtet*] itself.[28]

Once we encounter the artwork in this way, from out of the mode of being of the work and not from the standpoint of our subjective states, we are drawn into what Heidegger describes as the event character of the work, a "setting into work of truth" in which we participate. Heidegger calls this the "Open". The Open is liberated by the setting into work of truth that is the mode of being of the artwork itself. What he calls the "clearing of openness and establishment in the Open" belong together;[29] when this occurs, truth happens. Heidegger adds that this happening is historical, and truth happening in the work of art is one of several ways Heidegger mentions in this essay that truth establishes itself in beings.[30]

This is one way in which the mode of being of the work, the happening of truth and what it brings into the Open, is related to the community, as a founding event which, Heidegger writes, can occur in just this way, only once. So, as the work of art is an event which institutes a world, it is an unmediated source of something entirely new. In the example of a Greek temple, Heidegger sees the world opened up by the work in this strong sense of a foundation for a historical people: the temple "first gives to things their look and to men their outlook on themselves".[31] The world of the work is a meaningful order that gives to persons and things their proper place. In the example of the Greek temple, Heidegger sees a genuine response to the god that is present there, an integrating power. In the temple-work, the presence of the god is one which gathers humans into community, the world of this historical people: "Only from and in this expanse does the nation first return to itself for the fulfillment of its vocation".[32] This is the world of the temple-work of ancient Greece.

What in traditional aesthetics would be called spectators or viewers, Heidegger calls the "preservers" [*die Bewarhenden*] of the work. He does not refer to the subjective states of the preservers, but positions them as taking part in the event of the artwork: they are "standing-within" the truth that is happening in the work.[33] Towards the end of the essay, Heidegger states that all art is essentially poetry.[34] The nature of poetry, in turn, is the "founding of truth".[35] It is here that Heidegger links the bestowing and grounding of art-as-poetry with the idea of a beginning. And this is the move, or translation, by which he displaces the traditional modern subjectivist aesthetic conception of the artist/creator of the work, a masterful performance of genius, as the work's "beginning".[36] Rather, it is the grounding/bestowing of the work which itself has "the unmediated character of... a beginning".[37] Heidegger further describes the strife of truth in terms of "founding as beginning": "Always when that which is as a whole demands, as what is, itself, a grounding in openness, art attains to its historical nature as foundation".[38] This notion of the unmediated character of the work of art, that it is "a leap out of the unmediable",[39] is very different from the way Gadamer will see the work not only as *not* being a singular event which is foundational, but also as one which is mediated.

The enigma of truth, *a-letheia*, unconcealedness, to which Heidegger brings our attention, and the Open into which we are brought – as co-related preservers – through encounter with the artwork, is thought *from within the ontological status of the work itself.* It is here that we find the moment of self-transcendence, a transcendence of the subjectivizing ground of traditional aesthetics, which is itself an expression of the dominance of the modern *cogito*. And it is here, crucially, that Heidegger's own approach to the work of art locates the individual *taking part* in the event of truth [*a-letheia*]; we do not direct this encounter; rather, we are called to participate in the event of the work itself. Here, then, our taking part, being drawn into the Open, shares the movement in which we are similarly transported in the festival or carnival celebration.

From this brief look at Kant and Heidegger himself, we can see that the shadow cast on the essay on the nature of art is that of the question of the meaning of Being. I want to mention at this point three statements from the end of Heidegger's discussion in the *OWA* essay:

1) The origin of the work of art – that is, the origin of both the creators and preservers, which is to say of a people's historical existence, is art. That is so because art is by nature an origin: a distinctive way in which truth comes into being, that is, becomes historical....
2) But this reflective knowledge [reflection on art] is the preliminary and therefore indispensable preparation for the becoming of art. Only such knowledge prepares a space for art, their way for the creators, their location for the preservers....
3) Are we in our existence historically at the origin? Do we know, which means do we give heed to, the nature of the origin? Or, in our relation to art, do we still merely make appeal to a cultivated acquaintance with the past?

This question must be asked: if the work of art functions as a site where we are brought into an experience of Being, through the enigma of the artwork, no longer grounded in and directed by the experience of subjectivity, but rather having the character of an event in which we participate, how is this to be mediated with the everyday, which in Heidegger's view is under the sway of metaphysical/instrumental thinking? Yet, this event is also, we are told, an unmediated beginning. One response is to say that, in the above quotation, where Heidegger says that the origin of the work of art is the origin of a people's historical existence, the work of art institutes a foundation, which grounds a people historically. Just as the temple-work served as a founding event for the Greeks, allowing the god to be present, and creating a space for gathering in its presence, there is the suggestion at the end of the essay, where Heidegger says that all art is in essence poetry, that a new, founding event is possible.

Hence the question, "Are we historically at the origin?" One response, then, is to say that the artwork grounds the everyday/community to the extent that a particular artwork can serve as that historical situation's "origin": it can serve as an unmediated founding event which gathers a people who are "historically at the origin". Another suggestion is to see that Heidegger seems to be talking about – and inciting and inviting us to – an experience of Being and truth [*aletheia*] which we encounter in the event of the artwork, and suggests that, in what we encounter and participate in, something is transformed. We undergo a transformation. Yet, in the encounter with this event of the work of art, for Heidegger, what we take part in is something wholly new; it occurs as a founding event, and is thus singular and unmediated. What is less clear is this: if the work of art is to be reconceived as Heidegger reconceives it, without recourse to the traditional modes of modern aesthetics, how is this encounter with the enigma of the artwork supposed to, in turn, transform a people? I will return to elaborate on this in a moment.

At this juncture it is important to ask about two things: (i) the relation of the aesthetic and the "everyday", and (ii) the relation of the aesthetic and the political. These are related, but not the same. The first concerns the question of how the transformation we undergo in the encounter with the event of the work of art is connected to our everyday experience: are we to interpret Heidegger as saying that the encounter with the artwork is necessarily a moment outside of our usual comportment toward things and the world of the everyday, and yet that, once we return to the everyday, we ourselves are changed, having participated in the event of the artwork's truth and having thus become aware of the Open "to which we are ever subject"? This question is one of the mediation of the everyday. Yet the question of mediation with the everyday is also a question of the relation of the aesthetic to the political, in reference to the more specific question of how this trans-subjective encounter can function as a site of resistance. That is, how can an encounter with the work of art transform those deformed aspects of our modern age? A related issue concerns whether this experience of the work of art can be shared. Presumably transformation cannot occur at a merely private and individual level; there must be some communication involved.

With these questions in mind, let us look at one passage in the *OWA* essay where Heidegger refers explicitly to the relation of the encounter with the artwork and ordinary experience:

> The more solitarily the work, fixed in the figure, stands on its own and the more cleanly it seems *to cut all ties to human beings*, the more simply does the thrust come into the Open that such a work *is*, and the more essentially is the extraordinary thrust to the surface, and the long-familiar thrust down. But this multiple thrusting is nothing violent, for the more purely the work is itself transported into the openness of beings – an openness opened by itself – the more simply does it transport us into this openness and thus at the same time *transport us out of the realm of the ordinary*. To submit to this displacement means: to *transform* our accustomed ties to world and to earth and henceforth to restrain all usual doing and prizing, knowing and looking, in order *to stay within the truth that is happening in the work*.[40]

It is clear that our encounter with the artwork is transformative: there is a displacement from the everyday and we are "transported into the openness of beings". Our encounter with the work of art is one of participating in an event where the artwork takes us beyond ourselves and our subjective standpoint. In light of my questions above, what does "displacement" mean, and what is the relation of this experience to the everyday, which is in need of transformation? And secondly, is the encounter with the origin of the artwork – while super-subjective – nonetheless a *privative* experience? It is trans-subjective, but can it be genuinely intersubjective?[41]

In terms of my initial analogy of the work of art with the festival, reflecting on these issues *vis-à-vis* Heidegger leads one to return to ask about the relation of the distinctive space/time of the festival enactment and everyday life. One question raised here concerns the different senses of community of the everyday and of the festival: the everyday is an inter-subjective, public and social, domain, which is a community comprised of difference and plurality, and the community created in the sharing and participation in the festival, is a sharing in which differences recede and are elided. In this light it is important to look at the above difference and how what one experiences (a transformation?) while taking part in the festival gets mediated, played out, in one's continued existence in the everyday; i.e. is the trans-subjective experience of the work of art continuous, or discontinuous, with everyday life? Heidegger seems to suggest a discontinuity.

Further, we have seen that Heidegger views the artwork as a founding event which is an *unmediated* source of something entirely new. Recall the example of the Greek temple, in which the world opened up by the work serves as a foundation for a historical people: the temple "first gives to things their look and to men their outlook on themselves". What kind of community is it if one of its defining characteristics is to be unmediated? Heidegger's idea of an unmediated founding event, which occurs only once, could be applied to any society; it is precisely not rooted. It is, in this sense, grounding but not grounded.

It is here that one becomes wary of some of the political implications of Heidegger's account of the work of art. Because it is unmediated, and because Heidegger, in the context of a critique of the subjectivization of aesthetics, nonetheless gives no role to the pole of subjectivity, and to such aspects of encounter with an artwork as "meaning" or "significance" or "interpretation", Heidegger's rhetoric that the artwork founds a "historical people" suggests that the community instituted here is a collectivity in which individuality and difference are erased; the pole of subjectivity is lost.[42] In the context of Heidegger's discussion of the work of art as origin, addressed to and with its hope for instituting a "historical people", a collectivity first and foremost seems to supplant the pole of subjectivity and with this, individual judgment.

And with this loss of the subjective pole, the link between the aesthetic and the political becomes highly suspicious; the rhetoric of an unmediated founding event can so easily slide into a loss of self in the face of a mystical and autocratic authority, like National Socialism.[43] Further, because of this radical de-emphasis of the pole of subjectivity, any democratic monitoring, interpretation, or evaluation, call for justification or accountability of the founding event, are elided. For in order not to blindly accept what is doing the instituting or founding of a people, one needs recourse to such things as individual judgment and interpretation and adjudication, all of which in turn require access (i) to the dimension of assessing the founding event's/artwork's "significance" and "meaning", as well as (ii) to a rootedness in an intersubjective community, the ground where such activities take place.

I would now like to turn to Gadamer, who addresses these questions of mediation more directly.

GADAMER'S ACCOUNT OF THE WORK OF ART: PLAY AND FESTIVAL

Gadamer's account of the work of art also takes shape as a critique of traditional aesthetics, in the form of a critique of what Gadamer calls "aesthetic consciousness", the attitude to art which is a consequence of Kant's aesthetics. For Gadamer, Kant is the root of the modern subjectivization of aesthetics because he narrows the notion of aesthetic experience to the subject's state of mind (the free play of the cognitive faculties, understanding and imagination).

Gadamer sees a problem with aesthetic consciousness in the way the encounter with the work of art is related to the life context of the one experiencing the artwork; the aesthetic experience is one of abstraction from the life-context of the viewer. There is a quite subtle, but important, distinction Gadamer makes in which, in the experience of art from the standpoint of aesthetic consciousness, the type of experience of the work of art is one which wrenches the person experiencing it out of their life-context, yet relates them back to the whole of their existence. The type of aesthetic experience [*Erlebnis*], is a particular – and ultimately for Gadamer, deficient – mode of experience for two reasons: (i) it conceives experience as discontinuous; and (ii) it takes place in "immediacy". Gadamer says of both the status of the work, and our experience of it, from the standpoint of aesthetic consciousness: "As the work of art as such is a world for itself, so also what is experienced aesthetically is, as an *Erlebnis*, removed from all connections with actuality".[44] The sense of the whole to which aesthetic *Erlebnis* refers, again, isn't related back to the flow of experiences which make up the continuity of one's life, and this is problematic for Gadamer. Rather, aesthetic experience does not combine with other experiences to make one open experiential flow, but:

> An aesthetic *Erlebnis* always contains the experience of an infinite whole. *Precisely because it does not combine with other experiences to make one open experiential flow, but immediately represents the whole, its significance is infinite.*[45]

The effect of this is that art is conceived as being at a remove of some kind from "real being", an abstraction from everyday life and from reality itself. Given this, Gadamer wants to retrieve (from the Kantian subjectivization) the question of artistic truth against the notion of a separation of aesthetic experience from the whole, experiential flow, of one's life. He wants to emphasize in contrast a notion (from Hegel) of experience as *Erfahrung*, one in which we ourselves are changed through encounter with the work of art;

this also emphasizes a negative moment of experience. Gadamer locates the theoretical basis of the definition of art as separate from "real being" as being due to the epistemological dominance of scientific method, and the discrediting of all knowledge which isn't scientific.[46]

A further problem with the standpoint of aesthetic consciousness is its dependence on a notion of immediate aesthetic experience; it also explains the inexplicability of art through a momentary flash of genius producing the work. Aesthetic consciousness is the counterpart to a second abstraction which Gadamer perceives in traditional aesthetics: the abstraction of the work of art from its original life context. He calls this abstraction of the work, "aesthetic differentiation". Here, art is conceived as beautiful appearance (not reality) and is also characterized by abstraction, an *alienation* from reality. And here we see one major source for Gadamer's discontent: aesthetic consciousness isn't related to any unity of an ideal of taste, which he says, is a source of unity and community.[47] For Gadamer, aesthetic consciousness doesn't recognize any *belonging together* of the work of art and the world. Instead, and here is the real problem for Gadamer, aesthetic consciousness "no longer admits that the work of art and its world belong to each other, but on the contrary, aesthetic consciousness is the experiencing [*erlebende*] center from which everything considered art is measured".[48]

In contrast to Heidegger, then, Gadamer is expressly concerned with the relation of the work of art to the world of the everyday. With a subjectivized aesthetics, both the work of art and aesthetic experience depend on a process of alienation; here we have the "autonomy" of the work of art, which lifts the work of art from its rootedness in an original life-context, the removal Gadamer calls "aesthetic differentiation". The aesthetic differentiation of aesthetic consciousness preserves sites for simultaneity, like museums, and the "universal library".[49]

"Thus," Gadamer writes, "through 'aesthetic differentiation', the work loses its place and the world to which it belongs insofar as it belongs instead to aesthetic consciousness".[50]

To sum up, Gadamer says:

> Basing aesthetics on experience leads to an absolute series of points, which annihilates the unity of the work of art, the identity of the artist, with himself, and the identity of the person understanding or enjoying the work of art.[51]

There is, then, a task: to preserve the hermeneutic continuity which constitutes our being, despite the discontinuity intrinsic to aesthetic being and aesthetic experience.

In contrast to the abstraction of aesthetic consciousness and aesthetic differentiation, Gadamer's major contribution is to see our experience of the aesthetic *as a mode of self-understanding*.[52] Here again we come to the heart of the specific hermeneutical/phenomenological notion of self-transcendence of my title. It may seem counter-intuitive, i.e. the encounter with the work of art as a mode of self-understanding might seem to suggest a return to subjectivism. Rather, for Gadamer, all self-understanding requires an element of self-transcendence. In a characterization which is – unlike Heidegger's – explicitly concerned to thematize the relation of the work of art to the everyday, Gadamer says:

> Self-understanding always occurs through understanding something other than the self, and includes the unity and integrity of the other. Since we meet the artwork in the world and encounter a world in the individual artwork, the work of art is not some alien universe into which we are magically transported for a time. Rather, we learn to understand ourselves in and through it, and this means that we sublate [aufheben] the discontinuity and atomism of isolated experiences in the continuity of our own existence.[53]

It is this character of self-understanding that dictates Gadamer's more positive elaboration of the nature of the work of art and our encounter with it. He says that "for this reason, we must adopt a standpoint in relation to art and the beautiful that does not pretend to immediacy but corresponds to the historical nature of the human condition".[54]

Gadamer says that art is knowledge and experiencing an artwork means sharing in that knowledge, and he asks, "Is there to be no knowledge in art? Does not the experience of art contain a claim to truth which is certainly different from that of science, but just as certainly is not inferior to it?" And, "is not the task of aesthetics precisely to ground the fact that the experience [*Erfahrung*] of art is a mode of knowledge of a unique kind, certainly different from all moral rational knowledge, and indeed from all conceptual knowledge – but still knowledge, i.e. conveying truth?"[55] In order to conceive of our encounter with the work of art as an event of self-understanding, "understanding must be conceived as part of the event in which meaning occurs".[56] This unique mode of knowledge is mediated through the claim the work makes upon us.

In this account in *Truth and Method*, Gadamer is arguing that the experience of art contains an experience of truth which is unlike that of scientific truth, and that the individual is transformed through an encounter with the work and its claim. In terms of a comparison with Heidegger, here again we see on one hand that Gadamer is much more explicitly concerned with the idea of the mediation of the encounter with the artwork and one's experience of the "everyday". Yet, one must also ask at this point whether Gadamer is really clear about what this distinctive truth of the work is, and also whether he gives an adequate picture of the kind of transformation the individual/viewer undergoes in an encounter with the artwork.[57] Having raised these questions, it is nonetheless clear that Gadamer attempts to keep the "pole of subjectivity", his emphasis on the self-understanding of the viewer and its transformation, as a prominent moment of his account of the artwork. And this is one way in which mediation occurs in Gadamer's account. Further, unlike Heidegger's understanding of the event of the artwork as a foundational, instituting event, which occurs once, Gadamer sees the claim of the work as one which is repeatable, and I will return shortly to this important distinction. Here I want to emphasize the unique knowledge of the experience of art. To do this, Gadamer says, we must ask about the mode of being of the work itself.[58]

For him, the mode of being of the work of art is "play": play is the clue to the ontological explanation of art. The notion of play he has in mind, however, is free from the subjective meaning it has in Kant's *Critique of Judgment* and Schiller's *On the Aesthetic Education of Man*. In Schiller, play is to harmonize the form impulse and the matter impulse. Cultivating the play impulse is the end of aesthetic education.[59] And here, Gadamer points out, art becomes known in contrast to practical reality, in which we are alienated. Education by art, Gadamer says, in Schiller is thus education to art; not preparation for moral and political freedom, but an aesthetic state, a cultured society with an interest in art.

Play has a subjective sense in Kant, too. On Gadamer's reading, Kant's problem is a dualism of the world of sense and the world of morality: the dualism is overcome in aesthetic play and the experience of harmony in the work of art. But this reconciliation of ideal and life is only temporary. Beauty and art give reality a "fleeting sheen". The concern, then, is with how to mediate or integrate this with reality. The "freedom of spirit" gained in Kant, Gadamer says, is merely an aesthetic state and not reality. For him, beneath the dualism of Is and Ought which Kant reconciles aesthetically, there is a more profound dualism: "The poetry of aesthetic reconciliation must seek its own self-consciousness against the prose of alienated reality".[60]

Kant still has a concept of taste, given from common social life; it obeys a criterion. Schiller, however, precludes any criterion of content, and dissociates the work from the world. Gadamer writes, "Thus this is a specifically aesthetic kind of differentiation. This distinguishes the aesthetic quality of a work from all the elements of content that induce us to take up a moral or religious stance towards it, and presents itself solely in terms of aesthetic being".[61] Through aesthetic differentiation the work loses its place in the world to which it belongs,[62] and art and artist lose their place in the world.

Gadamer's own effort to think the mode of being of the artwork as play follows Heidegger in being oriented to a critique of a subjectivized aesthetics. Gadamer begins by distinguishing play as something that is not aimed at the player's subjective reflection. The players are not the subjects of play; instead play merely reaches presentation [*Darstellung*] through the players.[63] Further:

> The movement of play as such has, as it were, no substrate. It is the game that is played – it is irrelevant whether or not there is a subject who plays it. The play is the occurrence of the movement as such. Thus we speak of the play of colors and do not mean only that one color plays against another, but that there is one process or sight displaying a changing variety of colors.[64]

Central to the activity of play is that the "playful" subjective attitude of the players isn't what's decisive. And here, as I noted, we see Gadamer's own Heideggerian move to shift an approach to the work of art from beyond the standpoint of subjectivity, a move of self-transcendence. Instead of starting from the standpoint of subjectivity, "the primordial sense of playing is the medial one".[65] There is a constantly self-renewing, to and fro movement that constitutes play itself. Gadamer asserts the primacy of play over the consciousness of the player. A further general characteristic of the nature of play is reflected in playing: "All playing is a being-played".[66] And, in keeping with the moment Heidegger singles out in Kant, Gadamer also distinguishes play as not involving purposive ends; the end pursued is certainly a non-purposive activity, but this activity is itself intended; it is what the play intends. Although play is not about achieving a purposive, conceptual or useful goal, it nonetheless involves a commitment: play is something serious.

A further element, central to the work of art, is that play is limited to presenting itself.[67] *Its mode of being is self-presentation.* And where does the viewer/participant enter in? Gadamer states that "all presentation is potentially a representation for someone".[68] The intention of this possibility is the characteristic feature of art as play. In this view, the spectator is transformed into a player: "Artistic presentation, by its nature, exists for someone, even if there is no one there who merely listens or watches".[69] Gadamer calls play's consummation into art "transformation into structure". He proposes, further, that:

> Only through this change does play achieve ideality so that it can be intended and understood as play. Only now does it emerge as detached from the representing activity of the players and consist in the pure appearance (Erscheinung) of what they are playing. As such, the play – even

the unforeseen elements of improvisation – is in principle repeatable and hence permanent. It has the character of a work, of an ergon and not only of energiea. In this sense I call it a structure (Gebilde).[70]

Gadamer, then, makes the work of art autonomous, absolutely, while still linking it essentially to the "player". The work of art is dissociated from the representing activity of the player, but is still linked to representation. The link isn't a dependence because the artwork does not acquire a definite meaning only through the persons representing it, or through the original creator, the artist. In relation to them all, the play has an absolute autonomy, suggested by Gadamer's notion of the concept of the play's transformation into structure, through which it becomes a work.

Transformation is not gradual, but to clarify this Gadamer asks us to think of a person transformed; he/she becomes like another person. The play's transformation into structure means that what existed previously exists no longer, and that "what now exists, what represents itself in the play of art, is the lasting and true".[71] To start from subjectivity is to miss the point. What no longer exists is the players.

"But above all", Gadamer writes, "what no longer exists is the world in which we live as our own. Transformation into structure is not simply transposition into another world".[72] We know that Gadamer is highly critical of the idea of a diremption from reality in his critique of aesthetic consciousness. To clarify the nature of the change of transformation, Gadamer uses the example of a drama, which he says resembles a religious act in the sense that it exists as something resting absolutely within itself. For this reason, it doesn't permit a comparison with reality as the "real" measure of verisimilitude. Rather, Gadamer says, "It is raised above all comparisons – and hence also above the question of whether it is all real – because a superior truth speaks from it".[73] Gadamer seeks to further clarify the truth he speaks of here with reference to the way even the most radical critic of art, Plato, spoke of the comedy and tragedy of life and as presented on the stage, without distinguishing between them.[74] There is no longer a difference between the two if one looks to the meaning of the play that is unfolding, on one hand, on the stage, and on the other, in life. Gadamer says that this is "the joy of knowledge".[75]

To briefly continue the analogy, one who perceives the comedy and tragedy of life also recognizes the element of play that life shares with art: that we are not entirely the masters of our own fate (the standpoint of subjectivism) but that we are part of a game that is played with us. It is for this reason that Gadamer calls art's "transformation into structure" a "transformation into the true".[76] In terms of the issue of the encounter with the work and everyday reality, it is not enchantment, but "it is itself redemption and transformation back into true being. In being presented in play, what is emerges. It produces and brings to light what is otherwise constantly hidden and withdrawn".[77] And also: "The being of all play is always self-realization, sheer fulfillment, energeia with its telos within itself. The world of the work of art, in which play expresses itself fully in the unity of its course, is in fact a wholly transformed world. In and through it everyone recognizes that that is how things are".[78] The transformation into structure is central: "reality" is defined from this standpoint as what is untransformed, and art as the raising up [*Aufhebung*] of this reality into truth.[79] Looking to the mode of being of the artwork as "play" and its work of transformation, Gadamer, like Heidegger, avoids the starting point in subjective experience as a ground for the encounter with the work of art; yet, as I already noted above, unlike Heidegger, he also addresses the issue of the status of what we encounter and its relation to the everyday. Within the context of his critique of aesthetic subjectivization, one of his aims is to account for the ontology of the artwork in the way it nonetheless remains integrated into the experience [*Erfahrung*] in which we, in our self-understanding, are transformed. There is more emphasis than in Heidegger's account on the pole of subjectivity, although not a subjectivized experience.

As the concept of play accounts for the work as a representation, but one not dependent on the person's representing it, to further document the mode of being of the artwork, Gadamer also looks to the distinctive temporality of the work of art, which he refers to the temporal character of festivals.[80] This also signals a singular *departure* from Heidegger's view of the event of the work of art as a foundational, singular, event. By examining the temporal character of the work of art on the model of the festival, Gadamer allows an integral moment of mediation, which we do not find in Heidegger's account. The temporality of the festival has the character of repetition, but it is not a literal repetition. Every repetition is as original as the work itself. Here Gadamer draws further on the nature of the time of festivals, their periodicity. In describing festival temporality, Gadamer draws a contrast between what he calls "empty" and "fulfilled" time. The temporal character of celebration is not temporal *succession*. It doesn't dissolve into a series of successive moments. Gadamer distinguishes our normal, pragmatic experience of time, as what he calls "empty time", which is more subject to calculation. Two examples – extremes – of empty time are boredom, where one experiences time as a featureless and repetitive flow, and extreme busyness, where one is aware of never having enough time.[81] Here, too, an experience of reckoning time, of calculation, is prominent: "We fill our time with something or we have nothing to do. Either way time is not experienced in its own right, but as something to be 'spent'".[82]

The temporality of the festival – "fulfilled" or "autonomous" time – is entirely different; and it is also characteristic, according to Gadamer, of the work of art:

> We all know that the festival fulfills every moment of its duration. This fulfillment does not come about because someone has empty time to fill. On the contrary, the time only becomes festive with the arrival of the festival.[83]

Here we see how performance and participation in the work of art, its being as self-presentation, are central. Another feature of festival time – its periodicity – is significant here. The periodicity gives us insight into the way the festival can be both the same and yet different. We celebrate the ending of the old year – this is the same – but each year we do it differently. That is, the festival – what is celebrated – is different in each celebration. The experience of time of the festival, Gadamer says, "is rather its celebration, a present time sui generis".[84] Gadamer makes an interesting contrast between the fulfilled time of the festival and historicality in general: the festival isn't like a historical event, but neither is it determined by its origin so that we can say that there is an "original" festival and then that subsequent celebrations are merely derivative. Further, it is important for the comparison of the work of art that the nature of the festival is to be celebrated regularly. And this presents an important analogy with the work of art. The festival is, in its essence, the same, yet different.

That the festival exists always by different means, Gadamer says, suggests that it is temporal in a more radical sense than everything belonging to history. It has its being only in becoming and return: a festival only exists in being celebrated. Elucidating the mode of being of the artwork by this analogy with the festival enables Gadamer to conceive the historicity of interpretations which are constantly being made and elicited by the work in analogy with festival periodicity: the work is the same, yet different.

Finally, the "claim" of the work of art is also important; that the work of art makes a claim upon us is another way in which Gadamer overcomes subjectivized aesthetics, where the significance of the work is wholly decided by the viewer. "Claim" here means that what presents itself to the spectator as the play of art doesn't exhaust itself in the momentary transport, but "has a claim to permanence and the permanence of a claim".[85] Gadamer draws on Kierkegaard's thought that the claim is something lasting. Because a claim lasts, it can be enforced at any time. A claim isn't a fixed demand, but the ground for such a demand.[86] For this reason, Gadamer says that "contemporaneity" belongs to the being of the work of art. It constitutes the essence of its "being present".[87] This is not the simultaneity of modern subjective aesthetic experience, where several objects of aesthetic experience are held in consciousness at the same time, all indifferently with the same claim to validity, but this "means that in its presentation this particular thing that presents itself to us achieves full presence, however remote its origin may be".[88] Crucially, the contemporaneity of the work of art "is not a mode of givenness in consciousness, but a *task* for consciousness and an *achievement demanded* of it".[89]

In his discussion of contemporaneity in *Truth and Method*, Gadamer continues to emphasize that contemporaneity is like a religious ritual. What he calls here the "total mediation" in the experience of art is that, "neither the being that the creating artist is for himself – call it his biography – nor that of whoever is performing the work, nor that of the spectator watching the play, has any legitimacy of its own in the face of the being of the artwork itself".[90] What unfolds in the artwork is something Gadamer recognizes as "so much lifted out of the ordinary world and so much enclosed in its own autonomous circle of meaning that no one is prompted to seek some other future or reality behind it".[91] It is here that the spectator experiences a loss of self, an "ecstatic self-forgetfulness"; yet Gadamer maintains that this experience "also corresponds to his continuity with himself". His account here has more the character of assertion, when he says of this loss of self and its connection nonetheless to the self, that "precisely that in which one loses oneself as a spectator demands that one grasp the continuity of meaning. For it is the truth of our own world – the religious and moral world in which we live – that is presented before us and in which we recognize ourselves".[92] He continues:

> Just as the ontological mode of aesthetic being is marked by parousia, absolute presence, and just as an artwork is nevertheless self-identical in every moment where it achieves such a presence, so also the absolute moment in which a spectator stands is both one of self-forgetfulness and of mediation with himself. What rends him from himself at the same time gives him back the whole of his being.[93]

I would like to suggest a slightly different view of some of the implications of Gadamer's account of the artwork, particularly if one connects the account of the festival temporality of the work with the discussion of contemporaneity. In doing so, one can locate a prominent place given to the role of the viewer/interpreter of the artwork, and the fact of their specific historical situatedness, and set out some of the implications of Gadamer's account which differ markedly, and importantly, from Heidegger's. Recall that Gadamer, above, speaks of the task of *achieving contemporaneity*. This can be interpreted to mean that the "achievement" rests in the task of encountering the work in a way that encounters its truth, renews its claim, in the present act of engagement with the work, and weaves this into the fabric of one's life, one's self-understanding.

This understanding of the temporality of the artwork is very different from Heidegger's view of the artwork as a founding event, which occurs only once. For Gadamer, in contrast, the event of the artwork carries a claim that is permanent, yet not fixed. This has important implications for the issues raised at the outset of this paper, concerning the mediation of the artwork with the everyday, as well as the way that conceiving the work of art as an event like the festival, raises important questions concerning the role and nature of community. For, it is here with Gadamer's account of the contemporaneity of the artwork, and of its (temporal) mode of being the same, yet different, that it is possible to locate a space to recuperate some of the problematic aspects of Heidegger's account. For, although Gadamer, like Heidegger, elaborates his positive account of the work of art from an ontological starting point – and I have noted that perhaps he is not always clear about the type of knowledge and truth the viewer encounters there – nonetheless, in the account of the [festival] temporality of the artwork, where its being is to "be the same yet different", we can locate the hinge where the ontological joins the ontic. That is to say, where the ontological joins those ontic features missing from Heidegger's account. Gadamer instantiates a space not only for the pole of individual subjectivity – it is subjectivity on which the achievement of contemporaneity, and the enactment of the artwork's "difference" (that it is the same, yet different), is dependent – but also for the ontic dimensions of "meaning" and "interpretation" of the work.

Gadamer's own effort to think the mode of being of the artwork as play follows Heidegger in being oriented to a critique of a subjectivized aesthetics. Yet the nature of the event that constitutes our encounter with an artwork is understood quite differently and in a way more concerned with the mediation of artwork with the everyday. For, unlike Heidegger, Gadamer's critique of aesthetic consciousness and differentiation seeks to emphasize the degree to which art is never wholly separate from the world but always belongs to a world from which it emerges, and which it also helps to understand.

Moreover, if one pursues this significance of the temporality of the artwork, and that it can be "the same" while still being the subject of innumerable encounters with – and interpretations of – the artwork, this then locates the work in the ontic realm in which the work's meaning and significance can be discussed. Further, this means that the experience of the work of art can be shared. For it is precisely here that a work can be a pivot around which different historical and cultural interpretations get articulated. We have seen that it is a part of Gadamer's account of the work of art that it be "the same, yet different": that it is mediated by a shared, or cultural identity, both in belonging to the context and world in which the work was created, and in the world of the present act of encountering/interpreting it. This opens the space, too, for the possibility of different interpretations of a work and a discussion of possible conflicts of interpretation.

Gadamer does not himself draw these conclusions directly from his account of the work of art. Yet, as I have indicated above, with a different emphasis, one can make a link here between the subjective and the intersubjective dimensions at the point where an exchange concerning different interpretations of the meaning of a work takes place. Gadamer's account, then, fruitfully allows for a juncture of the subjective and intersubjective dimensions in our encounter with the artwork.[94] As I have suggested in this paper, this is one way of marking a major difference between Gadamer and Heidegger's account of the event of self-transcendence: Gadamer, unlike Heidegger, gives a more concrete sense to how we understand the trans-subjective experience of the work of art and its mediation with the everyday, and shows that this can be a shared, intersubjective experience, not merely a privative one.

But my point is not to dismiss Heidegger altogether. Rather, it is crucial to recall that Heidegger's overarching concern in his ontological approach to the work of art is also a practical one. It is a critique of modern aesthetics and its attending subjectivization, which is itself a manifestation of modern metaphysics and its problematical rootedness in subjective states. When this is linked to, for example, Heidegger's later analyses of the dangers of modern subjectivism, such as in the essay "The Question Concerning Technology", Heidegger keeps us focused on the gravity of various deformations of late-modernity resulting from the dominance of unbridled subjectivism. It is important to keep in mind that this critical impulse is also behind Heidegger's effort to transcend the self through the work of art.

Boston College, Chestnut Hill, Massachusetts

NOTES

1 This paper was first presented as "Art as Play and Festival: Transcending the Self through the Work of Art" at a conference on the festival (and the cosmos) at the American Society of Phenomenology, Aesthetics, and Fine Arts, held in Cambridge, Massachusetts in April, 1999.

2 I intend in a follow-up study to this paper to do a more detailed analysis of the festival itself, including such things as the loss of self it involves, ethical questions of the service into which the festival is placed, and the type of community generated. In terms of the different types of community, the festival, and the everyday, the everyday realm is a social and political sphere comprised of difference and plurality, and the community of the festival is one of a sharing participation, in which differences are elided, and recede.

3 Immanuel Kant, *Critique of Judgment*, translated by Werner S. Pluhar (Indianapolis: Hackett Publishing Co., 1987), pp. 53–54. Cf. "The Second Moment of a Judgment of Taste", S. 6, "The Beautiful Is What Is Presented without Concepts as the Object of a *Universal* Liking". Kant says, "It follows that, since a judgment of taste involves the consciousness that all interest is kept out of it, it must also involve a claim to being valid for everyone, but without having a universality based on concepts. In other words, a judgment of taste must involve a claim to subjective universality" (p. 54).

4 Due to the scope of this paper, and my concern to look at the theme of self-transcendence and its relation to the festival, I will not go into detail about the different readings Heidegger and Gadamer give of the tradition of aesthetics. Nor will I be able to examine Heidegger's and Gadamer's different readings of Kant. But note briefly here that Heidegger altogether ignores Kant's link of judgments of taste with *sensus communis*. Gadamer, in *Truth and Method*, does examine this link in Kant, but claims that in Kant there is a dissolution of the moral element of *sensus communis*. Because of this, Gadamer does not pursue the importance of *sensus communis* in Kant's discussion as a link of the aesthetic (subjective) and the political and ethical (intersubjective) dimensions. For an excellent elaboration of this, which links Kant's discussion of (reflective) judgments of taste to reflective judgment in general, see Rudolf Makkreel, *Imagination and Interpretation in Kant: The Hermeneutical Import of the Critique of Judgment* (Chicago: University of Chicago Press, 1990), especially pp. 157–158, 168–169.

5 A brief note on this expression "transcendence of the self": it is an achievement of the phenomenological approach to describe the alternative to a subject-centered position in terms of an alternative which presupposes neither an "inside" nor an "outside", a distinct subject versus object, but rather to conceive the connection as a co-relation. In the context of Heidegger's and Gadamer's account of the work of art, the notion of the individual's *participation* in something greater than oneself is a fitting description of this movement of "self-transcendence".

6 Martin Heidegger, "The Origin of the Work of Art", in *Poetry, Language, Thought*, translated by Albert Hofstadter (New York: Harper and Row Publishers, 1971). Hereafter I will refer to this essay as *OWA*.

7 Jacques Taminiaux discusses the shifts in Heidegger's discussion of art, where in the earlier, 1935, version of the 1936 "Origin of the Work of Art" essay, Heidegger isn't interested in the "enigma of art" but in *Dasein's* basic stand toward art. Jacques Taminiaux, *Poetics, Speculation and Judgment: The Shadow of the Work of Art from Kant to Phenomenology* (Albany: SUNY Press, 1993), p. 164.

8 *OWA*, p. 79.

9 *OWA*, p. 80.

10 *OWA*, p. 80.

11 A few months after his conferences on the origin of the work of art, Heidegger gave a lecture course in the winter semester of 1936–37 on "The Will to Power as Art". I will examine this in Heidegger's volume *Nietzsche: The Will to Power as Art* (vol. 1), translated by David Farrell Krell (New York: Harper Collins Publishers, 1979). Hereafter I will refer to this as *N., v. 1*.

12 *N., v. 1*, p. 83.

13 *N., v. 1*, p. 83. Note that this is Heidegger's account of modernity and it is a monolithic account to the degree that it does not allow for significant ruptures in the dominance of this emphasis on the *cogito*. I will return to question this, below, with reference to the position of Kant in the history of aesthetics.

14 Heidegger singles out Nietzsche's emphasis on art in terms of the artist as producer and creator. What is decisive for Heidegger is that Nietzsche is the last metaphysician, in that his notion of artist as creator is (simply) an expression of Nietzsche's view of the *will to power*. For Heidegger, this means that Nietzsche's thought about art still invokes a traditional-metaphysical notion of truth and of beings, and does not get to what, for Heidegger, is the central question, the question of Being; concomitantly, Nietzsche's approach does not allow us to think of the possibility of art beyond modern subjectivism.

15 *N., v. 1*, p. 84.

16 *N., v. 1*, p. 107.

17 *N., v. 1*, p. 70 (citing Nietzsche in *WM*, 811).

18 *N., v. 1*, p. 70.

19 I would like to note here that Kant's aesthetics is, for Heidegger, one of the only places he locates outside the modern expression of the Western tradition of a forgetfulness of Being, and its manifestation in modern subjectivized aesthetics. The question of the status of Kant's ability to see differently, beyond the subject-object dualism which dominates the modern period, raises interesting questions about the monolithic character of Heidegger's characterization of the history of the West as one of the metaphysics of presence and increasing forgetfulness of Being. I examine this issue in more detail in my *Gadamer: Between Heidegger and Habermas* (New York: Rowmann and Littlefield, 2000), see especially Chapter 5. On this theme, note the following remarks from Jacques Taminiaux: "When Kant characterizes the attitude by which we receive the beautiful thing as favor (*Gunst*), in a sense Kant causes the collapse of the modern correlation of subject and object, because such a favor stands beyond any possible conceptualization, any purposefulness, and any subjective complacency, and is rather the pure openness to the unconcealed as such. The surprising thing is that this profound reading of Kant is in no way integrated by Heidegger in his meditation on the history of art, insofar as the history is linked to the history of aesthetics" (p. 151). This raises the question: is the horizon of aesthetics a terminus, allied with modern subjectivity, or could it be a "new beginning" linked with our age? Jacques Taminiaux, *Poetics, Speculation, Judgment: The Shadow of the Work of Art from Kant to Phenomenology*, ed. and translated by Michael Gendre (Albany: State University of New York Press, 1993), p. 151. Finally, I noted in n. 13, above, the monolithic character of Heidegger's characterization of modernity, and would like to extend this to the question of whether there is, in fact, a "modern subjectivization of aesthetics"? Although Gadamer departs from Heidegger's monolithic conception of the history of the West in terms of a metaphysics of presence and forgetfulness of Being, Gadamer, in *Truth and Method*, seems to agree with Heidegger about a "modern subjectivization of aesthetics", although Gadamer gives a very different reading of Kant's role, namely, as one of the main causes of the subjectivization, rather than (in Heidegger) as a point of resistance.

20 *N., v. 1*, p. 108.

21 *N., v. 1*, p. 109.

22 One finds a particular resonance with this discussion of Kant in passages in "The Origin of the Work of Art", where Heidegger calls for a simple recognition of the work, "*that* it is". See *OWA*, pp. 65–66. Note, too, that there is a certain ambiguity in Heidegger's validation of this purity. For Heidegger himself is well known for his critique of a theory of pure perception, undertaken on the basis of pragmatic experience. Gadamer criticizes the notion of purity in *Truth and Method*, claiming that purity is never possible since even when encountering something like absolute music – pure in form – although there is no objective meaning, nonetheless, our experience of it "involves entering into a relation with what is meaningful"; *TM*, p. 91. Gadamer's statement, a bit later, that "pure seeing and pure hearing are dogmatic abstractions that artificially reduce phenomena. Perception always includes meaning", forms the basis of his critique of purity; *TM* p. 92. Gadamer's claim that "perception always includes meaning" signals a difference between Heidegger and Gadamer's approaches to art. His emphasis on meaning, which is a meaning for the subject/viewer, places greater emphasis on the pole of subjectivity than does Heidegger's account.

23 *N., v. 1*, p. 109.

24 *N., v. 1*, p. 110.

25 *OWA*, pp. 42, 59. Heidegger describes the example of a Greek temple and invokes Greek *phusis* when he says: "The temple's firm towering makes visible the invisible space of air. The steadfastness of the work contrasts with the surge of the surf, and its own repose brings out the raging of the sea. Tree and grass, eagle and bull, snake and cricket first enter into their distinctive shapes and thus come to appear as what they are. The Greeks early called this emerging and rising in itself and in all things, *phusis*. It clears and illuminates, also, that on which man bases his dwelling. We call this ground the *earth*" (p. 42). Heidegger refers to what comes into being through the work of art as occurring in the midst of "the being that grows out of its own accord, *phusis*" (p. 59).

26 *OWA*, p. 36.

27 *OWA*, p. 46.

28 *OWA*, p. 61 [German edition, *Die Ursprung des Kunstwerkes* (Stuttgart: Philipp Reclam, 1960), p. 61.]

29 *OWA*, p. 61.

30 Other examples Heidegger gives are: the act that founds a political state; truth shining forth as the nearness of the being that is most of all; essential sacrifice; the thinker's questioning, which names the question-worthiness of Being. *OWA*, pp. 61–62.

31 *OWA*, p. 43.

32 *OWA*, p. 43. Given the time when the essay on the work of art was written, 1935–1936, it is clear that Heidegger had in mind the question of the German nation and whether it was ready to stand poised for the fulfillment of its vocation. At the time Heidegger wrote this essay on the work of art, he made references to the political sphere of the present, and was also looking to the poet Hölderlin, as the one who might "ground" the German people historically. For a brief discussion which links Heidegger's *OWA* essay with his writings on Nietzsche and Hölderlin, and which connects his comments in the *OWA* essay with his views on the historical possibility of the German people, in the context of his membership in the National Socialist Party, see Kathleen Wright's entry, "Heidegger and Hölderlin", in the *Encyclopedia of Aesthetics*, ed. Michael Kelly (Oxford: Oxford University Press, 1998), pp. 383–386. For a general discussion of Heidegger's involvement with National Socialism, and his silence on this, there is much recent literature. See Richard Wolin (ed.), *The Heidegger Controversy* (Cambridge: MIT Press, 1993), and Rüdiger

Safranski, *Ein Meister Aus Deutschland: Heidegger und Seine Zeit* (Frankfurt am Main: Fischer Verlag, 1998), especially chapters 13–17, pp. 231–344. For a specific critique in a contemporary context of the concept of "dwelling", note Neil Leach, "The Dark Side of the *Domus*: The Redomestication of Central and Eastern Europe", in *Architecture and Revolution: Contemporary Perspectives on Central and Eastern Europe*, Neil Leach, ed. (London: Routledge, 1999), pp. 150–162.

33 *OWA*, pp. 67–68.

34 *OWA*, p. 72.

35 *OWA*, p. 75.

36 *OWA*, p. 76.

37 *OWA*, p. 76.

38 *OWA*, p. 76.

39 *OWA*, p. 76.

40 *OWA*, p. 66. Emphasis added.

41 It is at this juncture that a reading of Kant which gives weight to the account of common sense [*sensus communis*] provides a crucial link between the subjectivity of judgments of taste and their intersubjectivity. As noted above, Heidegger ignores this feature of Kant's aesthetics.

42 One cannot help but think here of the ideological function of festivals and the loss of self experienced in National Socialist glorification of the idea of a *volkisch* community. See, for example, George L. Mosse, *The Crisis of German Ideology: Intellectual Origins of the Third Reich* (New York: Howard Fertig, 1964, 1998).

43 This is not to say, as Jürgen Habermas does, that we should dismiss Heidegger's submissive rhetoric *in toto* as leading to and advocating a passive weakness in the face of autocratic authority. See my "Heidegger and the Rhetoric of Submission: Technology and Passivity", in *Rethinking Technologies*, ed. Verena Andermatt Conley (Minneapolis: University of Minnesota Press, 1993), pp. 115–139.

44 *TM*, p. 70.

45 *TM*, p. 70.

46 *TM*, p. 84.

47 *TM*, p. 84. Gadamer writes: "What is considered valid in a society, its ruling taste, receives its stamp from the commonalities of social life. Such a society chooses and knows what belongs to it and what does not. Even its artistic interests are not arbitrary or in principle universal, but what artists create and what the society values belong together in the unity of a style of life and an ideal of taste". *TM*, pp. 84–85.

48 *TM*, p. 85.

49 *TM*, p. 87.

50 *TM*, p. 87.

51 *TM*, p. 95.

52 *TM*, p. 97.

53 *TM*, p. 97.

54 *TM*, p. 97.

55 *TM*, pp. 97–98.

56 *TM*, p. 164.

57 I am examining Gadamer's account of the artwork in *Truth and Method*, where he perhaps does not elaborate clearly enough what this transformation of the viewers in their encounter with the truth of the artwork is like. For a discussion of Gadamer on this point, see Kate Hamburger, *Wahrheit und aesthetische Wahrheit* (Stuttgart, 1979). Gadamer does elaborate on the nature of

aesthetic experience in some of his more recent writings on aesthetics in a way that addresses the lacunae I have just noted. See Hans-Georg Gadamer, *Gesammelte Werke*, Band 8 & 9, entitled respectively, *Esthetic und Poetik* I und II (Tübingen: J. C. B. Mohr (Paul Siebeck), 1993). For a brief discussion of this, see Jean Grondin's entry, "Gadamer and the Truth of Art" in the *Encyclopedia of Aesthetics*, ed. Michael Kelly (New York, Oxford: Oxford University Press, 1998), pp. 267–271.

58 *TM*, p. 100.

59 *TM*, p. 82.

60 *TM*, p. 83.

61 *TM*, p. 85.

62 *TM*, p. 87.

63 *TM*, p. 102.

64 *TM*, p. 103.

65 *TM*, p. 103.

66 *TM*, p. 106.

67 *TM*, p. 108.

68 *TM*, p. 108.

69 *TM*, p. 110.

70 *TM*, p. 110.

71 *TM*, p. 111.

72 *TM*, p. 112.

73 *TM*, p. 112.

74 *TM*, p. 112. Note that Gadamer's reading of Plato's positive valuation of tragedy here goes against the usual reading which emphasizes Plato's criticism of the poets and tragedy. The role of tragedy is also significant in examining Heidegger's shift to a positive valuation of the artwork, *techne* and *poiesis*, in the *OWA* essay. Jacques Taminiaux's discussion, noting Heidegger's effort to positively value *techne* and *poiesis* as a production of the art of artists in the 1936 *OWA* essay, entails a shift from the 1927 *Being and Time*, which had evaluated these only negatively, as inauthentic. Writing of a Platonic bias of speculation in Heidegger's emphasis on the contemplation of Being, affirming the *bios theoretikos*, Taminiaux indicates that Heidegger would have found justification of the shift to a positive valuation of *techne* and *poiesis* were he to have looked to Aristotle's *Poetics*, which argues for a "clear distinction between the *techne* and *poiesis* of artisans and experts and the *techne* and *poiesis* of artists and poets specializing particularly in tragedy" (p. 381). That is, Aristotle's claim that there is a philosophical teaching in works of art, above all tragedy, goes against Heidegger's following of (the traditional) Plato, who values them negatively. See Taminiaux's discussion of Heidegger here in "Philosophical Heritage in Heidegger's Conception of Art", *Encyclopedia of Aesthetics*, ed. Michael Kelly (New York, Oxford: Oxford University Press, 1998), pp. 380–383. In his reference to Plato above, Gadamer does not cite the source of his reading of Plato's positive valuation of tragedy, and its ability, as we infer from his statements, to also provide philosophical teaching. This points to Gadamer's unconventional reading of Plato. For more on this, see Hans-Georg Gadamer, *The Idea of the Good in Platonic-Aristotelian Philosophy*, translated by P. Christopher Smith (New Haven: Yale University Press, 1986). Finally, Gadamer's very different readings of Plato and Aristotle, emphasizing Platonic dialogue and Aristotelian *phronesis*, stand at the center of his hermeneutical philosophy as a democratization of Heidegger's thought. I will develop this below, in specific relation to art, but for a fuller articulation, see my *Gadamer: Between Heidegger and Habermas* (New York: Rowman and Littlefield, 2000).

[75] *TM*, p. 112.
[76] *TM*, p. 112.
[77] *TM*, p. 112.
[78] *TM*, p. 113.
[79] *TM*, p. 113.
[80] In what follows, I will also draw on Gadamer's description of the festival in his essay, "The Relevance of the Beautiful: Art as Play, Symbol and Festival", in *The Relevance of the Beautiful and Other Essays*, ed. Robert Bernasconi (Cambridge: Cambridge University Press, 1986), pp. 3–53. Hereafter I will refer to this as *RB*. This essay originally appeared in German in 1977. Gadamer also discusses the festival, more briefly, in *Truth and Method*, pp. 122–123.
[81] *RB*, p. 42.
[82] *RB*, p. 42.
[83] *RB*, p. 42.
[84] *TM*, p. 123.
[85] *TM*, p. 126.
[86] *TM*, p. 127.
[87] *TM*, p. 127.
[88] *TM*, p. 127.
[89] *TM*, p. 127.
[90] *TM*, p. 128.
[91] *TM*, p. 128.
[92] *TM*, p. 128.
[93] *TM*, p. 128.
[94] It is here again that Kant becomes relevant in making judgments of subjective yet intersubjective taste through recourse to the *sensus communis*. Gadamer does not himself explicitly make this link, yet in later sections of *Truth and Method*, he emphasizes the importance of exercising practical judgment [*phronesis*] in the social and political domain. He does not, however, expressly link the political to the aesthetic. See also Makkreel, *Imagination and Interpretation in Kant*.

GOTTFRIED SCHOLZ

THE GREATEST OPERA EVENT OF THE EIGHTEENTH CENTURY:

Costanza e Fortezza *and Its Political and Religious Message to the Europe of 1723*

The theater of life and the stage of the world – this great spiritual concept was manifested through many examples throughout 3000 years of western civilization. One of these may demonstrate its intellectual message as well as its artistic value.

On August 28, 1723 a memorable and unique event took place at the Hradschin in Prague. The most important operatic production that the 18th century ever experienced was staged at an open-air theater on the grounds of the old riding school – a theater which had been constructed expressly for that purpose. At 8:00 PM, as was strictly and hierarchically regulated, Emperor Karl VI and his wife, Elisabeth Christina, met with an impressive number of aristocrats from the Hapsburg Empire and neighbouring courts, as well as higher and lower representatives of Bohemia. They were admitted into an ephemeral, palatial building of wood, linen and stucco, which was illuminated by a thousand wax candles and innumerable oil lamps.

One must ask what purpose this enormous expenditure could serve and what this event could mean for that time period. First of all, the occasion celebrated the thirty-second birthday of the Empress, an event which was already expected to be celebrated with the ritual of a festive operatic production, among other things. That it all took place in Prague rather than in Vienna this time is explained by the fact that on September 5th, Karl was to be crowned King of Bohemia and three days later Elisabeth Christina would be crowned Queen. Therefore this large festival was not only a celebration of a birthday and a preliminary celebration for the coronation, but it also represented a national political demonstration of the power of the *Domus Austriaca*. Certainly all of Europe had to take notice.

The entire coronation ceremony would not necessarily have been required, since Bohemia had lost its independence as a result of the Battle at the White Mountain, and was declared a Hapsburg *Erbreich* – in other words, a Province – through the *Vernewerte Landesordnung* (renewed state law) of 1627. It goes without saying that this tragic shadow existence sparked a

A.-T. Tymieniecka (ed), Analecta Husserliana LXXIII, 229–235.

fundamental bitterness among the Bohemian nobility, and this operatic event can therefore be seen as a friendly concession on the part of Karl. Likewise, his father before him created a festive ceremony when he was crowned in Bohemia more than twenty years after becoming a ruler there (in 1680, just in the face of the Turkish threat).

Often overlooked is the internal national consequence of a coronation in both the seventeenth and eighteenth centuries. Up until then the representative body (provincial diet) would only accept the oath of allegiance after having submitted their requests to the monarch. Of course Karl declared the Hapsburg Dominion to be indivisible and inseparable through the Pragmatic Sanction of 1713. This legal document was accepted by the Bohemians in 1720, so a *Wahlkapitulation* was hardly thinkable. Since Bohemia had lost its legal status as an electoral kingdom through the above-mentioned law, it also had another constitutional position comparable to the *Römischer Reich deutscher Nation* or the Kingdom of Hungary, both of which had elective constitutions. Therefore, immediately after the sudden death of his brother Joseph I, Karl VI had to make sure that the coronation took place in Frankfurt and Pressburg, as without a formal ceremony he would not have been able to make legally binding decisions in these regions.

Upon closer inspection of the political power situation within Europe, this grandiose operatic production of 1723 takes on a special importance. Just a few years before, the two daughters of the late Emperor Joseph I had married two electors who ruled in the immediate vicinity of Bohemia – Friedrich August II (August III) von Sachsen and Karl von Bayern. Despite these new kinsmanlike ties, this was a powerful threat because at that time Karl VI had no male heirs, and although the aforementioned electors had agreed to the Pragmatic Sanction, this did not necessarily mean that after the death of Karl they would also really feel bound to their oath. One could also argue that a daughter of the former monarch had earned the right to be the successor instead of Maria Theresa. In fact, such arguments eventually did lead to the Austrian war of succession in 1740. In 1723, it was certainly sensible to assert one's presence in Bohemia with a demonstration of power in Prague, and to embellish this with the typical pompous grandeur of courtly absolutism.

An opera is always a total work of art (*Gesamtkunstwerk*) for which representatives of all the various arts must unite, in order to realise an idea together. The opera, *Costanza e Fortezza*, was realised through the united efforts of the librettist Pietro Pariati, the imperial conductor and composer Johann Joseph Fux, the composer and arranger of the ballet which took place within the opera Nicolò Matheis, the theater engineer and stage setting architect Guiseppe Galli-Bibbiena, and finally, the conductor Antonio Caldara (who substituted for the fragile composer). The names of the famous Italian soloists of the time mean nothing to us now. However, the two women, four castrati and three male singers were among the most famous representatives of their milieu. The concentration of high voices in a ratio of 6:3 is explained by the acoustics of a outdoor production, as higher voices carry better when there is no closed resonant body. Approximately one hundred choral singers and two hundred instrumentalists assembled for this performance, among them important personalities of later music history – Johann Joachim Quantz (oboe) and Carl Heinrich Graun (cello). It is significant to note that Johann Sebastian Bach did not attend, as he was not allowed to leave his employer in Leipzig for a trip to Prague. Four thousand spectators were counted, and the cost of the entire enterprise was over 66,000 florins, which by today's standards is an inconceivably large sum.

The preparation of the performance is described in the protocol of the ceremony which is stored in the house, court and state archives (partially edited by Hans Leo Mikoletzky). The journey of the entire entourage to Prague was comparable to the preparation of a military campaign. It included a total of 321 light and 125 heavy wagons drawn by 1,788 horses. A construction plan was necessary in order to, among other things, improve the roads and strengthen the bridges for the weight of the vehicles. The accommodations in Prague are documented by the Billeting Acts (*Einquartierungs-Akten*) which have survived. Even today, local historians can find that the poet Pariati stayed in house Number 4 on the Kleinseitner Ring and that Fux lived in house Number 324. The trip from Vienna to Prague lasted 12 days; the whole imperial journey to and from Vienna lasted five months.

In Prague, the theater first had to be erected. Then the necessary machines had to be made functional for the transformation, and finally ripieni had to be found to reinforce the regular court musicians in the choir and orchestra. The safety report of the Bohemian Chancellor, Count Schlick, describes how carefully one proceeded with such a state ceremony, in order to prevent possible disturbances right from the beginning.

Karl VI chose "Constantia et Fortitudine" as his motto. These are two terms from Roman antiquity which were meant as instrumental ablatives to symbolise his personal involvement in the performance of his duties. It will never be known, however, how much they actually helped in the enforcement of his government. At any rate, he was not blessed with luck at the beginning of his reign. As the designated heir to the Spanish Hapsburg throne, he did his best to establish himself there. His opponents – the French Bourbons whose claim to power led to the Spanish war of succession – thwarted the realization of the dream of his youth. Nevertheless, up until his death he saw himself as the Spanish King. The Spanish crown adorned his coffin among other things, and the Castillian coat of arms glitters within his complete coat of arms over the Imperial Chancellery in Vienna, which he built.

The action of the opera libretto follows the model of the Jesuit allegory theater – the principle of *fabulae docent*. The reports found in Livius and Plutarch serve as a starting point for the text. They describe the victory of the young Roman Republic over the Kingdom of the Etruscans. These allegories would be misunderstood, if we thought that the Bohemian King identified with the Estrucan Kingdom and that the insurgent Roman Republic implied an association with the Czech rebellion in 1618. Karl VI did not see himself as Porsenna, King of Etruria, but rather as the *Imperator Romanorum* in the form of the Roman Consul, Publio Valerio. The principle of Rome (which won the victory in the libretto) is significant in the Roman *Reichsidee* of the Emperor. Franz Matsche notes in great detail the symbolic identification of Karl with Emperor Augustus, as well as Vienna being named *nova Roma* in the writings of the court second-hand book seller and numismatologist, Karl Gustav Heraeus.

In every Baroque opera, the relationships of the lovers and the intrigues to which their love is subjected are part of the action. But even here we must differentiate between the realistic and the symbolic levels of meaning. One can recognise the peace of Rastatt (1714), where the sphere of influence between the Bourbons and the Hapsburgs was separated, in the final peace of the plot that is reached between Rome and Etruria. It is therefore possible to assume, as Othmar Wessely has already noted, that in the Etruscan King one sees Ludwig XIV, that his grandchild, Philippe d'Anjou, can be found in the figure of Tito Tarquinio, and that both portray the opponents of the Roman party. It would certainly be audacious to see the nieces of Karl VI as both of the Roman maidens who found themselves in Etruscan power. The clearly national political statement that the responsibility of the Emperor is to stand as a *Reichsinstitution* over his familiar group, can be found in one part of the libretto. In the aria, *"Padre son,"* Publio Valerio sings (translated into modern English):

> I am a father, but at the same time a son of the Roman Empire, loyal and true to the death. I know, however, that my daughter and son are in shackles, but I have to respect the Roman throne more highly, as it must be kept free.

Despite all differences, it is not only the custom of diplomacy, but also part and parcel of the expressive possibilities of the arts, to remain ambiguous. Indeed, it is in the individual interpretation and the explanation of the metaphors used, that the fascinating approach to the artistic message is found. Ambiguity is then not negative, but rather actually a substantial element in the creation of art. During this time of courtly absolutism, more could be asserted through vague suggestion than in direct speech, which was subjugated to the obligation of stilted etiquette.

The libretto of Pariati offers an impressive example. If we recognise in the act of coronation a declaration of political power, then the symbolism which carries the plot of the opera becomes a political statement. The Emperor rejects the French King's claims to Spain for his grandson, Philippe, in the same way that the Roman Consul rejects the claims of the Etruscans for the young Tito Tarquino. But then, in the third act, as Tito wants to inflict violence on the imprisoned Clelia, Persenna turns away from him, whereby peace is made possible.

Certainly one can also interpret the action through Christian allegory, for the Roman protagonists stand for virtue (Clelia), stability (Valeria), bravery (Orazio) and readiness for martyrdom (Muzio), whereby the positive aspects are finally victorious over the intrigue, violence and brutality of the Etruscans. Such meaning coincides more with the counter-reformation tradition of the Jesuit theater, which had a great influence in Vienna and Prague in the early eighteenth century. Thus the key message has at least two meanings, but both are undoubtedly an expression of imperial power, integrity and its victory over all hostility.

The dramaturgy of this festive evening followed the usual sequence of three acts. Contemporary copper engravings have preserved for us in the minutest of details the extraordinarily rich stage settings: the effect of the magnificent symbolic power of the theater architecture on the eyes of the time must not be underestimated. When, for example, in the first scene, two palaces stand across from each other (indicating separation by the Tiber river) a closer examination shows on the left, the middle part of the Etruscan palace in ruins, whereby the equivalent on the right is a Roman palace presented in total contrast as a raised dome. This dome crowns the Temple of Vesta, through which again reference is made to the noblest woman in the production, namely, the Empress. Likewise, the left side which is falling apart can be connected with the very Spanish/Bourbon rule that one would have been happy to see crumble. The projection of Christian *Pietas Austriaca* in the pagan mythology can be seen in the river god rising out of the Tiber and promising the rescue of Rome, thus substituting for the intervention of God. The supernatural powers are also on the side of the Roman party, and so the Consul also remains steadfast despite the difficult political final situation. The act closes appropriately with a ballet symbolizing Costanza Romana.

The second act, which takes place in the military encampment of the Etruscans, is very impressive. This can be seen in the extensive details of the story, such as the confusion between lovers' promises and rejections, the heroic courage of Muzio and the characteristic strengths of the Roman party. The scene closes with a ballet symbolizing Fortezza Romana.

The third and final scene shows an ideal palatial garden – the garden of Tito Tarquino at Gianicolo. The conclusion leads, according the Viennese practice, to a *Licenza* – both an address of tribute to the Empress and a politically, cleverly incorporated tribute to the city of Prague, the location of the festive occasion and the coronation city within which both high and low society should have been more closely connected to their monarch and his house.

As the festive opera drew to a close, likewise the High Baroque stage spectacle slowly faded away from the European cultural landscape. No longer would such a large artistic and political extravaganza be produced for a musical dramatic performance. Although operatic events continue to deal with political content, *Costanza e Fortezza* stands out as a cultural-historical climax that can also be seen as the end of an evolution of national political presentations of courtly absolutism.

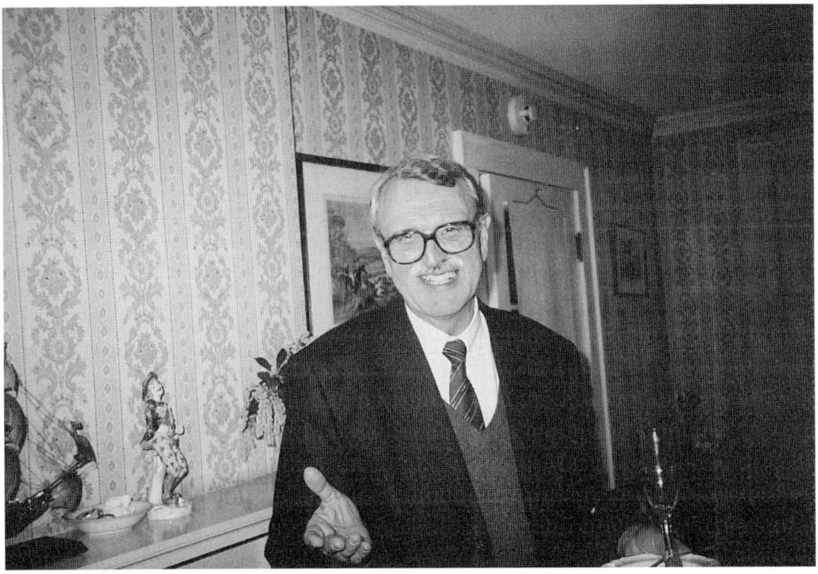

Gottfried Scholz

LEE F. WERTH

EUGENE O'NEILL'S DIVERSE USE OF FOG AS AN EXISTENTIAL METAPHOR

Eugene O'Neill had spent time at sea as a common sailor. Perhaps it was a "game of romance and adventure," as Tyrone claims in the play *Long Day's Journey into Night*, when refusing to acknowledge that young Edmund truly understood poverty.[1] Nevertheless, it is doubtful that life aboard a square-rigger sailing well into the twentieth century would be a game. O'Neill seems to have learned much from his sea adventures, however unromantic the drudgery; and he no doubt learned the significance of fog, which unlike darkness, is deceiving, not merely blinding.

Fog not only hides things, but makes them appear or seem to be what they are not. The reason for this is easily understood: sound travels more quickly in water or in saturated air. Only in dry air are we able to accurately sense the direction and origin of a sound, or its range. One may easily test this: while swimming, wait for an outboard powered boat to approach. When your head is underwater, the sound seems to be all around with virtually no indication of direction or distance. As to the optical properties of fog, staring into fog will sometimes produce a sort of after-image resembling a dark form, seemingly a ship bearing down on one. And like cloud formations, misty and smoky seas will take on the forms of fantasy.

Yet fog too can be beautiful. Tiny droplets create a sort of silver fur on everything. When shafts of light do penetrate the mists, an ethereal atmosphere is created in which one escapes the ordinary world, or perhaps seems transported beyond the mundane to something more profound and spiritual. Seaman O'Neill knew this well.

But to a sailor, fog is, first and foremost, a hazard to navigation. In the plays *Chris Christophersen* and *Anna Christie*,[2] We do not find Chris to be fond of fog. He is a sort of existential straight man who never forgets the real significance of fog to a sailor adrift in shipping lanes (prior to the existence of radar and when radio was uncommon).

The characters in *Long Day's Journey into Night* seek to hide in fog, either the literal or the chemically induced kind. Chris Christophersen, the sailor, knows that fog hides him when he *wants* to be seen for what he is; a sailor aboard an imperiled vessel at sea, a simple man engaged in honest and often dangerous work. His daughter, Anna, finds mystical embrace in the

A.-T. Tymieniecka (ed), Analecta Husserliana LXXIII, 237–246.

nothingness of fog, a rapture spoiled only by the raucous sound of the foghorn, which to her delight and Chris' horror, breaks down when most needed.[3]

As a sailor, O'Neill knew the literal meaning of fog, but as an alcoholic he also understood fog as an escape. As a poet, he understood fog as the possibility of mystical embrace and an intimation of some greater and ineffable meaning which he tried to capture, but was left feeling as though he had only stammered, for as Edmund says in *Long Day's Journey into Night*: "Stammering is the native eloquence of us fog people."[4]

Strangely, it is not in the sea plays that fog is most clearly used as a metaphor for existential bad faith and escape. In the sea plays, fog is more often the sailor's fog, Chris' fog is produced by "dat ole davil, sea." It is a medium between life and death and not without its occult significance; but whatever significance it has comes to it from the *literal* possibility of its causing shipwreck. And although "shipwreck" is itself a common existential metaphor, literal shipwreck is a cause of death, not an inquiry into the ultimate origin of values or the significance, if any, of life and the human condition. Sailors simply want to stay alive. And so it is perhaps not so strange that the sea plays portray fog quite differently than what we find in *Long Day's Journey into Night*.

With respect to fog being a metaphor for mystical enlightenment, the Buddhist Void, Nothingness, and the bliss consciousness which is a non-striving, non-particularizing awareness, Anna Christie does achieve mystical embrace when fogbound and at sea. But she is not a sailor, even if sailing is in her blood. Edmund, in *Long Day's Journey into Night*, is a kindred spirit to Anna Christie (in either of her incarnations, whether in the play, *Chris Christophersen*, or *Anna Christie*).[5] But Edmund finds his state of mystical embrace while lying under a bowsprit and looking aloft at the sails during a *clear* and moonlit night. Fog is no friend of Edmund's, at least not at sea.

We are to understand that what Anna and Edmund experience as an ineffable significance, which at best is revealed only "through a glass darkly," is genuine and real, and a reason for human hope. We can hardly expect Edmund, the sailor, to glimpse reality through fog; that is for a landswoman, Anna. But each in their respective mystical environments finds authenticity and a kind of salvation and redemption.

Let us now attend to passages from O'Neill's plays which include fog both literally, as a setting, and metaphorically.

In the tradition of ghost stories, O'Neill uses fog as an occult bridge between the dead and the living. This use is effective enough, but not particularly interesting in any philosophical sense, except, perhaps, to make us wonder why mist should be associated with shades and spirits. But fog and mist can assume many forms as well as being cold and clammy and, of course, disorienting. The ghost-fog relationship seems natural.

In his play *Fog*, O'Neill has the crying of a dead child alert a passing ship, thereby saving those who are fog-bound in a lifeboat. It is a "Twilight Zone" plot.

> The Officer: "Sounds like a kid balling, doesn't it?"... It kept getting plainer and plainer until there was no chance for a mistake – weird too it sounded with everything so quiet and the fog so heavy – I said to him again: "It's a kid sure enough, but how in the devil did it get out here?" And then we both remembered we had been ordered to keep a lookout for any of the survivors of the "Starland" who hadn't been picked up yet.... We could hear the kid crying all the time....[6]

The officer is told by the poet: "The child has been dead for twenty-four hours. He died at dawn yesterday."[7] The poet, who has shown greater courage and survival skills than the businessman with whom he has been trapped, expresses a desire to somehow commune with the dead mother and child:

> (*gently*) I think I will stay with the dead.
> (*He is sitting opposite the two rigid figures looking at their still white faces with eyes full of a great longing*.)[8]

The death-fog relationship is also illustrated in *Bound East for Cardiff*. Fog imperils the ship; Seaman Driscoll makes matters worse by saying:

> Twas just such a night as this the auld Dover wint down. Just about this toime ut was too, and we all sittin' round in the fo'castle, Yank beside me, whin all av a suddint we heard a great slitherin' crash, and the ship heeled over till we was all in a heap on wan side. What came afther I disremimber exactly.... [9]

Yank does not wish to die in a fog. As he lies dying, he sees fog inside the ship. He hallucinates and does indeed die before the fog lifts. The fog then lifts, indicating that the ship's crew will have a clearer view than we have as individuals during our lives, and perhaps the hope that Yank's questions are now answered. And then too, being a good shipmate is of import, even when we lose our bearings, which, being human, we are bound to do.

Whereas other writers have related fog to death, O'Neill's diametrically opposed uses of fog as an existential metaphor are quite innovative: fog is a means of escape and existential "bad faith"; yet it can also be a way of recognizing the inauthenticity and self-deceptions associated with our ordinary social, moral and religious platitudes, i.e., the false bearings of dogmas.

When cast adrift alone and estranged, we must come to terms with ourselves. We must find ourselves in order to find our way, not conversely.

Existentialists frequently use nautical metaphors, speaking of horizons, or being adrift, or off-soundings and at sea, of shipwreck, or at anchorage; but fog is perhaps the most profound. Consider Heidegger's comments:

> Real boredom comes when "one is bored." This profound boredom, drifting hither and thither in the abysses of existence like a mute fog, draws all things, all men and oneself along with them, together in a queer kind of indifference. This boredom reveals what-is in totality.[10]

There is that desire to have an *open horizon*, to be unhemmed in, to escape *Beyond the Horizon*.[11] But as O'Neill's characters discover in the play with that name, if one goes over a particular horizon, one takes oneself "overthere" in the escape into a disingenuous freedom. A horizon may open, but it is just another horizon. To go beyond the horizon is really only another form of hiding within the fog, as do Mary and Edmund in *Long Day's Journey into Night*. Perhaps Nietzsche is too glib when he says:

> At last the horizon seems open once more, granting even that it is not bright; our ships can at last put out to sea in the face of every danger; every hazard is again permitted to the discerner; the sea, *our* sea, again lies open before us; perhaps never before did such an "open sea" exist.[12]

Karl Jaspers observes about Nietzsche:

> Nietzsche was always conscious of moving on the sea of the infinite, of having given up land once and for all. He knew that, perhaps, neither Dante nor Spinoza knew his loneliness; somehow they had God for company.[13]

And yet if one is *always* on the open sea, any particular horizon is arbitrary and a matter of optics. If one is truly alone on an open sea and has given up land once and for all, then an open horizon is tantamount to fog; one has no "position," and in a sense can never be lost. Jaspers captures this with the "Encompassing":

> We always live and think within a horizon. But the very fact that it is a horizon indicates something further which again surrounds the given horizon. From this situation arises the question about the Encompassing. The Encompassing is not a horizon within which every determinate mode of Being and truth emerges for us, but rather that within which every particular horizon is enclosed as in something absolutely comprehensive which is no longer visible as a horizon at all.[14]

The foregoing quotations and enigmatic remarks are clarified by Anna Christie's discovery that she feels *at home* in the fog, i.e. the same sense of belonging that Edmund feels under the bowsprit on a moonlit night. The relevant passages clarify what might be called an existential anchorage or an ontological harbor.

The Anna of the play *Anna Christie* says:

> Funny! I do feel sort of – nutty, tonight. I feel old.
>
> Chris: Ole?
>
> Anna: Sure – like I'd been living a long, long time – out here in the fog. I don't know how to tell you just what I mean. It's like I'd come home after a long visit away some place. It all seems like I'd been here before lots of times – on boats – in this same fog. You must think I'm off my base.
>
> Chris: Anybody feel funny dat vay in fog.
>
> Anna: But why d'you s'pose I feel so – so – like I'd found something I'd missed and been looking for – 's if this was the right place for me to fit in? And I seem to have forgot – everything that's happened – like it didn't matter no more. And I feel clean somehow – like you feel just after you've took a bath. And I feel happy for once – yes, honest! – happier than I ever been anywhere before! It's nutty for me to feel that way, don't you think?[15]

The Anna of *Anna Christie* had been raised on a Minnesota farm, her virtue compromised, leading to employment as a prostitute. An examination of the play, *Chris Christophersen*, reveals that the elegant Anna raised in England feels similarly about fog. The similarities and differences highlight fog as "Encompassing" rather than as escape. American Anna feels cleansed and freed from her earlier and iniquitous life. The refined Anna of England also speaks of feeling old in a manner intimating prior incarnations,[16] and of being immensely happy and belonging. Any difference in their respective reactions to fog concerns the cleanliness theme: the Anna raised in England has no need for a metaphorical redemption. Rather, she has perfectly wholesome plans to employ her typing skills in an urban office, plans which although practical, relate to a form of life incapable of engaging her true self which finds its home in fog. (Although Chris has an identical persona in the plays *Anna Christie* and *Chris Christophersen*, his daughter Anna is a very different person; it might be argued that she is different only in those respects which have been shaped by respectively different environments. Poverty is unlikely to be ennobling.)

The following lines from *Chris Christophersen* reveal an enlargement of perspective which dwarfs mundane plans. Anna's self doesn't shrink into the fog out of frustration from goals unachieved. Rather, she transcends all trivial aspirations. (It is to be remembered that Anna is aboard a barge adrift in shipping lanes. She is alone with her father whom she virtually has only just met.)

> Chris: You act funny tonight, Anna. You ain't sick?
>
> Anna: No. How silly! I'm feeling better than I ever did.

Chris: You vas queer gel, Anna. Ay tank maybe Ay'm damn fool for bring you on dis voyage. Vell, you gat yob on shore – Ay gat yob, too – yust soon ve gat back to New York. Dat's best tang for you – yob in country, py yingo!

Anna: Please don't talk about – that – not until we do get back, Father. All those things, all my plans seem so far away now – and dead! I don't care to think of them. They're only the outside of me. They don't matter one way or the other. It's too – big – out here – to even consider them. I feel – something way down inside me – something I've never felt before – tonight. But I'm talking silly, aren't I, Father?[17]

Anna is aware of the danger of the situation, at least intellectually. Her father has clarified the situation. Yet she continues:

Anna: But I don't feel that the sea hates me – not tonight. – It's funny, Father. I do feel strange tonight. I feel – so old!

Chris: Ole?

Anna: Yes – as if I'd lived a long, long time – out here in the fog; as if I'd come back home after a long visit away someplace. It all seems so – familiar – as if I'd been here before many times – on boats – in this same fog. And yet of course I know that's silly.

Chris: Anybody feel funny dat vay in fog.

Anna: But why don't I feel afraid? I ought to, oughtn't I, a girl who's always lived inland? And you've told me what danger we're in. But I don't feel anything – but – restful – as if I'd found something I'd always been seeking – as if this were the place for me to be – and I feel happy! Yes – happier than I've ever been anywhere before! It's queer for me to feel that way, don't you think?[18]

In *Long Day's Journey into Night*, morphine-addicted Mary, Edmund's mother, finds solace and escape in fog. The intensity of her fondness for fog varies directly with her morphine usage. Throughout the play, the foghorn is a harbinger of Mary's hopeless addiction and Edmund's terminal tuberculosis. Mary loathes the foghorn; both Mary and Edmund seek to lose themselves in fog. Jamie, Edmund's brother, consumes alcohol to obtain a personal fog; however he seems oblivious to virtues of the atmospheric kind.

The family members blame their personal failures on one another, avoiding responsibility for their actions or inactions and for being who and what they are or have become. There is indeed much fog-alienation-guilt and despair, but also love and sometimes hope. Perhaps the dialogue seems outdated or melodramatic. Yet I suspect that for many, it brings back memories of having heard family fights forever involving the same accusations and allegations.

We do not find Anna Christie's fog in *Long Day's Journey into Night*. Here the fog is brutal and oppressive, and unlike Carl Sandburg's poem, this fog doesn't "move on"; it may briefly hide but always returns.

Mary (back on the habit) says:

	That foghorn! Isn't it awful, Cathleen?
Cathleen:	It is indeed, Ma'am. It's like a banshee.
Mary:	I don't mind it tonight. Last night it drove me crazy. I lay awake worrying until I couldn't stand it any more.
Cathleen:	(deleted)
Mary:	It wasn't the fog I minded, Cathleen. I really love fog.
Cathleen:	They say it's good for the complexion.
Mary:	It hides you from the world and the world from you. You feel that everything has changed, and nothing is what it seemed to be. No one can find or touch you any more.
Cathleen:	(deleted)
Mary:	It's the foghorn I hate. It won't let you alone. It keeps reminding you, and warning you, and calling you back. But it can't tonight. It's just an ugly sound. It doesn't remind me of anything....[19]

Edmund expresses similar sentiments:

The fog was where I wanted to be. Halfway down the path you can't see this house. You'd never know it was here. Or any of the other places down the avenue. I couldn't see but a few feet ahead. I didn't meet a soul. Everything looked and sounded unreal. Nothing was what it is. That's what I wanted – to be alone with myself in another world where truth is untrue and life can hide from itself. Out beyond the harbor, where the road runs along the beach, I even lost the feeling of being on land. The fog and the sea seemed part of each other. It was like walking on the bottom of the sea. As if I had drowned long ago. As if I was a ghost belonging to the fog, and the fog was the ghost of the sea. It felt damned peaceful to be nothing more than a ghost within a ghost.[20]

Edmund is hiding, renouncing life rather than affirming it as he does under the bowsprit on the moonlit night when he becomes the reality around him: a living universe and not a ghost.

And finally we come to those lines in which Edmund briefly sees the secret: "For a second there is meaning!"

... I was on the Squarehead square rigger, bound for Buenos Aires. Full moon in the Trades. The old hooker driving fourteen knots. I lay on the bowsprit, facing astern, with the water foaming into spume under me, the masts with every sail white in the moonlight, towering high above me. I became drunk with the beauty and singing rhythm of it, and for a moment I lost myself – actually lost my life. I was set free! I dissolved in the sea, became white sails and flying spray, became beauty and rhythm, became moonlight and the ship and the high dim-starred sky! I belonged, without past or future, within peace and unity and a wild joy, within something greater than my own life, or the life of Man, to Life itself! To God, if you want to put it that way. Then another time, on the American Line, when I was lookout on the crow's nest in the dawn watch. A calm sea, that time. Only a lazy ground swell and a slow drowsy roll of the ship. The passengers asleep and none of the crew in sight. No sound of man. Black smoke pouring from the funnels behind and beneath me. Dreaming, not keeping lookout, feeling alone, and above and apart, watching the dawn creep like a painted dream over the sky and sea which slept together. Then the moment of ecstatic freedom came. The peace, the end of the quest, the last harbor, the joy of belonging to a fulfillment beyond men's lousy, pitiful, greedy fears and hopes and dreams! And several other times in my life, when I was swimming far out, or lying alone on a beach, I have had the same experience. Became the sun, the hot sand, green seaweed anchored to a rock, swaying in the tide. Like a saint's vision of beatitude. Like the veil of things as they seem drawn back by an unseen hand. For a second you see – and seeing the secret, are the secret. For a second there is meaning! Then the hand lets the veil fall and you are alone, lost in the fog again, and you stumble on toward nowhere, for no good reason!

(*He grins wryly.*)

It was a great mistake, my being born a man, I would have been much more successful as a sea gull or a fish. As it is, I will always be a stranger who never feels at home, who does not really want and is not really wanted, who can never belong, who must always be a little in love with death![21]

Edmund may disagree, but I do not think he stammered, even if he does claim that "Stammering is the native eloquence of us fog people."[22]

The question arises as to the relationship between mystical experience which is usually associated with Asian philosophy, and existentialism. This would, of course, depend upon the respective Asian philosophies and versions of existentialism. However, some remarks may prove helpful even if the topic merits greater investigation and analysis.

What Edmund and Anna have found in their respective mystical experiences is a freedom *from* choice with its concomitant loss of the burden of personal selfhood, a burden described in terms of existentialist expressions such as "condemned to freedom." Edmund and Anna obtain a brief respite from this burden of particular selfhood. Neither can sustain the mystical experience, though how long it lasts or can last is not clear. Edmund tells us that the veil falls again. Anna senses that at some time she will again have to think about jobs and life ashore – but not now, please, not now.

A mystic of the Zen Buddhist variety would that claim the self or ego is always an illusion, and that Edmund and Anna may have experienced their true nature, particularly Edmund, who becomes the sea and starlit night. Anna seems to be in a sort of halfway state in which she senses her past incarnations; she feels old.

An existentialist such as Sartre would deny that our reality is identical with the nature around us. If Edmund becomes the sea and night sky, it is because for a brief time his being-for-itself has achieved coalescence with his being-in-itself, i.e., he becomes one with nature and facticity. He ceases striving (also a Buddhist theme), and in that cessation of striving and of having a project he lacks the "upsurge" which would allow for self-consciousness and the being-for-itself. Edmund slips back into selfhood because it is impossible for a being-for-itself to maintain an authentically disengaged state for very long. A crude analogy might be the satisfaction of a physical hunger. Briefly, we may say, "That hits the spot!" If Edmund is "a little in love with death," as he claims, it is, perhaps, because death is when our for-itself ceases. However, most existentialists do not see death as in any way desirable. Consider Camus, who claims that the quantity of life is important, more so than its quality. Or consider that Heidegger (who disdains being called an existentialist, preferring to be known as an existential philosopher) wishes us to understand that "da-sein" (our individual ontological relation to Being) ceases upon death.

In existentialist terms, Anna and Edmund must return to their freedom *of* choice and the responsibility it entails. The freedom *from* choice they briefly experience has revealed the reality of genuine beauty and therefore a reason for hope, but they remain who they are. However, they should remember that they also continually create who they are; and perhaps, even Edmund may one day look at his creation and say, "It is good!"

Cleveland State University

NOTES

1 Eugene O'Neill, *Long Day's Journey into Night* (New Haven: Yale University Press, 1989), p. 147. This play was first published in 1956, although written many years earlier.

2 Eugene O'Neill, *Complete Plays 1913–1920* (New York: Library Classics of the United States, Inc., 1988). *Chris Christophersen* (1919) antedates *Anna Christie* (1920).

3 *Chris Christophersen*, ibid., pp. 833–4.

4 *Long Day's Journey into Night*, p. 154.

5 *Anna Christie* was a revision of *Chris Christophersen*. Chris has the same persona in both plays. We shall subsequently see in what way the respective Annas are different.

6 *Fog*, Eugene O'Neill, *Complete Plays 1913–1920*, p. 111.

7 Ibid., p. 112.

8 Ibid., p. 112.

9 *Bound East for Cardiff*, Eugene O'Neill, *Complete Plays 1913–1920*, p. 191.

10 Martin Heidegger, "What is Metaphysics?" reprinted in Walter Kaufmann, *Existentialism from Dostoevsky to Sartre* (New York: New American Library, Inc. 1975), p. 247.

11 *Beyond the Horizon*, Eugene O'Neill, *Complete Plays 1913–1920*, pp. 571–653.

12 Friedrich Nietzsche, "The Joyful Wisdom (*The Gay Science*)," relevant excerpt appearing in Robert C. Solomon, *Existentialism* (New York: The Modern Library, Inc. 1975), p. 46.

13 Karl Jaspers, "Existenzphilosophy," relevant excerpt appearing in Walter Kaufmann, *Existentialism from Dostoevsky to Sartre* (New York: New American Library, Inc. 1975), p. 199.

14 Ibid., pp. 211–2.

15 "Anna Christie," Eugene O'Neill, *Complete Plays 1913–1920*, p. 982.

[16] Anna feels old, as though she had been in the same situation many times before. This could either be taken in the context of Nietzsche's eternal recurrence or in terms of the wheel of karma and the countless reincarnations we all are alleged to suffer. Since she claims to have been on boats in saying she was here before, eternal recurrence seems more accurate. The relevant passages are subsequently quoted.

[17] *Chris Christophersen*, pp. 836–7.

[18] Ibid., p. 840.

[19] *Long Day's Journey into Night*, pp. 98–9.

[20] Ibid., p. 131.

[21] Ibid., pp. 153–4.

[22] Ibid., p. 154. An examination of Edmund's prose-poem reveals interesting parallels with Albert Camus' novels. Edmund mentions feeling like a stranger and not belonging, a familiar enough theme in Camus. More interesting are the references to swimming which are closely allied to the swim which we find in *The Plague*. There we find the only happiness and belonging which appears in the novel, i.e., the experience shared by Dr. Rieux and Tarrou: "… they swam side by side … at last free of the town and of the plague…. Neither had said a word, but they were conscious of being perfectly at one, and the memory of this night would be cherished by them both." Albert Camus, *The Plague* (New York: Vintage Books, 1972), pp. 239–40, first published in 1948.

ALBERTO CARRILLO CANÁN

LIFE AS SELF-PRODUCTION IN KIERKEGAARD'S EARLY WORK

The one who does not understand that the hero dies any time before he dies, shall not come forth in understanding life.

With a religious aim, Kierkegaard develops in his early work what can be called a gallery of literary personae. The first of such personae is the ironic Socrates in Søren Kierkegaard's doctoral writing, *On the Concept of Irony* (1841). Straightaway one finds the aesthete and the ethicist in *Either-Or* (1843), and then one is confronted mainly with Abraham and Agamemnon in *Fear and Trembling* (also 1843). Up to this point Kierkegaard has forged a noticeable designation for his main personae: they are "*heroes*" or "*knights*", namely aesthetic, ethical and religious ones. Certainly the term "life" (liv) as such is not a concept created by Kierkegaard, but substituting it *to some extent*, he uses the term "existence" (eksistens)[1] that later became famous. The paradigm of "existence" is *heroic individual life* insofar as it is defined through *borderline situations*, which Kierkegaard sums up in *Fear and Trembling* under the title "resignation". Yet such a heroic life is not proper to special individuals, but rather very possible for anybody, for the "resignation" defining it is simply human.[2] The goal of this paper is to examine Kierkegaard's formal concept of heroic life as a general structure of self-production concerning everybody. But according to Kierkegaard's own claim in his third *Edifying Speech* of 1843, one must be "concerned with (...) oneself alone" (K5 134/III 303), and this therefore arouses the question of who is the spectator in such self-production in the theatre of existence, understood as heroic life?[3]

THE CATHARTIC CONCEPT OF HERO

In view of the undoubted and painful seriousness of Kierkegaard's concern with the religious, at first glance it may appear unsuitable or even frivolous to speak as if to him of "the *theatre* of existence". Nevertheless, he conceives of existence as a personal "task" of formation (K5 139n/III 166n) in a sense very different, even *contrary* to common philosophical ideas of formation, for according to Kierkegaard the principal issue in formation is by no means *knowledge*, but *passion*. So, he says in referring to Lessing that the "proper

A.-T. Tymieniecka (ed), Analecta Husserliana LXXIII, 247–257.

humane is passion" (K5 140/III 166). In fact, Kierkegaard's work as a whole draws inspiration from Lessing, especially from his interpretation of Aristotle's concept of *catharsis*.[4] Formation thus becomes in Kierkegaard a "transformation of existence" (K14 92/VII 335) that implies truly toilsome *passional* steps, actually passional "collisions" (K5 86f./III 126f.) and in the end includes gaining one's own "absolute (...) transparency" (K21 39/XI 154)[5] through something like an *inverted* Aristotelean catharsis. Such catharsis does not affect any spectator, but rather the performer.[6] This authorizes one to speak of self-production, at least provisionally.

Kierkegaard uses Gotthold Ephraim Lessing's interpretation of Aristotle's theory of catharsis in construing his own well-known theory of the three "stadia" or "spheres of existence" (K14 211/VII 436), as he calls them explicitly in *Stages on the Life's Way* (1845). In fact there are not three, but four main instances: first, quite customary life, which for Kierkegaard does not deserve to be called existence; and then there are the aesthetic, the ethical, and the religious stadia of existence. These instances are defined through *relationships* between three terms in Kierkegaard's model, namely 1) oneself, 2) God, 3) the world, where "world" means some other person or thing, in fact simply something other than oneself. The relationships among the mentioned terms are primarily one's own *passional* links to oneself, to God and to the world, and to be more precise, "pity and fear" in a very broad sense. And precisely the conjunction of "pity and fear" or "antipathy and sympathy"[7] grounds Kierkegaard's exploitation of the theory of catharsis. We need some preparation to understand Kierkegaard's exploitation of this theory; we especially need *three* main ideas of Kierkegaard's that he uses together. The *first* one is the model of what Kierkegaard calls "consciousness" in the posthumously published fragment *De omnibus dubitandum est* (ca. 1843). The *second* one is Kierkegaard's concept of "spirit" as "dialectic" structure, for example as he uses it in *Either-Or* and also in *The Concept of Fear* (1844). The *third* one is the idea of *multiple catharses* appearing towards the end of the writing *Stages on the Life's Way*. The religious use of multiple catharses supposes the "dialectic" concept of "consciousness". In fact, "consciousness" as "dialectic" consciousness is the structure that makes heroic life, that is, existence through "cathartic" transformations, possible, for this structure is the "spirit".

CONSCIOUSNESS AS CONTRADICTION

In the fragment *De omnibus dubitandum est*, Kierkegaard defines consciousness as the "trichotomous" structure consisting of 1) "me", 2) "this", the consciousness linking both. Kierkegaard, in fact, refers to consciousness as "the *relationship* of which the first form is *contradiction*. (...) The determinations of consciousness (...) are three-parted (trichotomous), as language also shows. For when I say: *I* become *conscious* of *this* sensory impression myself, I say a *trichotomy*" (K8 156/IV B1 147n). A suitable way to clear the meaning of this "trichotomy" lies in the traditional phenomenological model of the givenness (awareness) of oneself in the givenness (awareness) of something else. For example, I am conscious of me in seeing this (rose, etc.). But here one must stress the idea of the "contradiction" between "me" and "this". Especially illustrative of the "contradiction" is "the paradox of being in love", which means the "paradox of *self*-love as love to *another*" (K8 36/IV 206) mentioned in Kierkegaard's *Philosophical Fragments* (1844). The general structure involved in the "contradiction" can thus be formalized as the paradox of self-referentiality in relation to something.[8] Just this twofold relationship is a main formal *structure* one needs in order to understand the application of Lessing's theory of catharsis in Kierkegaard's model of the stadia of existence. The main issue is here some kind of "cleansing" undermining the relativity to something else and, thus, stressing self-referentiality.

SPIRIT AS "DIALECTIC" STRUCTURE

The other main *structure* Kierkegaard uses in exploiting Lessing's theory of catharsis is "spirit" understood as conjunction of "opposites".[9] This makes up the conceptual link to the conjunction "fear and pity". In *The Concept of Fear* Kierkegaard refers "fear" to "spirit" by saying: "one shall not find fear in animals, for (...) it is not determined as spirit" (K9 40/IV 313), and "the less fear, the less spirit" (K9 40/IV 314). But the point is the "amphiboly" of "fear", and thus, of "spirit". Actually Kierkegaard says: "When we wish to examine the <u>dialectical</u> determinations of fear, then it turns out that these indeed involve the <u>dialectical amphiboly</u>. Fear is *a sympathetic antipathy and an antipathetic sympathy*" (K9 40/313, i. a.). Kierkegaard understands the "amphibolous" structure combining "*sympathy*" and "*antipathy*" purely formally as a *conjunction* of "liking" and "disliking" or "attraction" and "repulsion", and for this reason Kierkegaard can speak of "sweet fear, sweet fearfulness" (K9 49/314), as when children are *at once* both attracted to and repelled by darkness. The point is, thus, "spirit" conceived as a conjunction of two opposite passions and the predicate "dia-lectic" meaning simply this "opposition".

This use of the term "dialectical" can be confirmed in *Either-Or*. Referring to the "medieval" "idea of representation", Kierkegaard says: "The wonderful *dialectic* of life is here always illustrated by individuals who in general stand *opposite in twos*; life is there always only under one form (*sub una specie*), and the great *dialectical unity*, which life possesses under *both* forms (*sub utraque*) *in unity*, is not imagined. The *opposites* stand thus mostly indifferent *each outside of the other*" (K1 93/I 69). "Dialectical" "unity" means therefore, purely formally, two that are not "indifferent outside of each other", but on the contrary joined together forming one "unity", and certainly a unity of "opposites". "Spirit" is such a "dialectical unity" of passions.

With regard to the concept of catharsis, up to this point we have two main structures: 1) The formal paradox of self-referentiality in relation to something else, 2) the dialectic conjunction of opposite passions. Both structures are by now *combined* in defining both the "stadia of existence" and the corresponding "cleanings".

THE FOURFOLD CATHARSIS

Towards the end of the *Stadia*, Kierkegaard explicitly uses the Aristotelian idea of catharsis; in referring to the "aesthetic", the "ethic", and the "religious hero" (K12 484/VI 423), he speaks of "aesthetic healing" or "cleansing" and of "religious healing" or "cleansing" of "fear and pity" (K12 491n/VI 429n). To understand the Kierkegaardian multiple catharsis one must, as was said above, refer to Lessing's well-known interpretation of Aristotle's concept of catharsis as used in writing tragedy. In Lessing's interpretation, there are at least four points that are exploited by Kierkegaard: 1) Lessing's assertion that the catharsis (cleansing) of passions should not be understood as cleansing of *all* passions but *only* of pity and fear.[10] This agrees with Kierkegaard's reduction of all passions to formal liking and disliking, and it endows his theory with unity. 2) Lessing's assertion that dramatic pity and fear must act *together*.[11] This point is important for Kierkegaard's idea of "dialectic" as a unity of opposites: in this case, attraction and repulsion. 3) Lessing's stressing of the self-referential aspect in fear.[12] This point is the cornerstone of the idea of cleansing as such; so, as we will see, cleansing purely formally means *self-referentiality*. 4) Lessing's assertion that pity arises *through* fear.[13] This last point is fundamental for the religious exploitation of the catharsis theory, that is, for obtaining a series of catharses resulting in the religious cleansing of the individual.

Kierkegaard's model demands the broadest sense of the terms "fear and pity" so that they become "sympathy and antipathy" or, in fact, so that they reach their uttermost formalization up to "attraction and repulsion". This is the condition for inserting the *material* twofold "fear and pity" into the "dialectic" structure Kierkegaard calls "spirit", understood as mere *formal* unity of "opposites". But "spirit" must for its part be conceived of as "consciousness". In fact, to the above-quoted passage in *De omnibus dubitandum est* about the "consciousness" as trichotomous "contradiction", Kierkegaard adds: "*Consciousness* is *spirit*, and what is unusual, is that when something in the world of spirit is divided it becomes three" (K8 156/IV B1 148). This declaration rounds out the above-mentioned, necessary (for Kierkegaard), *combination* of the twofold structure of self-referentiality in relation to something else on the one side, *and* of the "dialectic" twofold of "spirit" as unity of "opposites", on the other.

Attraction and repulsion can refer to oneself, to God and to the world. This is the frame of the "dialectic" relationships between the three terms: me, God, the world. Furthermore, these relationships are not only "dialectic", but they also belong to the realm of "consciousness". This means that attraction and repulsion must be ordered according to criteria of self-referentiality and of relationship to something else. The "dialectic" relationships are the *general* structure of customary life *and* of the three stadia of existence, but each one of these four instances receives its determination according to one version of "consciousness". Moreover, Kierkegaard conceives of each change from one form of "consciousness" to another as a "jump" corresponding to a special type of catharsis. There are three of them: from customary life to aesthetic existence, from aesthetic to ethical, and from ethical to religious existence.

Before I can explain this main point, I must now examine the *basic* use Kierkegaard makes of the four mentioned aspects in Lessing's interpretation of the Aristotelian idea of catharsis. In fact, this concerns mainly the conjunction "fear and pity".[14]

FEAR AND PITY

At *first*, Kierkegaard applies Lessing's interpretation according to which the catharsis or cleansing of passions concerns only "fear and pity" and, precisely, their cleansing through "fear and pity". As was said above, Kierkegaard needs the simplicity of the general thought model he calls "spirit", consisting, in a formalized sense, of opposite passions: sympathy and antipathy. Thus Lessing's *reduction* of the task of cleansing passion to the cleansing of only two opposite passions meets Kierkegaard's *formalization* of all passions into attraction and repulsion ("sympathy" and "antipathy") as well as his definition of "spirit" as a couple of opposites.

In the *second* place, Kierkegaard applies Lessing's interpretation according to which "fear and pity" act together. The *simultaneity involved here becomes, in Kierkegaard,* structural unity: a unified and thus "*dialectic*" effect of the opposites "sympathy" and "antipathy".

In the *third* place, Kierkegaard applies Lessing's interpretation according to which dramatic "fear" is necessarily self-referential. This point demands some careful consideration. I quote Lessing: "referring to somebody else there is nothing harmful which we do not fear for *ourselves*" (W4 326). Before that, he says: "it is the fear that (...) arouses for *ourselves*" (W4 323). Lessing affirms this kind of self-referentiality. For a better understanding of the involved problematic, one can remember Heidegger's concept of "mood" in *Being and Time*, whose *general structure* he illustrates even through "fear". He says: "There are three points of view from which the phenomenon of fear may be considered. We shall analyze the 'which' [das Wovor] of fear, the fearing and the 'for which' [das Worum] of fear" (SZ 140). By the way, one recognizes here the Kierkegaardian structure of "consciousness", a trichotomy consisting of "me", "this" and the "consciousness" linking them: 1) Me, if I feel fear for me, 2) "this", for which I feel fear, and 3) fear itself as the link between the previous two. The feeling for oneself is the self-referentiality Lessing stresses, and obviously it amounts to self-referentiality in relation to something else, in this case the "for which" of fearing. Obviously, *dramatic* fearing can now be subsumed without any problem under Kierkegaard's idea of "consciousness". Pity as such can also be subsumed under "consciousness", as we shall immediately see.

In the *fourth* place, Kierkegaard applies Lessing's interpretation according to which pity arises through fear. I must complete the following passage of the *Hamburgische Dramaturgie* quoted above, "referring to somebody else there is nothing harmful which we do not fear for *ourselves*" (W4 326). Lessing then refers to dramatic actions which refer to such a harmful thing, saying, "*only through* this fear they excite pity" (W4 326). There is no action "which can excite fear for us without exciting our pity at once" (W4 326). Certainly we are dealing with "pity for others" (W4 328) and not with self-pity. But we must consider here two additional ideas of Lessing's. The first one follows from the fact that pity arises through self-referential fear, for this means that "fear for ourselves" "is a *necessary ingredient* of pity" (W4 330). The second one is that "pity" truly *coincides* with "fear" in a very definite sense. So, in referring to self-referential fear, Lessing says: "In a word: this fear is *the* pity relative to ourselves" (W4 323). Thus, the fear *itself* is a conjunction of self-referential "fear and pity". One can therefore consider the original conjunction "fear and pity" as consisting of self-referential "fear and pity" *plus* "fear and pity" which are relative to something else – for example, pity for the hero and fear for his fate.

The idea of pity arising *only* through self-referential fear permits us to subsume the conjunction "fear and pity" *as such a conjunction* under the structure discussed above that Kierkegaard calls "consciousness". But this idea is especially important, for it implies that the whole conjunction becomes based upon *fear*.

EXISTENCE AND CATHARSIS

In the twenty-eighth part of his Hamburger Dramaturgy, Lessing briefly states the whole formal problem of cleansing *our* fear and pity through *dramatic* fear or pity. Obviously there are in principle four cases. But here Lessing reminds us that according to Aristotle's theory of catharsis, each of these four cases implies two cases. In fact, cleansing fear, for example, – through fear *or* through pity – implies, according to the Aristotelean thought-model, transforming too much fear *or* too little fear into pertinent fear, and thus into "virtue". In the following I shall only summarize the main cases of this transformation, and certainly grounding it on "resignation" as the main constituent of heroic existence as stated for the first time in Kierkegaard's early work. Furthermore, the detailed consideration of the relationships between the three terms, "me, God, and the world", especially in view of the problem of "consciousness" as self-referentiality in the relation to something else, must be omitted here. I must only mention that according to Kierkegaard, the term God does not imply relativity to something else as the individual, for God is not a thing external to the individual.[15]

Customary life, according to Kierkegaard, is characterized by too little fear in a precise theological sense. Kierkegaard rejects what theologically is considered the *praesumptio*, as *perversa securitas*. This presumptuous sureness of the individual is the lack of a "fear of God" (K6 128n/III 298n) and Kierkegaard calls it "happiness". We saw above that "the less fear, the less spirit". Consequently, he speaks of the "happiness of spiritlessness" (K9 96/IV 363). In fact: "In spiritlessness there is no fear, for that it is *too* happy and contented, and spiritless" (K9 97/IV 365). Perverse sureness means the lack of a fear of God, but this is nothing like a lack of fear for one's own sins. This implies, in fact, 1) too much sympathy for oneself, as well as *both* 2) too much fear, and 3) too much sympathy for the world. Fear and pity thus reach an *extreme* here and must be cleansed.

In fact, fear of and sympathy for *the world* characterize customary life. These passions become *virtues* as *love* for other individuals. But "love" must be "ideal". Love, as *relationship* to another person, becomes ideal in reference to one's own *relationship*, that is, to love itself. There are two main cases of this reference to one's own love. The first case is the seducer. In the second case, one holds on to love for the sake of itself. The first case is the ideality of the seducer; the second case is the ideality of love as such.[16] Kierkegaard refers, for example, to somebody who loves a princess, and says: "What the princess does cannot disturb him; only mean natures have the law for their comporting *in another*, the presuppositions for their comporting *outside of themselves*" (K5 45/III 95). *Keeping* love for the sake of love is the ideality of love, and in this ideality both fear for the beloved and sympathy for him become virtues. The aesthetic hero, when *loving* the other, must subsume himself for the sake of the other, thus sacrificing himself. Resignation requires, of course, that he *keep* the love, otherwise there is no resignation, no sacrifice, just forgetting. Keeping the love while suffering is the ideality of the aesthetic hero.

The aesthetic existence thus implies three cases: the ideality of the seducer, the mere ideality of love and the ideality of the aesthetic hero. The ideality of the seducer means too little sympathy for the beloved; the mere ideality of love means *too much* sympathy for the beloved, whereas heroic ideality means *too much* fear for him. One thus needs the cleansing which leads from the aesthetic into the ethical existence.

The transformation of the aesthetic existence into the ethical one demands a transformation of the aesthetic ideality of love. This occurs in referring the love to a principle, and in this way love for another becomes a *duty*; in fact, speaking more precisely, love becomes the *fulfillment* of a duty. The simplest instances of this love are, of course, "marital love", and love for one's own children. The subordination of love to duty amounts to cleansing love of too much sympathy for the beloved. But there are many duties, and a "collision" may occur between "duty" and "wish", in this case love. A typical case would be that of Agamemnon, who had to sacrifice Iphigenia and so fulfill a duty against his love for her. But this kind of hero must, of course, be ideal, too. That is, he acts on his love for Iphigenia by *keeping* it. To keep love while suffering for the sake not of the beloved but of a principle is the resignation proper to a tragic hero. To sacrifice Iphigenia, that is, to renounce her is *love* for the principle, again as *fulfillment* of a *duty* toward it.

Both the lucky and the unfortunate ideality of ethical love, that is, the lucky and the unfortunate fulfillment of duty, are the ennobled transformation of the early, merely aesthetic ideality of love. Nevertheless, the lucky ethical ideality of love is again *too much* sympathy, and certainly for the beloved individual as well as for the principle, whereas the heroic ethical ideality of love is *too much* sympathy for the principle as well as *too much* fear for it. A new cleansing is needed.

The highest ideality occurs in relating to God. The step from ethical to religious existence occurs when the ideal love for another *and* the ideal love for a principle are both sacrificed or "suspended" for the sake of believing in God. Once subordinated to faith, the other two forms of love become true "virtues". To be "ideal" means here the actual relationship to God; and the proper handling of this relationship is again to keep it. The religious hero succeeds in sacrificing the principle *and* the beloved person for the sake of God. So, the individual keeps his love for God. But a main difference between the religious existence and the aesthetic and ethical ones lies in the heroic. The truly religious existent is *always* a hero or knight of faith, for he *permanently* repeats the "movement of renouncing" the beloved. Kierkegaard illustrates this situation by quoting Lucas 14:26: "Who comes to me and does not hate his own father, mother, woman, children, brother, sister, moreover his own life, such a one cannot be my disciple" (K5 79/III 120n). Nevertheless, Kierkegaard thinks of a "second movement" beyond resignation, a "movement by virtue of the absurd". The religious hero truly experiences the suffering inherent to resignation, nevertheless he believes and seizes upon his circumstances and he enjoys the beloved, which he is prepared to sacrifice at any time. This permanent disposition to sacrificing the beloved is the permanent suffering of the religious, but this kind of suffering goes together with enjoying existence.

In the *Stadia on the Life's Way*, Kierkegaard remembers Pascal saying that "suffering is the natural state of a Christian" (K12 490/VI 428), whereas in *Fear and Trembling* he asserted that for the Christian hero, exemplified by Abraham, "his action is an absolute contradiction to his feeling" (K5 82/III 123). The self-production of the individual consists in the end in provoking Christian religious suffering. And this is an inverted catharsis. The object of fear is no longer the "stroke of fate", which strikes the tragic hero, but "guilt" (K12 492/VI 430), which means more specifically: "[to be] afraid because of one's own guilt. Fear and pity are different for the religious man [than for the aesthete] and are not cleansed when one turns outward, but rather when one turns inward. Aesthetic healing consists in the single person (...) disappearing before himself, like a mote of dust which is bought along with the goods, into that which is the common fate of all people, of humanity (...). *Inversely* as the former, religious healing consists in changing the world and centuries and generations and millions of contemporaries into something disappearing, (…) so that the single one who remains is the single person himself, yes, this certain single person who is held under the determination of guilty-not guilty in his relation of God" (K12 492n/VI 430n). Through faith, that is, believing in God *and* in one's own sin, the conjunction of "fear and pity" – formally the conjunction of liking and disliking – becomes the virtue of rightful love. The beloved is loved in the twofold action of renouncing him and enjoying him. This is joy for the other, in and through suffering for him, but above all through suffering for God, a suffering that is suffering for one' own sin.

CONCLUSION: DRAMA AND RELIGIOUS SPEECH

Kierkegaard's application of the catharsis theory in his early work becomes explicit at the end of the *Stages on the Life's Way*. In this writing, he says that the "religious cleansing" of "fear and pity" occurs at first for the religious speaker; in fact his speech, if correct, is always "merely a monologue" (K12 493/VI 431). The truly religious speech *transforms* the speaker's "fear and pity" into *self-referential* "fear and trembling" (K12 494/VI 432). But nevertheless he can, through his speech, act on the hearer. In this second aspect, Lessing's idea of dramatic "cleansing" becomes the famous Kierkegaardian "edifying" discourse (K12 495/VI 432). Thus, in the drama of religious speech there are in the end three spectators: the self-transforming speaker, the hearer of the religious speech, who in principle should be *induced* to self-transform too, and God. The religious speaker is "exclusively concerned with himself" (K12 493/VI 431), for he does not give his hearer (or reader) any indication, suggestion, etc., but only provokes and expresses his own suffering; nevertheless such expression must resonate with the hearer. This is an application of the *pietist* practice of communication still to be explored.

Universidad Autónoma de Puebla, Mexico

BIBLIOGRAPHY AND ABBREVIATIONS

K5 = Kierkegaard, S., Gesammelte Werke (GTB 604), Gütersloh 1986.
K6 = Kierkegaard, S., Gesammelte Werke (GTB 605), Gütersloh 1980.
K8 = Kierkegaard, S., Gesammelte Werke (GTB 607), Gütersloh 1985.
K9 = Kierkegaard, S., Gesammelte Werke (GTB 608), Gütersloh 1983.
K12 = Kierkegaard, S., Gesammelte Werke (GTB 611), Gütersloh 1982.
K13 = Kierkegaard, S., Gesammelte Werke (GTB 612), Gütersloh 1988.

K14 = Kierkegaard, S., Gesammelte Werke (GTB 613), Gütersloh 1982.
K21 = Kierkegaard, S., Gesammelte Werke (GTB 620), Gütersloh 1985.
W4 = Lessing, G.-E., Werke in sechs Bänden, vierter Band, Zurich 1965.
i.a. = Italics by author of the quoted text.

NOTES

1 Cf.: "My life becomes (...) the existence of a poet. Can somebody think something more unfortunate than this?" (K1 39/I 20). See the list of abbreviations at the end of this paper. The second reference in the parentheses is to the Danish edition, Søren Kierkegaards Samlede Værker, udgivnet af A. B. Drachman, J. L. Heiberg og H. O. Lange, vols. 1–14.

2 Cf.: "The *unusual* that makes up an aesthetic *heroine* should be here that she (...) keeps her love. A maiden of such quality could not be used in [my] experiment (...). So I chose a *customary* maiden. (...) Of course, 'she' has had, as *any* maiden, a possibility to become great" (K12 485/VI 424).

3 This paper rests on three early papers of mine: "Concerned with Oneself and God Alone. On Kierkegaard's Concept of Remorse as the Basis for his Literary Theory" (1997), forthcoming in Analecta Husserliana; "The Paradoxical Transformation of Existence. On Kierkegaard's Concept of Individuation" (1997), forthcoming in Analecta Husserliana; "Kierkegaard's Amphibolous Conjunction of Joy and Sorrow and His Literary Theory" (1998), forthcoming in Analecta Husserliana, too.

4 Cf. Lessing, G.-E., *Hamburgische Dramaturgie*. For another view of Lessing's influence on Kierkegaard see Rogan, J. "Keeping Silent through Speaking," in: G. Pattison, *Kierkegaard on Art and Communication*.

5 As Kierkegaard states it in *The Sickness onto Death* (1849).

6 Concerning the spectator, dramatic history becomes "edifying discourse". See the end of this paper.

7 As Kierkegaard formalizes it in *The Concept of Fear*. See below. For this problem, cf. my papers mentioned in note 3.

8 Cf. the second and the third of my papers referred to above.

9 Cf. the third of my papers referred to above.

10 Cf. *Hamburgische Dramaturgie*, parts 77 and 78.

11 Cf. *Hamburgische Dramaturgie*, parts 75 and 76.

12 Cf. *Hamburgische Dramaturgie*, parts 74–77. See also my first paper mentioned.

13 Cf. *Hamburgische Dramaturgie*, parts 75 and 76.

14 For a detailed treatment of this conjunction, see my first and third papers mentioned.

15 Cf., for example, "(...) God is nothing external, as a woman is, whom I can ask if she is content with me". (K13 153/VII 134)

16 For the problem of "ideality", see my third paper mentioned.

SECTION V

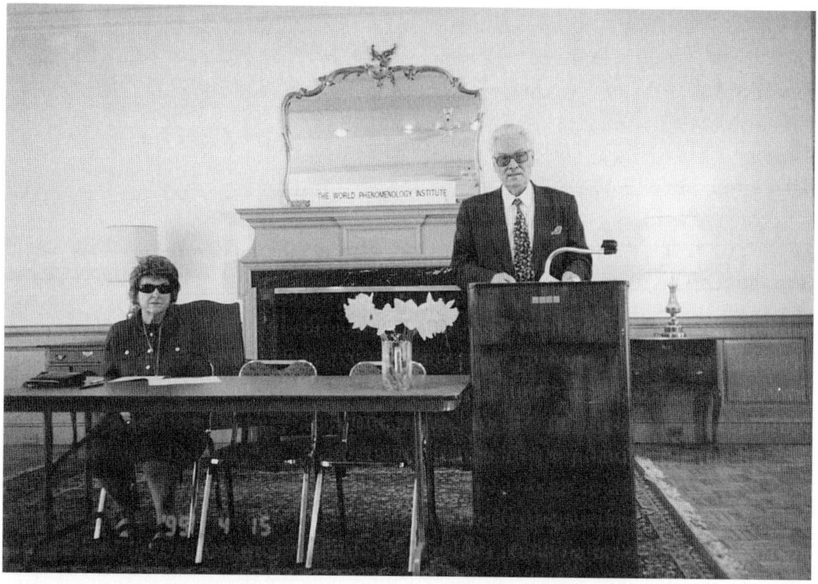

Constanin Dolgow lecturing, Marlies Kronegger presiding.

MAX STATKIEWICZ

THE IDEA OF CHAOS AND THE THEATER OF CRUELTY

Ce jeu n'est plus un jeu mais une vérité. – Genest[1]
Si le théâtre double la vie, la vie double le vrai théâtre.... – Antonin Artaud[2]

The expression "idea of chaos" seems as paradoxical, from the Platonic perspective, as the "idea of theater" does. Indeed, *ideai* are the principles of order in Plato's metaphysics, cosmology, and politics; they are the paradigms of the order of representation, excluding the possibility of chaos. Theater defies order, especially the order of the city (i.e., justice), by blurring the distinction between social types (*tupoi*), which are based on Ideas. By impersonating (*mimeisthai*) a person from another type or class, or by identifying with a different character on the stage, one risks really becoming (*einai*) such a person or character, at least according to Socrates in Plato's *Republic*. One risks crossing the boundary between appearance and reality, introducing confusion and undermining the basis of the political order.[3] Consequently, there is no place for theater in the ideal world of the *Republic*; the tenth book demands the banishment of the poet, principally the tragic poet, from the city – a gesture that has marked the whole history of both Western theater and Western philosophy of art.

Although Artaud's phrase "idea of chaos" and his idea of becoming and of the marvelous which appear in one of the programmatic essays for his "theater of cruelty," probably do not refer directly to Plato's metaphysical *Ideas*, they most certainly relate to his "idea" of theater.[4] After centuries dominated by the Aristotelian model of the "cathartic theater," Artaud returns to the Platonic notion of the theater as a threat to the metaphysical and political order. At the same time, he proposes to confer on the "theater of cruelty" the task reserved for philosophy in the Republic, that is, of apprehending the world and of providing guidance for the action in the world. The "theater of cruelty" ought to realize a "metaphysics-in-action" (*métaphysique en activité*).[5]

If the phrase "All the world is the stage" were not just a synecdoche for *theatrum mundi*, a "*lieu commun de l'humanisme*,"[6] if all the world were a stage *stricto sensu*, it would not be a world in the sense of *cosmos*. It would threaten to degenerate into confusion, just as theater does for Plato when it ignores the boundary between the illusory and the real. Order depends on, in fact arises from, the separation between heaven and earth,[7] between the

A.-T. Tymieniecka (ed), Analecta Husserliana LXXIII, 261–275.

divine and the human; between generations; between body and soul; and, in terms of theater, between actors and spectators. The latter is perhaps the most originary separation, generating the community of the city and its cosmological and political myths of foundation. Indeed, the sacrificial ritual is generally considered the *Ur-Theater*, and the function of such ritual is the rejection of the *pharmakos* – a double, an incest, a monster of confusion – that would threaten the harmonious existence of the human community.[8]

If the *pharmakos* is not fully exorcised, the display of the accompanying confusion might have a discordant effect. A stage of violence might provide, in Augusto Boal's view of the theater of action, a rehearsal for revolution.[9] From Aeschylus' *Prometheus Bound* and Sophocles' *Antigone* to Genet's *Balcony* and Weiss/Brooks' *Marat/Sade*, Western theater portrays, analyzes, and perhaps "rehearses" revolutionary situations. If other literary forms and arts engage in making and unmaking worlds, they do it less "dramatically" for they are further removed from direct action in the political community. The role of theater in this respect is privileged because of its inherently social character and its paradigmatic status of *représentation* (performance/representation). A particular form of representation symbolizes and governs a particular political, artistic, and scientific world order.

To see the world and human life in terms of theater is to see it as ephemeral, quickly passing, unstable. Such seem to be the *êlthes, eides, apêlthes* of the laughing philosopher or the "seven ages, seven roles" of Shakespeare's melancholy Jacques.[10] Pindar's "dream of a shadow" (*skias onar*) and Shakespeare's "shadow of a dream" make the same point.[11] In Plato, both the dream and the theater are common images of the ephemerality and instability of sensuous experience. Theater in particular epitomizes a "disorderliness" or injustice (*adikia*) caused by the lower part of the soul and the lower class in the State, both dominated by sensuous desires. If Plato does not explicitly equate the world with the theater or the dream, it is precisely because the world (*kosmos*) connotes order. Only the realm of Ideas and that of their faithful images (*eikones*) deserve to be called "world."[12] Plato's notion of order is based on representation and thus requires more than one world – at least two: the world of models and the world of resembling images or copies. Theater, however, just like a dream, produces not copies but simulacra (*phantasmata*), the effects of mimesis that counterfeit the model. Dreams and plays, therefore, are not to be trusted by philosophers, at least not by the philosophers of the ideal, orderly (*dikaia*) city. "There is an ancient dispute (*palaia diaphora*)," says Socrates in the *Republic*, "a quarrel between philosophy and poetry or mimesis."[13] In the *Republic*, the contention is promptly decided in favor of philosophy; theater is banned from the ideal world of the "philosophical" city.

Aristotle's intervention in the ancient quarrel between philosophy and theater is far from beneficial for the latter. To be sure, theater is recalled from exile, but on what terms? It is now integrated into the rational system of representation with well defined boundaries. The world is a world, and the theater is a theater; there is no danger of confusing the two. The relationship between the world of the stage and the world of the spectators is cast as resemblance (*homoiotês*), which is distinctly separated from identity. Even the most exact imitation cannot have the same effect as the real.[14] Aristotle's mimesis is not a play (*paidia*) with the limits of reality and illusion, as in Plato; rather it is the education (*paideia*) of adults already aware of their place, their status in the social structure of the city. Theater reinforces this structure. The characters on the stage should conform to established types, differentiated according to their virtue and power.[15] The queens and kings in a tragedy are portrayed as "better than ourselves," and characters in a comedy as worse. We should not completely identify with them, but only empathize with their vulnerability to *hamartiai* – mostly offenses against the political and religious laws – in order to become aware of our own vulnerability in the same respect.

Thus, for Aristotle, theater essentially plays a conservative role in the education ordained by the State, one from which an unlimited range of social metamorphoses should be excluded.[16] The ghost of the dangerous theater that Plato feared has been exorcised, or rather purified. It is, indeed, the institution of catharsis that plays the crucial part in the submission of theater to the philosophical rule.[17] Catharsis requires both a sense of solidarity with the protagonists of tragedy (they are *homoioi hêmin*, "like ourselves") and a sense of distance (they are also *beltiones ê kath'hêmas*, "better than we are"). Although of different status, both spectators and characters are similar in their propensity for disorder, even if this propensity is not always conscious. To put it into Platonic terms, both characters and spectators are potentially dissimilar from the established model that would guarantee their identity; they are potentially "dissimilar from themselves" (simulacra). The purpose of catharsis is to guard against potential social anarchy by strengthening the accepted order of representation. Aristotle's "purified" theater becomes the instrument of choice in the educative/ideological task of consecrating the ideal social types. The respect for the separation between world and theater and between stage and auditorium constitutes the beginning and principle (*arkhê*) of the order of catharsis.

The idea of world-as-theater should be understood against this model, which already began to impose itself with the reception of Aristotle's *Poetics* in the sixteenth century.[18] The Baroque theater (at the end of the sixteenth century and in the first half of the seventeenth century) limited, for a while, the expansion of the Aristotelian model by introducing the figure of *theatrum mundi* that contained the seeds of a dangerous theater. Also, drawing on the medieval tradition of the popular Corpus Christi cycle and of the Mystery plan – and in contradistinction to the Aristotelian model – it questioned the boundary between the "reality" of the spectator's belief and the stage performance.[19] The "theater within the theater" has often been used as a means of such questioning. Shakespeare's *Hamlet* is the most famous example. The young Hamlet abhors the deception of the Danish court and strives for the truth of passion that "passes show." Paradoxically, he finds it in theater. A troupe of comedians is visiting Elsinore and is made to perform *The Murther of Gonzago*, a play supposed to reflect the "reality" of the play *Hamlet*, that is, according to the general principle announced by Hamlet himself, "to hold as 'twere the mirror up to nature: to show virtue her feature, scorn her own image, and the very age and body of the time his form and pressure."[20] In the play, the performance is meant to reveal the king Claudius' guilt, just as the chief player's poetic description of Priam's death and of Hecuba's grief revealed Hamlet's need for *motivating* passion. The performance of *The Murther of Gonzago* should expose the usurper-king and arouse Hamlet's action, that is, it should intervene in the "reality" of the play *Hamlet*. This is not what happens. Claudius, the "real" king, although disturbed by the play, eventually succeeds in distinguishing himself from his counterpart in *The Murther of Gonzago*, and succeeds in asserting the distinction between his court and the troupe of comedians. The latter are dismissed and Hamlet – their director and prompter – is sent to England. However, the sense of order among the spectators of *The Murther of Gonzago* as well as among those of *Hamlet* has been undermined. They cannot help but question the reality of their world and the efficacy of their actions; they cannot help but fear, along with Hamlet, an imprisonment in an untrue world.[21]

The transition from theater to life is more successful in Jean de Rotrou's *Le véritable Saint Genest* (*The True Saint Genest*), which also contains a play within a play. The inner play takes place at the court of Diocletian, and it represents the martyrdom of Adrian, a Christian. Genest, the Roman actor playing the martyr, undergoes a true conversion to Christianity, even as he is acting on the stage. Testifying to the power of the prohibited faith, he becomes a martyr himself. Just like Hamlet, Genest finds truth in theater. He has succeeded in "d'une feinte en mourant faire une vérité."[22] However, as in *Hamlet*, the audience is not affected by the play. Genest's *action* is regarded as the continuation of his *acting*: "Qui vescut au Théatre expire dans la Scène."[23] This, of course, is true only on a superficial metaphorical level. In fact, Genest, even while having recourse to theatrical language, disregards the rules of the theater and ignores the prompter:

> ... Il n'en est plus besoin.
> Dedans cette action, où le Ciel s'intéresse,
> Un Ange tient la Pièce, un Ange me r'adresse;
> Un Ange par son ordre a comblé mes souhaits,
> Et de l'eau du Baptesme effacé mes forfaits;
> Ce monde périssable, & sa gloire frivole
> Est une Comédie où j'ignorais mon rôle....[24]

Henceforth, Genest is no longer an actor in the common sense; rather, he cannot but play himself under the prompting of Heaven. Here Genest anticipates a possible rejoinder to Derrida's characterization of Western theater as "the theater of the prompter," that is, as theater answering to an external master text and at the same time "inspiring" and "purloining" the words pronounced on the stage, according to the logic of the prompter (*souffleur* in the double sense of "prompter" and "thief").[25] Artaud's theater of cruelty will follow the path of Genest who, against the representational order of the prompter, proved theater capable of the passage from *mimeisthai* to *einai* – the passage that Plato's Socrates feared most. Both will eventually be condemned for having displaced the boundary between being and seeming, between truth and falsehood in theater and in the world, i.e., for having proven that "ce spectacle est une vérité."[26]

The transition between the world and the theater, life and dream, also plays an important role in Calderón's play *La vida es sueño* (Life Is a Dream), a play about theater and dream.[27] Were the world a theater, it might seem wise to play one's role just as the Stoic philosophers and Shakespeare's melancholic characters – Antonio of *The Merchant of Venice* and Jacques of *As You Like It* – recommend. This is the temptation of Segismundo, the hero of Calderón's play. Segismundo's role is that of a savage tyrant, the role assigned to him by the prophecy and by his father. Like Oedipus the king, King Basil does everything to avoid his fate, and like Oedipus, he runs headlong into it. Having imprisoned his son in a forsaken tower and thus having made a wild beast of him, he then reveals his son the prince's existence and identity to the world when the time comes for the transfer of power. The cruel treatment that he feared so much now seems inevitable. But the director of the *gran teatro del mundo*, evoked by Segismundo, is not subject to the rule of Necessity like Apollo, but rather like the all-powerful Christian God, offers free will (*albedrio*) to human beings. Segismundo's demand for freedom is based on the awareness of his free will: "And I, having more free will (*albedrio*)/would have less freedom (*libertad*)?"[28] The failure of the king leaves room for the abuse of freedom and the chaos of unrestrained license in a theater without a director or a privileged spectator who would be immune to violence. However, inspired by God, Segismundo recovers *in extremis* the sense of transcendence. He finds another, everlasting world, the rules of which he will strictly "follow." Thus, the *gran teatro del mundo* becomes the world in the full sense of the word: the image of an ordered universe governed by the eternal rules of justice. Eventually Segismundo's revolt against his father and his conversion undermine the "melancholy" view of the *theatrum mundi* and turn to the Platonic (and Christian) world of eternal substances. The danger of the world being a dream (*quiza es un sueño*, "perhaps it's a dream" – the leitmotif of the play) has been evaded by the introduction of the "philosophical" notion of the theater of representation.[29]

Plato's Socrates would be surprised to see the theater reveal the "essential structure" of things, a function he reserved in the *Republic* or philosophy. This form of philosophical theater unfolded during the neoclassical period. The "Cartesian" and "Academic" spirit of the seventeenth century eventually spelled out the rapprochement of philosophy and theater in the wake of Aristotle's *Poetics*. It produced the world of order and became even more philosophical than Aristotle the Philosopher wanted it to be. The rules of "the three unities," of verisimilitude, decorum, separation of genres, poetic justice, and catharsis, have made the neoclassical theater the model of an ordered universe. It was accompanied by the expulsion of privileged spectators from the stage and the establishment of the proscenium arch, or the "Italian stage," perfectly framing the spectacle.

Artaud's "theater of cruelty" challenges this model, which has dominated the Western stage since the second half of the seventeenth century and still dominates most of the *théâtre de boulevard* and Broadway theater. However, Artaud's challenge is not limited to the world of theater narrowly understood. It pertains to the whole Western philosophical *Weltanschauung*, to its contemplative, purely intellectual kind of philosophy. Artaud restored to the idea of world-as-theater the philosophical disquiet of certain Baroque plays. In Artaud's project, *theatrum mundi* definitively lost the security of a rhetorical figure, of a "lieu commun de l'humanisme." If the world is a "theater of cruelty," as Artaud suggests, it becomes an anarchic, evil, and cruel world, resistant to the philosophical "cosmodicy":

> It is cruelty that cements matter together, cruelty that molds the features of the created world. Good is always upon the outer face, but the face within is evil. Evil which will eventually be reduced, but at the supreme instant when everything that was form will be on the point of returning to chaos.[30]

Against the Aristotelian model of philosophical theater, Artaud recovers both the Platonic idea of the theater as danger – danger of anarchy, of confusion, of *adikia* or *Unfug*,[31] of setting out of joint the ideal world of justice – and the Baroque idea of world-as-theater, of theater becoming life. Artaud's project of the dangerous theater that plays with boundaries responds to the major challenge of the modern theater. When, in the late thirties, Walter Benjamin pointed out that the main task of the modern theater was "the filling-in of the orchestra pit" (*die Verschüttung der Orchestra*), he had the epic theater particularly in mind.[32] However, Benjamin's statement pertains even more to Artaud's "theater of cruelty."

The question of the boundary between the stage and the auditorium is crucial to the "vision" of theater. Its supposed fragility excluded theater from Plato's *Weltanschauung*, and yet its supposed firmness brought theater back into the philosophical orbit, in Aristotle's view. The latter, i.e., the Aristotelian "philosophic theater," appears in the rapprochement between the philosophical contemplation (*theôria*) and a theater's auditorium (*theatron*). The comfortable and passive position of the spectator in the neoclassical and bourgeois theater gave to this comparison the sense exploited by Marx in his *Theses on Feuerbach*: instead of philosophy of action, the contemplative attitude (*theôria*) fosters philosophy of interpretation.[33] The line of separation between the spectator (*theatês*) and the spectacle (*thea*) is important to the notion of contemplative philosophy both in Plato and in Aristotle. If philosophy is to guarantee the stability of the ideal Republic, the object of its contemplation cannot be the fleeting spectacle of sensuous impressions, but the reality of eternal Ideas perceived by the "eye of the soul" (*omma tês psukhês*).[34] In Plato, theater is rejected as a threat to the purity of the ideal contemplation. In Aristotle, theater is rehabilitated because it is shown to be able to uphold the principle of the philosophical contemplation of the real and the boundary of the sensible and the intelligible. Theater is philosophical to the extent that it extracts the general (*ta katholou*) from the particular (*ta kath'hekaston*). Sensuous spectacle is dispensable when the intelligible reality of *ta katholou* is presented through plot (*muthos* or *logos* – the "soul

(with the help of Aristotle) in his *Hamburgische Dramaturgie*.[53] Both acceptance and rejection confirm the moral order shared by the characters and the audience. In the theater of cruelty, on the other hand, even when particular characters are maintained, the order is threatened. Indeed, the identification with a Cenci, a Giovanni, a Heliogabalus, produces a disarray rather than a comfortable sense of order.[54] They are all "crowned anarchists," although only the latter officially bears this denomination. Heliogabalus is not even a consequent anarchist.

A strange rhythm intervenes in Heliogabalus' cruelty; this initiate does everything with art and everything in double. I mean everything on two planes. Each of his gestures is double-edged.

> Order, Disorder,
> Unity, Anarchy,
> Poetry, Dissonance,
> Rhythm, Discordance,
> Grandeur, Puerility,
> Generosity, Cruelty.[55]

Far from producing an effect of harmony, as did the Pythagorean binary oppositions cited by Aristotle, the doubles of the typical Artaudian hero lead to the questioning of any order and to the collapse of meaning. Artaud himself recognizes the risk of an audience's total identification with the anarchic characters. He anticipates some of the criticism directed against the theater of cruelty: "It will be claimed," he writes in the essay "No More Masterpieces," "that example breeds example, that if the attitude of cure induces cure, the attitude of murder will induce murder." The language of Artaud's answer to this criticism might provoke some misunderstandings if not referred to his clearer statements in other texts:

> ... let it not be forgotten that though a theatrical gesture is violent, it is disinterested; and that the theater teaches precisely the uselessness of the action which, once done, is not to be done (*qui une fois faite n'est plus à faire*), and the superior use of the state unused by the action and which, *restored*, produces a purification (*sublimation*).[56]

Artaud's view should be understood neither in the sense of aestheticism, which he often repudiates elsewhere,[57] nor in the sense of the classical Aristotelian catharsis.[58] Whereas for Aristotle "[t]he world of tragic events must ... be rational," and "the world [*tout court*] remains a rational, meaningful place...,"[59] for Artaud, as for Plato, the two worlds are in danger of confusion with one another.

Yet, for Artaud, this "idea of chaos" – intoxicating and dangerous – gives the theater a certain epistemological, in the sense of "apocalyptic," quality. The task of the "theater of cruelty" is to impart to the audience a sense of anarchy, both liberating and perilous. At the same time, it produces enough energy for challenging the established order of the world and a revelation (*apokalupsis*) of the truth of the human condition, namely, that "*nous ne sommes pas libres. Et le ciel peut encore nous tomber sur la tête*" (we are not free. And the sky can still fall on our heads).[60] The "truth," the idea of chaos, cannot be conveyed in a rational way because it consists precisely in the collapse of rationality. Only in a sensuous, aesthetic manner is it possible to experience the total loss of boundaries. Artaud's identification of cruelty with life – "I have ... said 'cruelty' as I might have said 'life'"[61] – should be viewed in this apocalyptic context. The principle that allows for the highest achievements of the human spirit reveals itself in the utter dissolution of "all bounds":

> Life cannot help exercising some blind rigor that carries with it all its conditions, otherwise it would not be life; but this rigor, this life that exceeds all bounds ... is what cruelty is.[62]

Artaud is confident that such an experience of theatrical cruelty will produce a magic transfiguration that will eventually lead to the restoration of the sense of order and to the idea of a "cosmic rigor."[63] Artaud's "*sublimation*" designates such a transformation of the spectator in the sense of an absolute generosity toward the world and toward its fragility. This generosity suggests less a Kantian passive contemplation with no interest in the existence of its object than a mimetic identification in the sense of acting-play-action, which is also a form of (aesthetic) knowing/sensing oneself in the world. Such experience profoundly marks the audience. Artaud explicitly "defies" the spectator of the theater of cruelty who has witnessed the violent spectacle "to give himself up, once *outside* theater, to ideas of war, riot, and blatant murder."[64] Thus, Artaud sees the effect of his theater in cathartic terms after all – in the modern sense of catharsis associated with sublimation: the violence of blood is eventually transformed into the "violence of thought" (*violence de la pensée*). In the very terms of his challenge, Artaud reestablishes the essential boundary between theater and the "outside" world, which thus becomes again a world in the full sense of the word – a cosmos.

State University of New York, Stony Brook

NOTES

[1] In Jean de Rotrou, *Le véritable Saint Genest*, IV. VII, 1325: "This play is no longer a play but truth." Where not specified, translations are mine.

[2] "If theater doubles life, life doubles true theater…," Antonin Artaud, letter to Jean Paulhan on 25 January 1936, in *Oeuvres Complètes* [henceforth abbreviated as *Oe.C.*], vol. V (Paris: Gallimard, 1964, 1979), 196.

[3] Plato, *Republic*, 377ab, 379a, 395c-d, 398b, 400d-402d, and *passim*.

[4] The consecrated English phrase "idea of chaos," if not strictly equivalent to the French "*idée sur le chaos*," is certainly evocative; it is not inappropriate, in view of the explicit reference to Plato's *Ideas* in the same essay. See Antonin Artaud, "*La Mise en scène et la métaphysique*," in *Oe.C.*, vol. IV (Paris: Gallimard, 1978), 35; "Metaphysics and the *Mise en Scène*," in *The Theater and Its Double* [henceforth: *Th.D.*], trans. Mary Caroline Richards (New York: Grove Wedenfeld, 1958), 36.

[5] *Oe.C.* IV, 43; *Th.D.*, 44.

[6] In Jean Jacquot's words; see his "Le Théâtre du monde: de Shakespeare à Calderón," in *Revue de Littérature Comparée*, vol. 31, no. 3 (July–September, 1957), p. 341.

[7] "In the beginning God created the heaven and the earth," says *Genesis*, and Hesiod's "first of all Chaos came to be" (ê toi men prôtista Khaos genet') implies, according to Cornford, that "sky and earth separated." F. M. Cornford, *Principium Sapientiae* (Cambridge: Cambridge University Press, 1952), 194f.; cf. G. S. Kirk, J. E. Raven, and M. Schofield, *The Presocratic Philosophers*, 2nd revised and augmented edition (Cambridge: Cambridge University Press, 1983), 34ff.

[8] See René Girard, *La Violence et le sacré* (1972); *Violence and the Sacred*, trans. Patrick Gregory (Baltimore: The Johns Hopkins University Press, 1977) and Victor Turner, *From Ritual to Theater: The Human Seriousness of Play* (New York: PAJ Publications, 1982).

[9] Augusto Boal in his *Teatro de Oprimido* (1974); *Theater of the Oppressed*, trans. Charles A. and Maria-Odilia Leal McBride (New York: Theatre Communications Group, 1985); cf. Johannes Birringer, *Theater, Theory, Postmodernism* (Bloomington, Indiana University Press, 1991), 146–168.

[10] "[The world is a stage, life is a passage (*parodos*),] you come, you see, you exit." Democritus, fr. 115, in Hermann Diels and Walther Kranz, *Die Fragmente der Vorsokratiker*, vol. II (1952; Zürich: Hildesheim, 1984), 165; Shakespeare, *As You Like It*, II.Vii, 139ff.; cf. *Merchant of Venice*, I.i, 77–9.

[11] Pindar, 8th *Pythic Ode*, v. 136; Shakespeare, *Hamlet*, II.ii, 258.

[12] See, e.g., Plato, *The Republic*, 500c-d and *Timaeus*, 27a, 28b, and *passim*.

[13] Plato, *The Republic*, 607b.

[14] Aristotle, *Poetics*, 1448b9–12.

[15] Aristotle, *Poetics*, chapter XV; in his famous speech of Shakespeare's *Troilus and Cressida* (I.iii, 75ff.), Ulysses calls these distinctions "degrees" and deplores the anarchy caused by their loss.

[16] This view of "Aristotle's coercive system of tragedy" has been developed by Augusto Boal in his *Theater of the Oppressed*, chapter 1.

[17] Theater is "*philosophôteron historias*" (more philosophical than history), according to Aristotle, *Poetics*, 1451b5–6; although he speaks of poetry in general, the immediately following remarks refer explicitly to comedy and tragedy.

[18] The Greek text of the *Poetics* was edited by Aldus in 1508, preceded shortly by the Latin translation of Giorgio Valla (1498); the famous *Explanationes*, a commentary by Maggi and

Lombardi, was edited in its definitive form in the middle of the sixteenth century. Vincenzo Maggi and Bartolomeo Lombardi, *In Aristotelis Librum De Poetica Communes Explanationes* (München: Wilhelm Fink Verlag, 1969).

[19] For the analysis of the baroque *theatrum mundi* topos, in the literal (architectural) and figurative sense, see, in addition to the Jaquot's essay cited in note 2, R. Alewyn, *Das grosse Welttheater: Die Epoche der höfischen Feste in Dokument und Deutung*, second edition, revised and expanded (München: C. H. Beck, 1985), F. A. Yates, *Theatre of the World* (London: Routledge and Kegan Paul, 1969), Ross Chambers, *La Comédie au château* (Paris: Librairie José Corti, 1971), and Didier Souiller, *La Dialectique de l'ordre et de l'anarchie dans les oeuvres de Shakespeare et de Calderón* (Bern: Peter Land, 1985) and *La Littérature baroque en Europe* (Paris: Presses Universitaires de France, 1988).

[20] William Shakespeare, *Hamlet*, III.ii, 21–24.

[21] The fragility of boundaries is well marked in the recent cinematographic production of *Hamlet* by Kenneth Branagh, where Hamlet literally jumps on and off the stage of *The Murther of Gonzago*.

[22] He succeeded "in dying to make his art (deception) true." Jean de Rotrou, *Le véritable Saint Genest*, V.vii, 1750.

[23] "Whoever lived in theater dies on the stage"; ibid., IV.vi, 1388.

[24] "There is no longer need [for a prompter].
In this action, watched by Heaven,
An Angel is the prompter, an Angel keeps me straight;
An Angel by his order gave me full satisfaction,
And erased my crimes with holy water;
This perishable world and its vain glory,
Is a play in which I did not know my part"
(Ibid., IV.vii, 1298ff.).

[25] Jacques Derrida, "La Parole soufflée," in *L'Écriture et la différence* (Paris: Éditions du Seuil, 1967), 253–292; cf. "Le Théâtre de la cruauté et la clôture de la représentation," ibid., 341–368; *Writing and Difference*, trans. Alan Bass (Chicago: The University of Chicago Press, 1978), 169–95 and 232–50, respectively.

[26] "That this spectacle/performance is the truth."

[27] The phrase "*gran teatro del mundo*" is pronounced by Segismundo in his "real" dream. Pedro Calderón de la Barca, *La vida es sueño*, Second Day, 1086; *Life is a Dream*, trans. Edwin Honig (New York: Hill & Wang, 1970).

[28] Calderón, *La vida es sueño*, First Day, 151–2.

[29] For the problem of transcendence in Calderón, see Walter Benjamin, *The Origin of German Tragic Drama*, trans. John Osborne (London: NLB, 1977), 81; William Entwistle clearly reads Calderón's plays in Platonic terms: "The world of phenomena is, admittedly, a dream, but there is a network of realities immediately underneath the surface and embracing the correlated postulates of all sciences. It is this world which Calderón reveals in a number of his great plays." William J. Entwistle, "Calderón et le théâtre symbolique," in *Bulletin Hispanique*, LII (1950), 41–54. The religious character of the idea of "world-as-theater" is even more explicit in Calderón's auto-sacramental *El Gran Teatro del Mundo*, with its transcendent author and privileged spectator – God; see Pedro Calderón de la Barca, *El Gran Teatro del Mundo*, 36–66; *The Great Stage of the World*, trans. George W. Brandt (Manchester: Manchester University Press, 1976), 2.

[30] Artaud, *Oe.C.* IV, 100; *Th.D.*, 104.

31 Heidegger's translation of *adikia*.

32 Walter Benjamin, "Was ist das epische Theater? (2)" in *Gesammelte Schriften* (Frankfurt am Main: Suhrkamp Verlag, 1977), 539; *Understanding Brecht*, trans. Anna Bostock (London: New Left Books, 1973), 22; cf. ibid., the first version of the "What Is Epic Theatre?," p. 1 ff.; cf. also Harry Zohn's translation in Walter Benjamin, *Illuminations: Essays and Reflections*, ed. Hannah Arendt (1968; New York: Schocken, 1969), 154. I have analyzed the Brechtian answer to the challenge of the modern theater in my essay "Theater and/of Ideology," in *Rethinking Marxism*, 10, no. 3 (1998): 38–50.

33 On the theatrical side of this traditional Pythagorean comparison, Artaud deplores the triumph of a voyeuristic theater of "Peeping Toms." See *Oe.C.* IV, 82; *Th.D.*, 84; in Arthur Miller's *The Archbishop's Ceiling*, the model of passive *theôria* is the result of the forced *theatrum civitatis*: "Our country is now a theater from which nobody is allowed to walk out and everyone is obliged to applaud."

34 Plato, *The Republic*, 527d-e.

35 Aristotle, *Poetics*, 1450b17–20.

36 Artaud, *Oe.C.* IV, 114; *Th.D.*, 117f.

37 Not "just… as great," as the English translation has it.

38 *Logos* is identified with *muthos* by Aristotle, e.g. *Poetics*, 1454b35, 1455a17.

39 "We can perfectly well continue to conceive of a theater based upon the authority of the text (*la prépondérance du texte*) more and more wordy, diffuse, and boring, to which the aesthetics of the stage would be subject. But this conception of the theater… is… certainly its perversion" (*Oe.C.* IV, 102; *Th.D.*, 106).

40 *Oe.C.* IV, 115; *Th.D.*, 118.

41 *Oe.C.* IV, 116; *Th.D.*, 119.

42 *Oe.C.* IV, 91; *Th.D.*, 94.

43 *Oe.C.* IV, 88; *Th.D.*, 91.

44 Letter to Roger Blin from 25 March 1946 in *Oe.C.* XI (Paris: Gallimard, 1974), 215.

45 *Oe.C.* IV, 79; *Th.D.*, 81; cf. Jerzy Grotowski, "Il n'était pas entièrement lui-même," in *Temps Modernes* (April 1967): 1887.

46 *Oe.C.* IV, 84; *Th.D.*, 86; cf. 93 and 96, respectively.

47 *Oe.C.* IV, 71; *Th.D.*, 73.

48 *Oe.C.* IV, 114; *Th.D.*, 116.

49 *Oe.C.* IV, 64; *Th.D.*, 67.

50 *Oe.C.* IV, 121; *Th.D.*, 124.

51 *Oe.C.* IV, 84; *Th.D.*, 86.

52 Bertolt Brecht, "On the Use of Music in an Epic Theatre," in *Gesammelte Werke, vol. 15, 476;* in *Brecht on Theatre: The Development of an Aesthetic*, trans. John Willet (New York: Hill and Wang, 1964), 87.

53 Richard III, an utter villain, even viler than the hero of Shakespeare's historical play, eliciting disgust (Aristotle's *miaros*) rather than sympathy (*Mitleid*); see G. E. Lessing, *Hamburgische Dramaturgie: Kritisch Durchgesehene Gesamtausgabe mit Einleitung und Kommentar von Otto Mann* (Stuttgart: Alfred Kröner Verlag, 1963), section 73 ff., p. 287 ff.; English translation of Helen Zimmern: G. E. Lessing, *Hamburg Dramaturgy* (1890, reprint, New York: Dover Publications, 1962), 173 ff.

54 *Oe.C.* IV, 27–30 (*Th.D.*, 28–30), *Oe.C.* IV, 147–210; *Oe.C.* VII, 9–137.

55 *Héliogabale ou l'anarchiste couronné*, in *Oe.C.* VII, 127f.

56 *Oe.C.* IV, 80; *Th.D.*, 82.

57 See, e.g., Artaud's letter to Jean Paulhan from 25 January 1936, in *Oe.C.* V, 272f.

58 The latter interpretation might be suggested by the standard translation of the French "*sublimation*" by "purification."

59 J. Lear, "Katharsis," in Amélie Oksenberg Rorty (ed.), *Aristotle's Poetics* (Princeton: Princeton University Press, 1992), 334f.

60 *Oe.C.* IV, 77; *Th.D.*, 79, Artaud's emphasis.

61 *Oe.C.* IV, 110; *Th.D.*, 114; cf. *Oe.C.* IV, 99; *Th.D.*, 103: "It seems to me that creation and life itself are defined only by a kind of rigor, hence a fundamental cruelty, which leads things to their ineluctable end at whatever cost."

62 *Oe.C.* IV, 110; *Th.D.*, 114.

63 *Oe.C.* IV, 98; *Th.D.*, 102.

64 "Quels que soient les conflicts qui hantent la tête d'une époque, je défie bien un spectateur à qui les scènes violentes auront passé leur sang, qui aura senti en lui le passage d'une action supérieure, qui aura vu un éclair dans les faits extraordinaires et essentiels de sa pensée – la violence et le sang ayant été mis au service de la *violence de la pensée – je le défie de se livrer au dehors à des idées de guerre, d'émeute et d'assassinat hasardeux"* (Oe.C. IV, 80; *Th.D.*, 82; my emphasis).

KRISTIN O'ROURKE

RITUAL AND PERFORMANCE IN THE THEATER OF ROMANTICISM: DELACROIX'S SELF-STAGING AT THE PARIS SALON

The connections between the art world and the theater during the nineteenth century were overt and multiple. From the manner of exhibition of art works, to their subject matter, to the poses and postures of the figures on the canvas, paintings overlapped with and reiterated the main principles of French theatrical productions. The Salon, the state-sponsored, Academy-dominated annual or biennial art exhibition, in the late eighteenth and early nineteenth centuries was itself a theater space, a space of spectacle, an anticipated social event, as much as it was an exhibition of contemporary art. The multi-leveled relationship between the theater and the specific institution of high art known as "history painting" began in this exhibition, in the recently developed public sphere, and in the growing critical realm of print journalism and pamphleteering.

The major history paintings at the time not only were derived from classical or modern theater and literature, but were themselves often perceived to be self-consciously theatrical, containing highly charged gestural expressions, stage-like settings, historical costumes and (usually) tragic story lines. According to Academic doctrine, this form of painting was the highest of the genres, the most public, most noticeable by virtue of its massive scale, and by extension the most ambitious. It was the kind of painting sought out first by Salon visitors, and discussed most frequently by the critics. It was also the most likely to make the artist famous, or at least written about, especially as the nineteenth century progressed and the Salons grew to overwhelming proportions. In the newly developing and highly competitive art market, artists became performers staging themselves (and building reputations and careers) through their art works for the public and the critics. Eugène Delacroix capitalized on this opportunity, forging his professional and artistic identity in this public arena.

The reflexivity of the theater and Neoclassical painting is often restricted to subject matter, composition, and mode of display. Neoclassical (or as it was called at the time, classical) history painting, promoted by the teachings and example of the Academy and the School of Fine Arts in Paris, can be clearly linked with the classical French theater of the time. The main features

A.-T. Tymieniecka (ed), Analecta Husserliana LXXIII, 277–288.

of both of these art forms were a concern with unity of action, decorum, the integrity of characterizations, the representation of the passions, faithfulness to the original text, and most importantly, clearly conveying the narrative. The difficulty of painting, of course, as opposed to all forms of literature, was the need to express all of those things in one moment on the canvas, to communicate all of that information to the viewer so that he or she could take it in at one glance. Diderot, in his *Notes on Painting*, written originally to accompany his Salon review of 1765, requests from a composition simplicity above all else, and democratically invokes the heterogeneous makeup of the average crowd of Salon viewers. Diderot cautions: "A composition intended for display to the eyes of a motley crowd of spectators will be defective if a man can't figure it out through plain good sense. It should be simple and clear. Consequently, no pointless figures, no superfluous accessories. Treat the subject all of a piece...." Diderot continues: "The painter has but one instant at his disposal, and it's no more permissible for him to encompass two instants than two actions."[1] It is this necessarily unified compositional structure of both drama and the visual arts, as well as the tight-fisted control over expression which it reveals, which was the subject of attack by Romantic writers and painters.

The increasing publicness of the Salon exhibition in the late eighteenth and early nineteenth centuries had both a positive and a negative effect. The exhibition was originally reserved for the members of the Academy to demonstrate their skill and learning to the public. The exhibitions were free and open, and were meant to instruct the public as well as the younger generations of artists through their ideal example.[2] What the Academy did not foresee was that their works were then offered up for criticism to the public and a new breed of cultural journalists, most of whom had no art background whatsoever.[3] The Revolution of 1789 opened the Salon to all artists wishing to exhibit, and while restrictions returned eventually, the absolute control of the Academy over the Salon was permanently dissolved. The Salon in the nineteenth century became a battleground for the preservation of tradition and of the French school, as typified by Jacques-Louis David and his followers, against the desire of independent-minded artists for reform. The Salon for the most part was a tool of the Academy and the School of Fine Arts, an institution which created and maintained a standard of high art practice that offered the young artist the correct route[4] to success. The ritual function of this art exhibition, which took place at prescribed intervals, was to preserve the sanctity of the Academy and its principles of high-art production through visual examples that served to normalize its imposed standards of both style and subject matter. As the continuing success of the French school was a reflection on the Academy, and by extension on the monarchy or other political regime, the artistic and propagandistic goals of the Salon can be understood as beneficial to the French state.

While classicism in the visual arts could exist in the theoretical realm, dependent on ideal principles of composition and aesthetic production, Romanticism was a movement conceived in the public sphere and supported by it. The subversive strategy of the Romantic school, which developed in reaction to the dominance of (Neo)classicism in the early nineteenth-century art world, was to use the very public space of the Salon to overturn these hierarchies, to counter the rational, beautiful, edifying paintings of the Academic school with their "Romantic" works which were perceived as grotesque, fantastic and anti-heroic. The Romantic painters used the Salon to stage their rebellion, the so-called "Battle of the Romantics and Classicists." Delacroix and his fellow Romantic painters offered large-scale history paintings to the Salon in the 1820s, monumental works that broadcast their agenda and that can be read as both technical manifestations of the new school of art-making and theoretical statements of their hard-to-define artistic principles. The furor surrounding these paintings, particularly Delacroix's 1827–28 *Death of Sardanapalus*, both in the critical arena of Salon reviews and in the Academic and governmental realms, worked to foster Delacroix's growing reputation as the leader of the "school of the ugly" in contemporary painting, and to cement his persona in the history of art. Delacroix is here singled out as the leader in this public struggle of Romanticism because he was perceived as such by critics and the public alike, during and after his lifetime. This characteristic continues to define his work and career in the cultural history of the nineteenth century.

In light of these artistic and ideological struggles, how are we to interpret the relation of the theater (and theatricality) to history painting in the late eighteenth and the nineteenth centuries? Theatricality in painting, in art historian Michael Fried's study of eighteenth-century Salon criticism,[5] is defined as the awareness of being looked at, when the subject of the painting is aware of itself as the object of the spectator's gaze. He opposes this to what he calls the quality of "absorption" in painting, whereby the object or rather the subject of the painting is absorbed in his/her own subjectivity, unaware of the spectator, and is therefore more "natural" in Enlightenment terms, less mannered and superficial, less like the Rococo-courtly style of painting associated with the Old Regime. Yet if we look at two very famous works of the late eighteenth century, one a Revolutionary icon actually painted before the revolution, *The Oath of the Horatii* by Jacques-Louis David (1784), and the other a proto-Romantic and apolitical work by his student Anne-Louis Girodet, the *Sleep of Endymion* (1791), do we really read these history paintings as non-theatrical simply because the main characters are unaware of our presence? Is it not in fact impossible to ignore the inherent theatricality of monumental painting, or really any publicly displayed art works at all in this time period?

In Girodet's *Endymion*, the main character is absorbed in his eternal sleep. If the spectator is completely discounted in a work such as this, then why is the body of the main figure so conveniently tipped towards the picture plane, privileging the viewpoint of the spectator in such an artificial manner? The viewer is further reminded of his/her voyeuristic position by the presence of the smirking Cupid holding back the branches so that the moonbeams of the goddess Diana can light up the main figure's body.

David's *Oath of the Horatii* offers the same possibility for this non-theatricality, as all of the characters are engaged either in preparing for the upcoming battle (the men on the left of the canvas) or anticipating their mourning after the battle (as exemplified by the sisters on the right). However, the debts that this painting owes to theatrical production appear obvious. Outside of the subject itself which derives from, but does not formally replicate, Corneille's drama *Horace*, the figures are disposed across the canvas in the foreground, parallel to the picture plane, their bodies posed in such a way as to generate the maximum emotional effect on the viewer and the maximum understanding as well. The background is summarily indicated, and contemporary criticism considered it to be more like a theatrical backdrop than a true recession into space. This highly theatricalized, highly gestural image reads like a film still. We see the whole story in one instant (or rather all of the story is encompassed in this one condensed action which we comprehend in time as our eyes travel over the canvas): in other words, Lessing's "pregnant moment." It would appear then that theatricality is an unavoidable condition of monumental public art works.

In order to clarify the relation between history painting and the theater, I would like to turn now to a discussion of three specific works: David's *The Lictors Returning to Brutus the Bodies of His Sons*; Guérin's *Phaedra and Hippolytus*; and finally Delacroix's *Death of Sardanapalus*. I would like to explore the differences among the circumstances and motivations of these three paintings, and suggest how Romantic history painting differs from its predecessors. While the *Brutus* took on a life of its own in relation to the politically motivated theater of the day, and Guérin's *Phaedra* continues to be seen only in conjunction with Racine's play, Delacroix's *Sardanapalus* represents a break from the equation of history painting with the theater. The relationship of Romantic history painting to Romantic theater builds upon the historical connections between these two art forms, but Romantic history painting succeeds in representing something beyond a reflection of the other medium. Romantic history painting presents its own struggle against the establishment, echoing those of the Romantic theater, but also thematizing a new performative mode of art wherein the painting does not represent or look like theatrical performances, and in fact it cannot do so, but rather *is* a performance piece. The painting becomes the stage on which the artist both presents his theoretical standpoint, and constructs himself for the viewer. The art work embodies the principles of the Romantic movement, as well as offering a visual example, a marker, of the now-mythical "Romantic struggle." The art exhibition itself becomes a theater, a microcosm of the realm of cultural production as a whole wherein ideological struggles are indexically represented.

Two particular history paintings exemplify the close relationship of Neoclassicism to the French theater in the decades before Romanticism. The first is one of the major early works by the artist credited with the rejuvenation of the French school in the immediately pre-Revolutionary era, Jacques-Louis David's *The Lictors Returning to Brutus the Bodies of His Sons*, from 1789. This painting is relevant not only for its visual qualities, but more interestingly for its intimate relation to the play *Brutus* by Voltaire, staged in Paris in 1790, at the National Theater. In a foreshadowing of the so-called "Battle of *Hernani*" pitting the Classicists against the Romantics forty years later over the premier of Victor Hugo's drama *Hernani*, this revival production of Voltaire's play was marked by the vocal presence of rivalrous political factions in the audience (Royalists and Republicans)[6] who would cheer or boo at moments appropriate to their political cause.[7] The story of Brutus, originally from Plutarch and Livy, presents a central protagonist who symbolizes patriotic duty as taking precedence over familial and emotional ties, as well as the founding moment of the Roman Republic. With the revival of Voltaire in the Revolutionary period, his play, written much earlier, acquired Revolutionary overtones; David's painting was likewise produced before the outbreak of the Revolution, yet also soon became a representative icon of the Republican spirit.

Voltaire's play was put on for two nights in November of 1790. At the end of the first night, a spectator requested that the bust of Voltaire by the Neoclassical sculptor Houdon be placed on stage. At the second performance, the marble bust of Voltaire was again on one side of the stage, and a bronze bust of Brutus (a work that had been bought back to Paris from Rome by David himself) was displayed on the other side. The visual equation of Brutus, the founder of the Roman Republic, with Voltaire, the philosopher of the Enlightenment, was purported in this double display. The interesting feature of the production however is not this; rather it is that at the very end of the play, the actors deliberately enacted a *tableau vivant* of David's picture. As the curtain fell, the main character seated himself in a chair in the shadows while the lictors could be seen carrying the bodies. As one spectator wrote: "Every Parisian knew David's painting, each recognized right away the intention of thus honoring the artist in public, in front of the nation. It was like a national holiday, heightened by unanimous applause."[8] All subsequent printed editions of the play incorporated this *tableau vivant* into its ending, giving stage directions on carrying out the painting's action as the curtain falls. David's painting, like Voltaire's play, was reappropriated to fulfill a Revolutionary ideal, galvanizing the majority of the public in their enthusiastic reception of the *tableau*, both painted and staged. In this way, the painting served as a visual representation of political motivations, and demonstrated the political potency of works of art, and of their public display in the realm of high art, the Salon exhibition.

The 1802 painting of *Phaedra and Hippolytus* by David's younger rival Pierre-Narcisse Guérin was itself based on a classical French drama, the play *Phèdre* by Racine staged in Paris in 1802, and this perceived closeness of the painting's composition to an actual scene from the play spurred much discussion at the time, garnering this work much praise as a sublime example of expressive painting. James Rubin reads in this painting's huge success by virtue of its connection to Racine's work, that Racine represented a post-Revolutionary turn back towards "Frenchness" after the Revolutionary barrage of Roman morality tales in the works of Voltaire and Corneille. Rubin describes Guérin's Racinian work as what he terms "an antidote to that of David."[9] Yet the overt theatrical connection, according to later biographers and art historians, has worked to the detriment of Guérin's reputation as an artist, which, following his death in 1833, began a steady decline. He was accused of coldness of execution, of lack of imagination for his too strict copying, and was denigrated for what came to be seen as a highly mannered and artificial style.

The painting does not actually represent a scene either from the written or the performed version of the play; in fact there is no scene in the play where all of these four characters are on stage together, nor in this particular scene, where Hippolytus is confronted by his father over his supposed illicit advances towards his stepmother Phaedra. In comparing the final sketch of the painting with the finished version, one can notice that in the process of finishing the painting, Guérin has, among other things, elongated his lines, and through his use of light, enhanced the drama by taking the focus off of Hippolytus, and placing it on Phaedra herself. Guérin has heightened the tension by placing all of the characters together in a tightly woven composition which links all four of them in a powerful moment of accusation, betrayal, and guilt. This painting represents both the strictness of the connection between Neoclassical theater and history paintings, and the tension of the slippage between the two. While the painting looks like it could be an exact portrayal of a contemporary play, the variations and adjustments made by the artist to enhance the one painted moment are elided or unseen, to the detriment of the artist. The very representability of Neoclassical artistic ideals and political precepts, as seen here in the works of David and Guérin, allows for this misleading sense of equality between the art work and its referent.

David, Guérin and others, however much they modernized, naturalized or revised their sources, still essentially reproduced works of the great masters of the French theater. What we see with Romantic history painting is a series of breaks from this tradition, first seen in the adherence in the strict symbolic division explicitly set up by Stendhal between Racine and Shakespeare in his 1823 work by that name. In his chapter entitled "What Romanticism Is," he begins:

"*Romanticism* is the art of presenting people with literary works which… are susceptible of arousing the most possible pleasure. *Classicism*, on the contrary, offers them that literature which gave the greatest possible pleasure to their great-grandfathers."[10] Delacroix's *Death of Sardanapalus*, based on the 1821 play by Lord Byron,[11] can be contrasted with the *Apotheosis of Homer*, a representative work by his rival J. A. D. Ingres, a pupil of David and upholder of the classical tradition throughout his career. Both paintings were exhibited in the Salon of 1827–28, but are worlds apart conceptually.

The early Salon paintings of Delacroix emphatically demonstrate this overturning of the classical rules. In them, clear gesture was replaced by expressive color, classical subject matters were passed over in favor of contemporary literary works and political events. The influence of contemporary Romantic literature and theater (of Byron and Walter Scott, among others) on painting was recognized and heartily supported, especially by the critics who reviewed these art works, most of whom were themselves poets, novelists, or playwrights. It is in the highly public arena of the Salon that the mingling of the spheres of the literary, theatrical and visual arts at this time was most evident.

What we see in Romantic history painting ultimately is a historically successful iteration of the "new rules," in other words, that "there are no rules." Delacroix's *Death of Sardanapalus* serves as a visual counterpart to these Romantic literary treatises of the 1820s. As opposed to Neoclassical theater, what we can see here is that Romantic theater was ultimately unproducible, and in fact very few Romantic plays were successfully staged. However, a painting is able to visually manifest those precepts; Romantic painting can represent that which is unrepresentable. Victor Hugo in his stridently Romantic preface to *Cromwell* (published in 1827–28), wanted to dispense with all of the "unities" of classical theater except for one, the unity of action, which he said functioned in drama as the viewpoint of the work. Tying this concept directly to painting, he believed that the "unity of action is the law of perspective in the theater."[12] Yet, by comparison, the unities of composition, action, time, and space are broken down in a painting such as Delacroix's *Death of Sardanapalus*, with its self-conscious disregard for logical space, proportion and perspective. This painting was described at the time as "an incomprehensible amalgam of men, women, dogs, horses, logs, vases, instruments of all types, of enormous columns, an immeasurable bed, thrown here and there, without effect, without perspective, and standing on nothing!"[13] This painting, which strains the tolerance of the viewer today, let alone in 1828, is the first step towards abstraction and the ultimate breakdown of form and three-dimensionality in modern art. This painting as an icon of the avant-garde, as a signifier for Delacroix himself and his new modern (i.e., bold) genius, appealed to future generations of artists on that level.

But the painting itself could be considered a failure in terms of the general critical opinion of the time, and it nearly destroyed Delacroix's career. The head administrator of the government's arts agency, after the closing of the Salon, told Delacroix that if he ever painted a work like that again, he would never receive another commission from the government, a serious threat as these government commissions were the main source of income for many painters. This threat, combined with the fact that the government did not purchase the painting after the Salon, as well as the generally negative critical response to the work, led Delacroix to retreat from his increasingly radical position as head of the new school of painters. This is perhaps the most anti-classical painting of his career, and it is at this point that his reputation was in essence sealed.

If the painting was such a disaster artistically and financially, why then did a talented and shrewd artist like Delacroix paint it? He called this painting his "Massacre no. 2,"[14] meaning both the second work of a similar theme following his first massacre, the 1824 *Massacre at Chios*, but also referring to the critical reception of both works, which was very unfavorable. What was his personal and artistic investment in what most considered at the time a professional blunder? Delacroix would never again so radically invoke the anti-classicistic principles of this signal painting, but it is this image in particular which continues to exert influence over his historical reputation. Delacroix's major Romantic Salon paintings (the *Sardanapalus*, as well as the 1824 *Massacre at Chios*, and the 1831 *Liberty Leading the People*) were indeed "succès de scandale": ill-received by most critics (except the most ardent Romantic partisans), they were nevertheless noticed at the time, and they have endured in the history of art. These early paintings formed the cornerstones of his reputation during his lifetime and in the many retrospective exhibitions following his death. Delacroix himself became an icon of Romanticism, a symbol to subsequent generations of avant-garde painters (Van Gogh, Matisse, Cézanne, Picasso, among others). It was not only his painting style and choice of subject matter that was emulated by these artists, but also his persona, as interpreted through his art works.

Delacroix's painting should be interpreted not as his sincere effort to create a Romantic school or to reform the French school,[15] but rather as a public gesture, a posturing in the camp of the anti-classicism intended to provoke, disturb and subvert. Delacroix talks in his journal and letters in this early stage of his career of wanting to make himself known, to be noticed among the throng of painters and their works. Delacroix's interest in creating the *Sardanapalus*, which he must have known would be highly controversial, was to align himself with other innovators of the time, and to make sure that he, as well as his painting, would be noticed. This painting is both a demonstration of the principles of Romantic art-making, and a theoretical statement in its own right, a painterly manifesto of the Romantic school. Delacroix's Salon paintings were more than works of art, they were performance pieces on the stage of the exhibition.

It is in this way that we can begin to see the Salon as a theater, as the mise-en-scène of cultural battles over the purpose and future not only of history painting, but of the arts in general, as well as over the institutional structures of the Academy and the museum. This was a later, and more public, development of the literary reform movements already taking place. It was also more indexical, as there were actual art products which could be directly compared. Today we see the galleries of the Louvre full of Delacroix's paintings, but at the Salon, all styles, formats, and genres were placed side by side, leading to clear visual comparisons on the part of the viewer and critic. The different styles of painting were subsequently organized by the critics (and the artists themselves) into opposed camps, and the discussion that developed in critical discourse throughout the 1820s became the Romantic/Classic dialectic. These ideological battles, as exemplified through the art works, were on display for three to six months on average at every Salon,[16] while the discursive battles were waged for years, the threads of which resurfaced from exhibition to exhibition.

Two popular representations illustrate the tensions inherent in this battle to break down or maintain the old institutional stranglehold. In one image, we see Delacroix dueling with Ingres in front of the Institute, and the other is a cartoon of the so-called "Battle of Hernani," with a caption which reads: "The Romans [or classicists] disordered by the first showing of *Hernani*."

The exhibition space was connected to the theater as a form of popular entertainment, yet the Salon exhibition was not only a site of spectacle, as it offered a site for the (self)performance of the artist, critic, and more importantly for Romanticism as a movement. The increasingly public Salon exhibition in the first half of the nineteenth century coincided historically with the development of the figure of the artist as a stereotype in public consciousness. Through the growth of the illustrated press and the proliferation of artistic and literary journals in the newly modern public sphere, the "artist" became a new character in the pantheon of societal figures, leading to a new public awareness of "schools" of artists with divisions among critical and artistic camps. The "battles" of the Romantics and the Classicists were fought and won in the public arena, and they are deeply connected to the public personalities of the artists, critics, and writers on both sides of the fence. Delacroix became the hero of the Romantic generation of artists in part because of his perceived difference from the Academy and its traditional principles of art making. This anti-institutional characteristic, although ambiguous in the case of Delacroix, has continued to define his work and career in the cultural history of the nineteenth century. The importance, even today, of the figure of the young, rebellious, anti-establishment Delacroix (however much of a myth it may be), cannot be overstated.

In the title of this paper, I refer to the "theater of Romanticism" as the very public and performative nature of Romanticism. Romanticism itself is a theater in which power plays, gestures, poses of the artists, their art works, the critics, etc., are enacted. Not only did Delacroix's performance on the artistic and discursive stage of the Salon function to define Romanticism at the time, but his paintings continue to raise the question posed by Baudelaire in his Salon review of 1846, in his chapter entitled, "What Is Romanticism?" He says that the answer lies not in the subject matter or in truthful representation, but rather in the "manner of feeling." Baudelaire goes on to say "For me, romanticism is the most recent expression, the most modern beauty.... He who says 'romanticism' says 'modern art.'"[17]

University of California at Santa Barbara

NOTES

1 Denis Diderot, "Notes on Painting," in John Goodman, ed. and trans., *Diderot on Art, Volume 1: "The Salon of 1765" and "Notes on Painting"* (New Haven: Yale University Press, 1995), 220.

2 Patricia Mainardi, *The End of the Salon: Art and the State in the Early Third Republic* (Cambridge, UK: Cambridge University Press, 1993); see chapter 1, "Pictures to See and Pictures to Sell," regarding the ideological struggles of which the Salon exhibition was the subject. "Pictures to see" implies the edifying principles of the free display of pictures to the public with no intent or need to sell them, "pictures to sell" refers to the Salon as an art marketplace, a notion of art as a trade and paintings as commodities, which therefore reduces the painter's social status.

3 Richard Wrigley, *The Origins of French Art Criticism: From the Ancien Régime to the Restoration* (Oxford: Clarendon Press, 1993); see chapters 5, "Censorship and Diffusion of Criticism during the Ancien Régime," and 6, "The Status of Criticism," for a discussion of the tensions between artists, critics, and government interests in the eighteenth and early nineteenth centuries.

4 The vocabulary of much Salon criticism of the 1820s discusses the Davidian or classical school as demonstrating the "bonne" or "vraie route" of art-making which would lead to success, and to the glory of the French School, whereas the Romantic school circling around the work of Delacroix offered a bad example to follow, that of the "fausse" or "mauvaise route" which would lead the young artist astray.

5 Michael Fried, *Absorption and Theatricality: Painting and Beholder in the Age of Diderot* (Chicago: University of Chicago Press, 1980); see chapters 1, "The Primacy of Absorption," and 2, "Toward a Supreme Fiction," for Fried's discussion of these terms as well as visual examples to support his argument.

6 Robert Herbert, *David, Voltaire,* Brutus *and the French Revolution: An Essay in Art and Politics* (London: Penguin, 1972), 15–16.

7 Herbert, 74–75.

8 G. A. Von Halem, *Paris en 1790*, trans. and ed. A. Chuquet (Paris, 1896), quoted in Herbert, 77–78.

9 James Henry Rubin, "Guérin's Painting of *Phèdre* and the Post-Revolutionary Revival of Racine," *Art Bulletin* 59, n. 4 (December 1977): 601. The article describes the contemporary and later critical reception of this work, and of Guérin's work in general, as well as the specific relation of this painting to Racine's play.

[10] Stendhal, *Racine and Shakespeare*, 1823, in Barry V. Daniels, *Revolution in the Theatre: French Romantic Theories of Drama* (Westport, CT and London: Greenwood Press, 1983), 126.
[11] Byron, *Sardanapalus: A Tragedy*, 1821. This play was not intended to be performed, but it was staged not long after its publication.
[12] Victor Hugo, Preface to *Cromwell* (1827) in Daniels, op. cit. 170.
[13] Antony Béraud, *Annales de l'Ecole française des beaux-arts* (Paris: Pillet aîné, 1827), 73–74: "… que de défauts choquans, et qui pis est, de défauts systématiques! quel dessin que celui de la plupart de ces figures! que d'efforts perdus pour arriver à l'expression, pour atteindre un *faire* original, et ne tomber que dans le bizarre! Donnerons-nous le tître de composition à cet amalgame incompréhensible d'hommes, de femmes, de chiens, de chevaux, de bûches, de vases, d'instrumens de toute espèce, de colonnes énormes, de lit démesuré, jetés pêle-mêle, sans effet, sans perspective, et ne posant sur rien!…"
[14] Eugène Delacroix, *Correspondance générale*, ed. André Joubin, vol. 1 (Paris: Plon, 1936), 211, and fn. 1; letter to Soulier dated Feb. 6, 1828, right after the exhibition of this work at the Salon.
[15] This notion of reform is the most common interpretation of a work such as David's *Oath of the Horatii*, 1784.
[16] In the later years of the Restoration, the Salon was intentionally limited in frequency. During the Restoration, the Salon took place in 1815, 1817, 1819, 1822, 1824, 1827–28. In the July Monarchy, this issue was straightforwardly addressed. In the 1830s, starting in 1831, the Salon took place nearly every year.
[17] Charles Baudelaire, *Ecrits sur l'art*, ed. Francis Moulinat (Paris: Librairie Générale Française, 1992): 77. "Le romantisme n'est précisément ni dans le choix des sujets ni dans la vérité exacte, mais dans la manière de sentir.… Qui dit romantisme dit art moderne.…"

BERNADETTE MEYLER

LINGUISTIC WORKS OF ART AT THE BORDERLINES: ONTOLOGICAL EXCLUSION IN INGARDEN AND GADAMER

In recent years, an increasingly pronounced estrangement has beset the discourses surrounding drama and the literary work of art. While literature departments now express a certain disregard for theatrical texts, viewing such compositions as irrelevant to their larger projects, those involved in the creation and production of plays have in turn relegated the very concept of "text" to the status of an antiquarian oddity and instead consider "performance" the only relevant aspect of their art. This schism may appear on first inspection to be occasioned by purely pragmatic factors; departments of drama, which, like schools of music, at least spatially attempt to unify the field's various aspects, have asserted the necessity for their discrete role within the university. When one examines instead the theoretical contexts surrounding the divide, however, it begins to seem less plausible that such institutional factors arose spontaneously, and one instead discovers their source within critical contexts themselves. Attempts to separate narrative from dramatic modes of expression and assert the priority of one over the other hardly commenced in the twentieth century; even Aristotle already distinguished epic from tragedy in his *Poetics*, and claimed the superiority of the latter in his culminating chapter. The rationale behind such a disjunction nevertheless alters with the historical situation. As a result, it was reinstituted in new terms by modern "theories of art," engaged in the project of dislodging aesthetics from its relatively secure position within post-Kantian philosophy. In particular, Roman Ingarden, writing from a phenomenological perspective, and Hans-Georg Gadamer, espousing a hermeneutic approach, essayed to undo certain dichotomies generated within philosophy through recourse to an account of narrative or dramatic representation. The quite disparate "solutions" Gadamer and Ingarden posited to the originally philosophical problem led, in both cases, only to the displacement of the divide into the literary sphere, where it emerged between narrative and dramatic modes of composition.

Although Ingarden, in *The Literary Work of Art*, and later Gadamer, in *Truth and Method*, rarely provide specific literary examples, and infrequently name particular texts, the objects most appropriate to each theorist's inves-

A.-T. Tymieniecka (ed), Analecta Husserliana LXXIII, 289–313.

tigations become clear during the course of his discussion. Janus-faced, both theories point in two directions, first, back toward a more general philosophical project, and, secondly, forward toward a method of application. This structure manifests itself less equivocally in Ingarden's study than in Gadamer's, since the latter reveals objections to the very concept of "method" and the strictures it implies as a scientific term. As J. S. Smith claims in an essay intended to elaborate possible extensions of Ingarden's theory of the literary work to the stage play,

> Roman Ingarden's literary theory analyzes the literary work of art with a rigidly intrinsic methodology of purely ontological description, but the ultimate results of this analysis are intended not only to help with the solution of the fundamental philosophical problems (e.g., the realism-idealism controversy), but also to provide certain practical methods and tools of approaching different kinds of art.[1]

According to Smith's understanding, Ingarden's theoretical description leads down a fairly straightforward path to certain methodological outcomes. Even in the case of *The Literary Work of Art*, however, not only do Ingarden's more categorical postulates prove amenable to specific types of analyses, but it transpires that solely by excluding some materials from full participation in "the literary work of art" can Ingarden accomplish his philosophical objectives. Ironically, the most concrete moments of Ingarden's text concern themselves with elaborating the liminal condition of other art forms in relation to the literary work; the principal among these "Borderline Cases," is identified as "the stage play," a genre discussed both in Part III of *The Literary Work of Art* and in the Appendix on "The Functions of Language in the Theater." As these passages reveal, even within the realm of dramatic compositions, Ingarden's approach can countenance some works much more than others; several aspects of naturalistic plays fit into his discussion of the literary work of art, while symbolic works seem not to exist for him. It is in this discussion of "borderline cases" that "method" intrudes, as Ingarden attempts to extend his theory to cover even those elements of the field to which it does not seem most appropriate, and in so doing reveals the assumptions that underlie his theses. Through such a collection of negative examples, the primary object of Ingarden's study, the realistic novel, emerges by counterposition. Here method itself, or the practicality of application, feeds back into the overarching theory, as the difficulties Ingarden encounters in analyzing the stage play demonstrate the tangible postulates he must adopt in order to reach the appropriate philosophical conclusions.

While Gadamer expresses distinct reservations about the concept of "method" itself within *Truth and Method*, his work too implies a certain relationship between theory and method, and relegates some objects of investigation to the "borderlines" of his study; one of the final subsections in Part One of *Truth and Method*, "The Question of Truth as It Emerges in the Experience of Art," is entitled "The Borderline Position of Literature." Whereas Ingarden's project required the priority of the "literary work of art" – or, in other words, the written text – Gadamer's goals necessitate instead the elevation of dramatic genres over compositions destined for personal perusal. While he eventually manages to reclaim "literature" for his method, Gadamer can only accomplish this conquest after first having recourse to the metaphor of "play." The nature of this metaphor, like the attributes of method itself in Gadamer's text, appears quite ambiguous. According to Mihai Spariosu,

> At first sight, play in general seems to be incidental to Gadamer's project of elaborating a 'philosophical hermeneutics'.... Yet a second look at Gadamer's project will reveal that here, no less than in Heidegger and Fink, play has an essential function: it serves both as a metaphor for describing the way in which the truth of Being occurs through man's hermeneutical activity and as the groundless grounding *(Ab-gründung)* of Gadamer's own thought.[2]

With this explanation, Spariosu describes a dynamic similar to that manifested by method in Ingarden's text: "play" considered as a concrete description of the hermeneutic process is brought into relation with "play" as a philosophical concept, just as the realistic novel simultaneously provided Ingarden with material on which his theory could be applied, and certain crucial philosophical substrates. As Gadamer's own text reveals, "play" labels both the literally dramatic text and the process of interaction between the human individual and all art forms. The explicit ambiguity of "play" within *Truth and Method*, and the continual interweaving of usages – "play" alternately naming an activity and a specific object – actually seems to be occasioned by Gadamer's understanding of method.

Before examining Gadamer's account, however, it is possible to elaborate a preliminary connection between metaphor and method. As the idea of the "borderline case" suggests, for each theory, certain objects seem somehow natural, while others require that the initial principles of investigation be extended in order to include them. Between the theory and this apparently appropriate application, a metaphorical connection is constituted, through which the theory universalizes the characteristics of the entity in question; principles of analysis derived from this generalization must then be extended toward other objects. The peril inherent in such a process, one that the current split between narrative and dramatic texts clearly demonstrates, resides in the possibility that the metaphor could be literalized, leading one to conclude that, for example, stage plays actually do partake in a significantly different mode of being. Michael Fried's essay "Art and Objecthood," published only seven years after the first edition of *Truth and Method*, embodies an even more extreme version of this effect, as he derides the "theatrical" impulse in contemporary "literalist" art, refusing to distinguish at all between dramatic texts and a "theatrical" experience. As we will see, Gadamer's 1975 text "The Relevance of the Beautiful: Art as Play, Symbol, and Festival," which employs twentieth-century examples of the artwork, sheds considerable light on Fried's thesis, and to some extent corroborates its generative impulse.

It is through Gadamer's understanding of "method" that the dramatic contents of his analysis become inextricably intertwined with his play theory during the course of *Truth and Method*. Towards the commencement of this work, Gadamer already expresses dissatisfaction with the concept of "method" that the humanistic tradition has derived from the hard sciences and suggests possible alternatives. He claims that,

> However strongly Dilthey defended the epistemological independence of the human sciences, what is called "method" in modern science remains the same everywhere and is only displayed in an especially exemplary form in the natural sciences. The human sciences have no method of their own. Yet one might well ask, with Helmholz, to what extend method is significant in this case and whether the other logical presuppositions of the human sciences are not perhaps far more important than inductive logic....[3]

Following this assertion, Gadamer endorses Helmholz's idea that "memory and authority" (p. 8) prove of more importance to humanistic studies than "method." Subsequently, while discussing play, Gadamer again comments on the relationship between Dilthey, "method," and the "human sciences," writing that,

> [T]he important thing, in my view, is to correct the philosophical interpretation of the modern human sciences, which even in Dilthey proves to be too dominated by the one-sided method-ological thinking of the exact natural sciences.... If we are reflecting on the experience of art – as opposed to the subjectivization of philosophical aesthetics – we are not aiming simply at a question of aesthetics, but at an adequate self-interpretation of modern thought in general, which has more in it than the modern concept of method recognizes (p. 125).

According to this statement, "the experience of art" itself interferes with the possibility of "method," or more generally, the theory-method distinction. The "experience of art" that Gadamer names is not to be conflated with the subjectivized "experience" ("Erlebnis") that he earlier criticized, but instead conceived as an encounter with the ontological reality of the artwork. Placing a certain priority on the experience of a specific and individualized entity encourages one not to create an overly general method. Since method lies in the province of the practical, intimately related to the experience of the artistic object, it will become inextricably linked to theoretical assertions about the nature of the work in a philosophy that considers the experience of the artwork indissociable from its being. Thus, for Gadamer, the work of art must dictate its own terms of understanding; whether Gadamer himself can adequately intuit those terms is a question that cannot yet be answered.

Despite the fact that Ingarden gives priority of place to the realistic novel while Gadamer insists upon the importance of "play," both theorists begin from comparable starting points and create accounts of the linguistic work of art[4] based on principles of ontological description. Since each presents his treatise as an intervention within the continental philosophical tradition, the question arises as to what problem Ingarden, and then Gadamer, attempts to solve. Positing a preliminary hypothesis, we might speculate that both writers, dissatisfied with the models of truth, and the concomitant comprehension of ontology that post-Kantian philosophy has proved capable of providing, use the work of art to introduce other notions of meaning into the extant discourse. In order to accomplish this objective, Gadamer and Ingarden must first establish that aesthetic questions cannot be considered in isolation from cognitive concerns, and then elaborate a viable ontology of the literary or dramatic work, particularly in relation to its mode of representation.

Although Kant's *Critique of Judgment* separates the aesthetic from the cognitive spheres, the entire work is framed in terms of an epistemological endeavor. Aesthetic judgment appears to assume a negative position in relation to this project, but is no less defined by it. Indeed, inherent within the "Third Critique" lies the paradox that the judgment of taste involves the cognitive faculties, yet the beautiful avoids all involvement with cognitive content. As Kant writes of aesthetic judgments in his "Preface," "Although they do not by themselves contribute to the knowledge of things, yet they belong to the cognitive faculty alone and point to an immediate reference of this faculty to the feeling of pleasure or pain according to some principle *a priori*...."[5] Formally, the judgment relies on cognition, but Kant cannot discern anything within the aesthetic object that would allow one to grasp it as knowledge; in other words, Kant opines that an understanding of "truth" cannot constitute a goal of aesthetic judgment. This stance can be formulated in terms of "concepts" as well, since the judgment of taste, unlike the determinative judgment, fails to subsume its object under a particular "concept," or purpose, although it must display "purposiveness." As Kant states in §5,

> The judgment of taste is merely *contemplative;* i.e., it is a judgment which, indifferent as regards the existence of an object, compares its character with the feeling of pleasure and pain. But this contemplation itself is not directed to concepts; for the judgment of taste is not a cognitive judgment (either theoretical or practical), and this is not *based* on concepts, nor has it concepts as its *purpose* (pp. 43–44).

Within this remark, the assertion that the judgment of taste is "indifferent as regards the existence of an object" contrasts with Kant's comments about the objects of practical, or moral, judgment; regarding these, he maintains that "it is not merely the object that pleases, but also its existence" (p. 43). Despite being characterized in cognitive terms, the aesthetic judgment lacks cognitive content, and the existence itself of the aesthetic object remains inessential. It would not, then, be inaccurate to conclude that Kant's *Critique of Judgment* considers the relation between the judging subject and the aesthetic object from an epistemological rather than an ontological perspective.

Kant himself unfolds the implications of this epistemological stance for his entire philosophy, positing even of the natural concept, treated in the *Critique of Pure Reason*, and the concept of freedom, the subject of the *Critique of Practical Reason*, that "neither of them can furnish a theoretical knowledge of its object (or even of the thinking subject) as a thing in itself; this would be the supersensible, the idea of which we must indeed make the basis of the possibility of all these objects of experience, but which we can never extend or elevate into a cognition" (p. 11). This very "theoretical knowledge of the object" constitutes the goal of both Ingarden's and Gadamer's texts, which adopt ontological approaches. Since Kant deems such understanding impossible not only for aesthetic entities but for all objects, Ingarden and Gadamer can place the former on the same plane as the latter through executing an ontological examination. Both attempt to comprehend ontologically – or in terms of their essential meaning – what is constitutive of the work of art, elements that they respectively view as embodied in the realistic novel and the dramatic performance.

As Ingarden announces somewhat disingenuously at the outset of his treatise, "The goal we have set for ourselves is basically a modest one. For the present we would like to provide only an 'essential anatomy' of the literary work; its main conclusions will only then open the way to an aesthetic consideration of the work."[6] This sentence emphasizes Ingarden's focus on the "essence" of the work instead of the aesthetic experience of the reader, an orientation that separates ontology from epistemology. From the commencement, however, he acknowledges that concentrating on "essence" occasions certain problems related to the parameters of the investigation; one must in a certain sense at first make a definitional claim that one then elaborates. Ingarden writes,

> To begin with, we shall provisionally determine the range of objects we wish to investigate by means of a series of examples. We do this "provisionally" – that is, from the outset we are always ready to change such a definition of the range of objects if in the course of the investigation this appears to be necessary.... A final definition of the range of literary works presupposes grasping and conceptually defining their essence. This will be possible only at the end of this study (p. 7).

The dilemma that already appears in this statement is that of filling in the meaning of "necessary" as it applies to the necessity of re-delimiting the study's borders. Although Ingarden does wait until "the end of [his] study" to relegate certain objects to "Borderline Cases," one might wonder whether these examples were not already considered marginal at the beginning of the text. Gadamer likewise diverts attention from questions of knowledge towards those of essence. Indeed, his efforts in the section of *Truth and Method* entitled "The Ontology of the Work of Art and Its Hermeneutic Significance" seem to be focused on substituting an ontological for an epistemological conception of the aesthetic. Explaining his notion of "play," Gadamer postulates that, "when we speak of play in reference to the experience of art, this means neither the orientation nor even the state of mind of the creator or of those enjoying the work of art, nor the freedom of a subjectivity engaged in play, but the mode of being of the work of art itself" (p. 101). Although the structure of Gadamer's project – much more than Ingarden's – significantly resembles that of Kant's, this distinction illuminates the difference between the two, a division between an epistemological or an ontological orientation.

We may ask, however, how the pronounced relation between Kant and Gadamer manifests itself in concrete terms, and what impact it has on a comparison between Gadamer and Ingarden. As Spariosu demonstrates in *Dionysus Reborn*, the term "play" has a lengthy history in the tradition derived from Kant, including Schiller's concept of the "play-drive." Gadamer himself expresses the desire to reclaim "play" from these subjectivist clutches, accomplishing this aim not by denying the structures that Kant identifies, but instead through viewing them from a different standpoint. Within the context of *Truth and Method*, Gadamer describes the relationship between play and purpose in terms similar to those Kant uses to explain "purposiveness without purpose." He argues that

> [P]lay has a special relation to what is serious. It is not only that the latter gives it its "purpose": we play "for the sake of recreation," as Aristotle says. More important, play itself contains its

own, even sacred, seriousness. Yet, in playing, all those purposive relations that determine active and caring existence have not simply disappeared, but are curiously suspended (p. 102).

However, in this text, Gadamer still envisions the "meaning" of art partly in terms of a certain content; he criticizes the Schillerian tradition by saying that "the idea of aesthetic cultivation – as we derived it from Schiller – consists precisely in precluding any criterion of content and in dissociating the work of art from its world" (p. 85).

When Gadamer subsequently confronts the problem of abstract art in "The Relevance of the Beautiful," he partially reconceives the significance of the work of art and proceeds to endorse more enthusiastically the Kantian conception of the encounter with the aesthetic object. After alluding to the idea that "the problem is that we cannot talk about great art as simply belonging to the past, any more than we can talk about modern art only becoming 'pure' art through the rejection of significant content,"[7] Gadamer writes of the "meaning" of the work of art that "Kant did not want to define this additional something as a content. And indeed, as we shall see, there are good reasons why it is actually impossible to do so" (pp. 20–21). Gadamer's enterprise, then, derives the structure of "play" from the post-Kantian tradition, although he transforms the concept by elaborating on it in ontological terms. Here it emerges that the ontological enterprise serves quite disparate goals in the texts of Gadamer and Ingarden: while Ingarden focuses more on the specific components of the work of art, Gadamer purports to direct his attention toward the question of structure adapted from Kant. It will be impossible to evaluate Gadamer's success in dissociating himself from the regime of content, however, until we have examined how Ingarden and Gadamer ontologically classify the work of art.

Although both Gadamer and Ingarden insist on examining the work of art from the standpoint of ontology, it is crucial to each theorist's project that he does not deem the aesthetic ontologically autonomous – or, at least, in reinstating ontological autonomy he ineluctably alters the meaning of that phrase.[8] Two aspects of this claim must be considered. As theories of art deviating from the tradition of philosophical aesthetics, Gadamer's and Ingarden's texts attempt to countervene the autonomy of an aesthetic sphere set off as a separate area of philosophy, necessary for the unification of the system, yet relegated to an independent study. However, the two thinkers still employ the linguistic work of art to solve certain philosophical dichotomies, particularly those related to cognition. This approach entails the possible consequence – one that is indeed present in *The Literary Work of Art* and *Truth and Method* – that the artistic object itself might no longer retain an essentially independent existence; in order to account for the presence of a "meaning" within the work of art distinct from that obtainable within the empirical sciences or logic, some mediation must be possible so that an individual can constitute or construe it.[9]

In *The Fate of Art; Aesthetic Alienation from Kant to Derrida and Adorno*, J. M. Bernstein provides an illuminating account of the historical reasons for Ingarden and Gadamer to reject the autonomy of the aesthetic.[10] Accurately describing theories of art related to theirs as post-aesthetic, he defines them through their negation of the tradition following Kant. Bernstein asserts that,

> If "aesthetics" in its narrow sense refers to the understanding of art as an object of taste outside truth and morality, then "post-aesthetic" theories of art are themselves critiques of truth-only cognition insofar as their going beyond aesthetics implies a denial of the rigid distinctions separating the claims of taste from the claims of knowing or right action.... According to post-aesthetic theories, art works must be understood in nonaesthetic terms because the very idea of aesthetics is based upon a series of exclusions which themselves assume a conception of truth in terms of its isolation from normative and "aesthetic" values....[11]

According to Bernstein, Kant, in establishing the autonomy of the aesthetic sphere, concomitantly created the conditions necessary for divorcing it from the cognitive realm, and, hence, helped to institute a distinct separation between art and meaning. Post-aesthetic theories then attempt to remedy this situation by reintroducing signification into the consideration of the artwork. Gadamer substantiates Bernstein's argument in "The Relevance of the Beautiful," where he first attributes to Kant the development of aesthetic autonomy, then denies the validity of Kant's claim. Initially maintaining that "Kant, who worked out most clearly the autonomy of aesthetics, was primarily oriented toward natural beauty" (p. 30), Gadamer proceeds to claim that "a deeper analysis of this aesthetic experience of natural beauty teaches us that, in a certain sense, this is an illusion and that in fact we can only see nature with the eyes of men experienced and educated in art" (p. 30). These statements demonstrate the connections tying the first connotation of "aesthetic autonomy" with the second; separating the aesthetic sphere from the cognitive in Kant also entailed – at least according to Gadamer's interpretation – denying the heteronomous construction of the artistic object.

This second connotation, that of the independence of the linguistic work of art itself, Ingarden rejects much more definitively in *The Literary Work of Art* than does Gadamer in *Truth and Method*. As Anna-Teresa Tymieniecka writes, echoing Gadamer's comment about Kant, "It is well known that Ingarden's primary interest in aesthetics – as he himself has repeatedly stated

– was to clarify by the comparison and opposition of the nature of the work of Art as a purely intentional artifact with the work of nature itself, the existential status of both."[12] Since the philosophical urgency of Ingarden's enterprise lies in his attempt to resolve the relationship between "real" and "ideal" objects, he employs the literary work to mediate between them. Ingarden insists on the ontic autonomy of these elements, an attribute that they would seem to share with the purely empirical and logical sciences in which Gadamer attempts to intervene; as he explains, "We are speaking here of real and ideal objects only as of something which in itself is ontically autonomous and at the same time ontically independent of any cognitive act directed at it" (p. 10). By considering texts as "intentional objects" that are ontically heteronomous yet depend for their existence on both real and ideal entities, co-existing in a modified form, Ingarden bridges the gulf. As he himself describes this intermediate term,

> By a purely intentional objectivity we understand an objectivity that is in a figurative sense "created" by an act of consciousness or a manifold of acts or, finally, by a formation (e.g., a word meaning, a sentence) exclusively on the basis of an immanent, original, or only conferred intentionality and has, in the given objectivities, the source of its existence and its total essence (p. 117).

Thus the "intentional objectivity" that can be identified as the literary work of art is ontologically constituted by the transmogrification of certain "objectivities" – both real and ideal – through an "intentional" act.

The concatenation of real and ideal elements within the purely intentional text is effected in Ingarden's view by their participation in the various "strata" of the literary work. Through his construal of these levels, Ingarden revises the conventionally conceived structure of "representation," demonstrating the imbrication and absolute transformation of real and ideal objectivities within the fictional composition. Although the strata may involve both real and ideal elements, these must be "concretized" by the reader before they attain completion. To summarize briefly, the stratum of meaning units partakes to some degree in the nature of an ideal objectivity. Ingarden sets out this situation in his "Initial Problems":

> Thus, this attempt to save the unity and identity of the literary work also fails. There remains only one way out of this difficult situation, namely, to admit the existence of *ideal* meaning units and yet not incorporate them into the literary work – so as to avoid the difficulties presented above – but invoke their aid only for the purpose of securing the identity and unity of the literary work (pp. 18–19).

After his investigation has almost concluded, Ingarden provides a synopsis of the procedure whereby what he has described occurs; although the meaning of a word can be identified as an ideal object, the restrictions that a sentence places upon it actualize it in a purely intentional ontic guise (pp. 361–364). Ingarden describes the function of the "ideal meaning units" as operating to enforce the "intersubjective identity" of the literary work of art (p. 364).

If one were to fault Ingarden's account at any particular point, his hypothesis about "ideal meaning units" could provide an appropriate point of ingress, but such criticism would extend beyond the scope of this treatment. It must be noted, however, that, like Gadamer, Ingarden does not merely attempt to establish the identity of meaning units, but the coherence of all "correct" concretizations of the literary work, which, although they might incorporate divergent features at points of indeterminacy, must somehow be seen as "the same." As we will see, one can compare Ingarden's claim that the unity and identity of the literary work is established through the constancy of these "ideal meaning units" with Gadamer's quite disparate assertions about what ensures the identity of a representation amidst its reproductions. Since he is also conducting an ontological investigation directed towards the essential nature of the linguistic work of art, he too must provide an account of this object's persistent "identity."

Returning to Ingarden and his description of the "stratum of meaning units," we can glean that this level serves to depict "states of affairs," and through these, the "stratum of schematized aspects"; both constitute the attributes possessed by the "stratum of represented objectivities," a stratum the reader comprehends only after synthesizing the particularities that the schematized aspects supply. Through this model of representation, Ingarden undoes the conventional understanding of mimesis, by which "real objects"are directly depicted within the literary work of art. Since the "stratum of represented objectivities" cannot be considered a mere reflection of the existing world, but is instead constructed out of "states of affairs," the Platonic model of imitation is displaced. Ingarden attributes the conventional account of mimesis to an artistic illusion, claiming that "it is no accident that our *gaze* is usually directed straight at the represented objects, scarcely touching the corresponding states of affairs. For it is in the nature of that which represents as such, that in the performance of the representation function it disappears to a certain degree from our *field of vision* in order to facilitate, above all, the illumination of what it represents" (p. 191; my italics). The visual imagery that Ingarden employs here is not unimportant, as it emphasizes the relationship between perception and both "states of affairs" and "represented objectivities," a connection that Gadamer himself supports in "Poetry and Mimesis," where he laments that "it is of course true that the ancient doctrine of imitation dominates poetic theory above all. Yet it seemed to justify itself most convincingly in the visual arts, for it is here that all talk of image and original most forcibly suggests itself...."[13] The link is additionally clarified when Ingarden subsequently discusses the misapprehensions of "the average reader" in terms of the "represented objectivities," arguing that,

> [T]he average reader, thanks to the natural functions of the meaning intentions understood and effected by him in the course of his reading, is interested only in the material makeup of the *contents* of the objects that come into question. As a result, the structures and features of real objects are transferred without further ado, and as a matter of course, onto represented objects, with the peculiarities of the latter being *overlooked*. In order to bring these into *full light*, a different, investigatory, and not *aesthetically perceptual* attitude is required (pp. 217–218; my italics).

This statement, which appears to describe the effects of illusionistic creation, suggests that the reader tends to forget the mechanics by which the appearance of actuality is produced. These are the invisible supports that Ingarden highlights with his analytic pen.

The disparity that Ingarden identifies between a conventionally conceived depiction of objects and his own account of representation contrasts distinctly with that of Gadamer and bears significant implications for his evaluation of the stage play. When discussing the nature of aesthetic concretizations, Ingarden asserts that "the most radical difference between a literary work and its concretization appears in the aspect stratum" (p. 339). This impinges upon the relationship between reading and theatrical representation; whereas a reader first concretizes these "schematized aspects" and then attains a mental projection of the "represented objectivities," the same "represented objectivities" are materially and perceptually[14] given for the spectator at a dramatic performance. It is this interference of the "real" – and at the same time the perceptual – in the experience of the play that alters the intentional structure of the literary work as Ingarden has conceived it and thereby relegates it to a "borderline case." As Ingarden opines,

> the stage play differs from a *purely* literary work... in that entirely new means of representation, precluded by the essential nature of a purely literary work, appear in it: 1) real objects engaged in performing the function of reproduction and representation and 2) aspects appropriately formed and predetermined by the properties of these real objects, in which represented objectivities are to appear... (p. 320).

For Ingarden, the problem appears to be that in the dramatic production one can no longer divorce "schematized aspects" from those "represented objectivities" they help to construct; instead of allowing the ontological reality of empirical entities to be given through a partially ideal formation of sentences, thereby reconciling ideal and real through the intentional structure of literature, the stage play supplies tangible actualities in their immediacy. It is these "realities," although somewhat reconceived, and the spectator's confrontation with them, that, on the other hand, underlie Gadamer's conception of the work of art. Not distinguishing between the reconstituted concretizations of the reader and the apperceptions of the viewer proves central for his project. This fact becomes particularly evident when one examines how the two theorists treat represented individuals; whereas Ingarden considers the relationship between two interlocutors on stage quite different from the actor's link with the audience member and opines that the former pair will display much greater emotional involvement, Gadamer sees the represented person directly – ontically – confronting the individual spectator. This subject can only be treated adequately when we compare Gadamer's account of the meaning produced by the work of art with that of Ingarden.

Another – irremediable – difficulty suffusing Ingarden's conception of the linguistic work of art, however, underlies this one, a point that the nature of "real" objects within the dramatic work can illuminate. In his desire to constitute the literary work in terms of real and ideal objectivities, Ingarden places stress on the aspects of the real entity; when considering actual stage props within dramatic texts, it becomes clear that many function not in their particular empirical capacity but in various symbolic guises. For example, the mound in which Winnie finds herself ensconced in Samuel Beckett's *Happy Days* points away from its physical qualities towards what it signifies about Winnie's own state. This supposedly "real" element of the dramatic production then appears to assume a role no more ontically independent than that of the words themselves; since it does not operate naturalistically within the work, it casts larger doubts upon the status of "represented objectivities" throughout even the purely literary text. Indeed, the intensity of Ingarden's focus on the stratum of schematized aspects and that of represented objectivities could distort the nature of the linguistic work of art. Although Ingarden does identify certain "metaphysical qualities" that the object stratum entails, these – as we will see – do not modify the "real" nature of the entities in question but instead encompass those lineaments of the literary work that after Northrop Frye we would call "mode," and that, as Ingarden himself writes, constitute "*situations or events*" (p. 291). Gadamer does not necessarily supply a more satisfying solution, however, and even though he does introduce the concept of the "symbol" into his consideration of artworks, the sense in which he employs this word diverges from its more technical literary connotations.

In order to comprehend the discrepancies between Ingarden's and Gadamer's conceptions of the identity and ontological determination of the artwork, let us examine more closely the divergent relations among schematized aspects, represented objectivities, and the experience of the recipient in Ingarden's accounts of the literary and dramatic works. According to *The Literary Work of Art*, the individual who produces a play, rather than the onlookers themselves, performs the task of concretization. For the reader, the experience is constituted through his own activity of concretization, whereas the viewer merely absorbs a post-concretized product. Ingarden, however, is unwilling even to allow that the "stage play" and the text to which it somehow corresponds can be considered two aspects of the same entity. In other words, Ingarden disputes the ontological identity of the written drama and the manifold of acted versions, asserting that, "in view of what we have said, it would be a mistake to claim that the stage play is – as we ourselves, following common practice, expressed it at the outset – a *realization* of a corresponding purely literary work" (p. 321). Indeed, he would ultimately contend that productions do not concretize the written text but merely a hypothetical model of the performance, finally declaring that "there appear in the stage play the structural differences we have discussed, which make it a *new* work with respect to the corresponding purely literary work" (p. 322).

Ingarden's perspective differs significantly from that of Gadamer, who localizes the identity of the dramatic text in its repeated presentation and extrapolates from this back to the literary work of art. As he writes in "The Relevance of the Beautiful": "The artistic experience is constituted precisely by the fact that we do not distinguish between the particular way the work is realized and the identity of the work itself" (p. 29). This statement echoes Gadamer's elaboration of the nature of play, presentation, and the picture within *Truth and Method*, a discussion perhaps best encapsulated by the assertion that "starting from the universal significance of play, we saw that the ontological significance of representation lies in the fact that 'reproduction' is the original mode of being of the original artwork itself" (p. 159). Gadamer's account of "reading" itself depends upon this understanding of the artwork. Initially, he hypothesizes that reading might fall outside the purview of "presentation," since, as he phrases it, "Literature, the written word, seems to be poetry alienated from its ontological valence" (p. 160), or, as Ingarden states the problem, the literary work of art appears ontologically heteronomous. Soon, however, Gadamer asserts that "reading with understanding is always a kind of reproduction, performance, and interpretation.... Thus the reading of a book would still remain an event in which the content comes to presentation" (p. 160). Whereas Ingarden secures the identity of the literary work through the ideality of word meanings and the possibility of an accurate concretization, Gadamer claims that it is within each event of "presentation" that the text achieves ontological existence and becomes an identifiable entity.

This divergence from Ingarden occurs because Gadamer does not reject the structure of mimesis, but instead reclaims it for his purposes. The nature of Gadamer's own peculiar philosophical problem occasions this outcome; rather than – like Ingarden – postulating real and ideal entities that can be unified through the literary work of art, Gadamer questions the value of the purely empirical from the standpoint of "truth," which he attempts to redefine in terms of being. Comparing the "real" of the world and of the dramatic situation, Gadamer maintains that

> "Reality" always stands in a horizon of desired or feared or, at any rate, still undecided future possibilities. Hence it is always the case that mutually exclusive expectations are aroused, not all of which can be fulfilled. The undecidedness of the future permits such a superfluity of expectations that reality necessarily lags behind them. Now if, in a particular case, a context of meaning closes and completes itself in reality, such that no lines of meaning scatter in the void, then this reality is itself like a drama. Likewise, someone who can see the whole of reality as a closed circle of meaning in which everything is fulfilled will speak of the comedy and tragedy of life. In these cases, where reality is understood as a play, emerges the reality of play, which we call the play of art (*Truth and Method*, pp. 112–113).

While Ingarden accepted the "reality" of objects outside the purview of the text, then proceeded to demonstrate why the literary text does not simply mirror these, Gadamer asserts that mimesis indeed occurs, but the value of the work of art's supposed "imitation" actually exceeds that of the original. "Real" entities no longer possess an independent existence for him, but are instead endowed with being through the process of "play" and the structure of "presentation" in an experience that no longer "belongs" to the subject, according to Gadamer, and in fact transcends the subject-object dichotomy.

Although the seemingly immutable appearance of visual art might be taken to oppose Gadamer's assertions about play, he assimilates the picture to an ontological status akin to that of drama. As Gadamer writes of the picture, "Every such presentation is an ontological event and occupies the same ontological level as what is represented. By being presented it experiences, as it were, an *increase in being*. The content of the picture itself is ontologically defined as an emanation of the original" (*Truth and Method*, p. 140). The metaphor of the object's "experience" is revealing in this instance, since it suggests a personification of the work of art. The way that Gadamer depicts the visual image would indeed support such a claim; he describes as revealing the fact that the ancient Greeks designated the picture by the same word as a "living being," then proceeds to attribute elements of an individual's existence to the ontic essence of the artwork. In an essay on "Image and Gesture," Gadamer first discusses ancient drama, then inquires, "What is the position of contemporary art with respect to this truth which has lost none of its validity?"[15] He responds with another question, wondering, "How can the anthropomorphic Greek religion of art return in the context of contemporary art? It certainly cannot produce recognition of the familiar, least of all the recognition of familiar figures.... The only thing that is universally familiar to us today is unfamiliarity itself.... But how can we express that in human form?" (p. 79). Although post-representational art can no longer be assimilated to conventional structures of mimesis, Gadamer still attempts to view it in human terms. Even in this context, truth remains, for Gadamer, a manifestation through performance by which the viewer or spectator is ineluctably altered; in other words, Gadamer reclaims the scene of *anagnoresis* or "recognition" from ancient tragedy and writes it into the relationship between a visual as well as a dramatic presentation and its recipient.

Gerald Bruns, partially assimilating the lessons of *Truth and Method* to Stanley Cavell's views on tragic "acknowledgment," emphasizes this "tragic" aspect of Gadamerian hermeneutics in his essay "On the Tragedy of Hermeneutical Experience,"[16] but examples already proliferate within Gadamer's own texts. Not merely employing the term "recognition" in a metaphorical sense, Gadamer often first places it back in its ancient Greek context, discussing the ramifications of "recognition" through the operations of the "symbol" before extrapolating the word's significance to the interaction between observer and aesthetic entity. Most notably, in "The Relevance of the Beautiful," Gadamer speaks of the "*tessera hospitalis*" – "token of hospitality" – that consisted of two halves broken then subsequently reunited, then invokes Aristophanes' discussion of the "*symbolon tou anthropou*" – "token of man" – in Plato's *Symposium*. The operations of the "*tessera hospitalis*" are outlined in numerous dramatic scenes, but Gadamer deems this type of symbol detrimentally affected by the fact that it can only function for purposes of recognition within a pre-established context; as he insists in *Truth and Method*, the picture differs from the symbol since it cannot depend on any previous act of "institution" to be "recognized" (pp. 153–5). The "*symbolon tou anthropou*," however, a phrase with which the character Aristophanes, recounting an allegory of love, designates the half of a being left wandering in the world after the whole "*anthropos*" has been sundered, increasingly approximates Gadamer's understanding of the artwork. In this case, "being" itself, in its existential totality, results from reuniting the symbol (including the dramatic production) with its other half (the audience member). Gadamer presents this situation in "The Relevance of the Beautiful," stating that

> [F]or our experience of the symbolic in general, the particular represents itself as a fragment of being that promises to complete and make whole whatever corresponds to it. Or, indeed, the symbol is that other fragment that has always been sought in order to complete and make whole our own fragmentary life. The "meaning" of art in this sense does not seem to me to be tied to special social conditions.... On the contrary, the experience of the beautiful, and particularly the beautiful in art, is the invocation of a potentially whole and holy order of things, wherever it may be found.... (p. 32).

"Being" here manifests the anthropological valence that extends throughout Gadamer's discussions of contemporary art. Although the classical concept of the symbol undergoes significant modifications in Gadamer's account, the dynamics of recognition persist. As his discussions demonstrate, the recognition that someone like Electra in Aeschylus' *Libation Bearers* experiences when encountering Orestes, the symbol-bearing individual, does not differ qualitatively from that which characterizes the spectator's relation to play. Thus, rather than placing "reality" outside the scope of investigation into the linguistic work of art, Gadamer opines that the presentation itself consistently embodies a certain "reality."

As this configuration of subject and object already begins to suggest, Gadamer does not envision the work of art as a purely independent entity, according to the customary conception of aesthetic autonomy, but he also cannot concur with Ingarden's views on the ontic heteronomy of the literary text. Indeed, Ingarden's doctrines appear to adhere excessively to prescribed notions of what actually constitutes existential autonomy, whereas Gadamer intends to revise this very idea. For Gadamer as well as Ingarden, the work requires some activation, but it is the advantage of "play" for the former philosopher that it "presents" something in its immediacy, allowing the recipient to "belong" to the existence of the work, as she cannot for Ingarden. The subject of one of Ingarden's preliminary sections in *The Literary Work of Art* is posed in terms of the question, "What does not belong to the literary work?" (p. 22). Although he only discusses psychologistic or epistemological misunderstandings of the text here, Ingarden dismisses the idea that either the author's or reader's experiences, or even "the whole sphere of *objects* and *states of affairs* which constitutes... the model of the objects and states of affairs 'appearing' in the work..." (p. 25), would "belong" to the written composition. Gadamer instead reclaims the viewer for the play itself, writing that "in being played the play speaks to the spectator through its presentation; and it does so in such a way that, despite the distance between it and himself, the spectator still belongs to play" (*Truth and Method*, p. 116). The nature of this "distance" must be elaborated at greater length, but Gadamer also indicates his desire not to fall victim to the subjectivizing impulse that Ingarden attributes to those who would characterize aspects of the recipient as "belonging" to the linguistic work of art; again in *Truth and Method*, Gadamer asserts: "We started by saying that the true being of the spectator, who belongs to the play of art, cannot be adequately understood in terms of subjectivity, as a way that aesthetic consciousness conducts itself. But this does not mean that the nature of the spectator cannot be described in terms of being present at something..." (p. 125). The spectator's – necessary – "presence" and receptivity designate him as a participant; through this ontological description of the dramatic presentation, play has undergone the requisite "transformation into structure" (pp. 116–117), turning a seemingly endless contingency and alteration into a type of constancy.

This "transformation into structure" has interesting consequences, since it partially eliminates the difference between the seemingly static nature of Ingarden's approach, based on the realization of objects, and the apparent dynamism of an aesthetic sphere based on "play." Gadamer commences by claiming that "the movement of play as such has, as it were, no substrate. It is the game that is played – it is irrelevant whether or not there is a subject who plays it. The play is the occurrence of the movement as such" (p. 103). He proceeds partially to negate this assertion first by articulating the belief that play actually consists in "self-presentation," then by inscribing a type of constancy in the unalterable "presence" of the work of art despite change. While it seems initially that the confrontation between the spectacle and the beholder actualizes a certain dynamism, the persistence of presence is demonstrated by the priority that visual images gradually assume.[17]

The results of Gadamer's ontological investigation into the particular type of "play" inherent to the artwork can be summarized as redefining the concept of autonomy, then attributing this quality to the dramatic or visual creation. As Gadamer states his conclusions:

> The fact that aesthetic being depends on being presented, then, does not imply some deficiency, some lack of autonomous meaning. Rather, it belongs to its very essence. The spectator is an

essential element in the kind of play we call aesthetic. I want to turn now to the famous definition of tragedy in Aristotle's *Poetics*. There the spectator's frame of mind figures expressly in the definition of tragedy's essential nature (*Truth and Method*, pp. 128–129).

Through thus invoking "tragedy" – as Fried appealed to "theatricality" – Gadamer avails himself of the critical tradition that, since Aristotle, has focused most on the relationship of the recipient to the work of art. Indeed, although Gadamer appeals to an initially Platonic version of mimesis, he attempts to reclaim the worth of the linguistic entity by adding Aristotle to his account. Gadamer's debt to Aristotle, and Ingarden's critical stance towards the Greek philosopher's views, are manifested quite lucidly in their divergent estimations of "distance," a concept that appears within both writers' texts.

The first point in *Truth and Method* at which Gadamer notes his qualified approval of Ingarden,[18] and concomitant deviations from the other's opinions, can provide a context for this discussion. As a footnote to the remark that "one fails to appreciate the obligatoriness of the work of art if one regards the variations possible in the presentation as free and arbitrary. In fact they are all subject to the supreme criterion of 'right' representation" (p. 118), Gadamer writes that,

Although I think his analyses on the "schematism" of the literary work of art have been too little noted, I cannot agree when Roman Ingarden… sees in the process of the concretization of an 'aesthetic object' the area of the aesthetic evaluation of the work of art. The aesthetic object is not constituted in the aesthetic experience of grasping it, but the work of art itself is experienced in its aesthetic quality through the process of its concretization and creation. In this I agree fully with Luigi Pareyson's aesthetic of "*formativita*."

Here we finally see elaborated the concrete differences that Ingarden's and Gadamer's conceptions of ontological autonomy entail. Whereas Gadamer conceives of the work of art as existing when realized through "concretization and creation," a process that unifies the work of art with the resultant aesthetic object,[19] Ingarden maintains that an inescapable "distance" divides the spectator from the dramatic occurrence, a separation occasioned by the ontological disparity between the recipient's mode of existence and that of the artwork.

Establishing this "distance" becomes intimately related to Ingarden's understanding of the stage play, and forces him to provide an extremely subdued interpretation of Aristotelian *catharsis*. As George Grabowicz, the translator of *The Literary Work of Art*, suggests in his "Introduction," supporting the relationship between any analysis of "distance" and drama:

A centrally important theoretical issue that Ingarden touches on is that of distance…. as a structural feature of quasi-judgments and the quasi-reality of the represented world…. For

> Ingarden, as for the above critics, the test case and most interesting proving ground is drama.... In each case they provide a carrier for distance as an aesthetic principle – a conscious, intersubjective bracketing of experience (lx-lxi).

Linking the ontological heteronomy of the work of art with the stress Ingarden places on distance, Grabowicz at the same time undercuts the idea of experience, one that Gadamer still endorses after transfiguring the term's received meaning. When Ingarden actually mentions the subject of "distance" during his discussion of the "metaphysical qualities" conjured by the stratum of represented objectivities, he references Aristotelian *catharsis*, but interprets the "*catharsis* of pity and fear" that Aristotle describes as a removed aesthetic consideration rather than a more intimate involvement of the spectator himself with the production. Ingarden writes first of this "distance" that it is not a distance of cognitive apprehension, but rather that "the 'distance' of which we speak here rests only on the unique phenomenon of 'not belonging to the same world' and brings with it the impossibility of genuine participation in the represented situation, a genuine transposition from our life-situation into the one represented in the literary work" (p. 295). Such an assertion directly contravenes Gadamer's comments on the viewer's "participation" in the dramatic scene.

Ingarden insists on this "distance" in his appendix on "The Functions of Language in the Theater" as well, attributing to it the difference between the reception of a character by another represented person and by an audience member. On his account, the "pity and fear" depicted within the play can be experienced only by fictional characters, and must be transformed into the binary pleasure/displeasure when they enter the universe of the spectator. Although in other respects he deviates from the post-Kantian tradition of aesthetics, Ingarden appears to have internalized the priority placed on the pleasure and displeasure related to the play of the faculties. Since certain types of drama cannot be assimilated to such a model of disinterest, he must eliminate these from his purview, not only giving priority to the textual work but additionally excluding plays that cannot conform to his theory. Not surprisingly, "mystery plays" (p. 394) are the target of Ingarden's exception, the same type of drama that Gadamer exalts in his essay "The Festive Character of Theater." Likewise, Ingarden attempts to secure his argument through appeal to plays with non-naturalistic dialogue, although his primary object, as we have seen, must be the illusionist novel or drama; Ingarden insists that productions featuring poetic speech on the part of the characters make apparent the purely representational nature of the spectacle.

In order to contrast Gadamer's understanding of Aristotle with Ingarden's, let us examine Ingarden's specific statement about *catharsis;* as he argues,

> In closest connection with this aesthetic manner of observing metaphysical qualities is what Aristotle had in mind when he spoke of "catharsis." Viewing them in an aesthetic attitude not only fills us with pleasure and bliss but also gives us that specific relief which we experience after all difficult events requiring the exertion of all our powers. It appears that precisely this relief and inner calm after an aesthetic apprehension of a metaphysical quality is what Aristotle meant by catharsis (p. 295).

Gadamer, of course, interprets *catharsis* as an entirely different phenomenon. Gerald Bruns' characterization of his position seems to possess at least some validity. As Bruns – creating another link between Aristotle and the idea of distance – reminds us in "On the Tragedy of Hermeneutical Experience": "Gadamer said in *Truth and Method...* that tragedy (on Aristotle's analysis) appropriates its audience and exposes it to the tragic experience. I mean the idea that the tragic audience is not kept at a distance by tragic representation but is, on the contrary, overwhelmed and transformed by it" (p. 81). Bruns proceeds to characterize "hermeneutical experience" as participatory rather than aesthetic; as he writes, "Cavell says that my relationship to the characters on the tragic stage is not (or not just) literary or aesthetic, much less cognitive.... I am outside the distance between the stage and my self-possession, facing the characters not so much as if they were real but as if my separateness from them no longer derives from their ontological peculiarity as aesthetic objects of fictional representations" (p. 85). However, Bruns' account does not exhaust Gadamer's comments on "distance," since the author of *Truth and Method*, like the "literalist" artists who provide a subject for Michael Fried, endorses some types of "distance." Even in "The Festive Character of Theater," Gadamer maintains that "theater is a product of Greece, both in name and nature. It is in essence a play produced to be looked upon, and the unification that it effects – in which we are all onlookers of one and the same event – is a unification at a distance" (p. 58). How can we reconcile the abnegation of "distance" throughout Gadamer's *oeuvre* with this partial acknowledgment of its existence? The answer to this comes in Gadamer's insistence on structure, and his attempt to escape a naive view of experience. Instead of exalting a complete internalization of the spectacle through the viewer's subjectivity, Gadamer insists on the audience as a group, and on the ontological rather than personal nature of the unity he describes.

Given Gadamer's and Ingarden's accounts of the linguistic work of art, one might draw one further historical extrapolation. The current – practical –

split between the theatrical and narrative modes is paralleled by a theoretical division between performative and representational models of the artwork. One could, given the priority Gadamer places on the experience of the dramatic performance, combined with Ingarden's insistence on a peculiarly mediated depiction of the real within the narrative form, speculate that the two theorists provide examples of this second distinction in its nascent state. Although this split between performative and representational derives from certain ontological "solutions" to philosophical problems, like those "discovered" by Ingarden and Gadamer, it has found itself generally mapped onto narrative and dramatic texts. If, aside from evaluating the theoretical distinction's validity, such an alignment were entirely accurate, it would serve both types of artworks quite well. However, proceeding from an *a priori* definition of what constitutes certain genres, it instead ignores the manifold aberrations within each mode and imposes extraneous criteria on the linguistic entity – precisely what Gadamer and Ingarden intended to avoid.

NOTES

1 Jadwiga S. Smith, "A Theory of Drama and Theatre: A Continuing Investigation of the Aesthetics of Roman Ingarden," in *Ingardiana III: Roman Ingarden's Aesthetics in a New Key and the Independent Approaches of Others: The Performing Arts, the Fine Arts, and Literature* (Analecta Husserliana 33 [1991]: pp. 3–62), p. 7. I would like to thank Wolfgang Iser, Joan Meyler, and Ben Heller for their encouragement and assistance at various stages in the composition of this essay.

2 Mihai Spariosu, *Dionysus Reborn: Play and the Aesthetic Dimension in Modern Philosophical and Scientific Discourse* (Ithaca: Cornell UP, 1989), p. 133.

3 Hans-Georg Gadamer, *Truth and Method*, revised edition, trans. Joel Weinsheimer and Donald G. Marshall (New York: Continuum, 1994), pp. 7–8.

4 In this context, I will employ the phrase "linguistic work of art" to encompass both the "literary work of art" discussed by Ingarden and the non-textual "play" to which Gadamer alludes.

5 Immanuel Kant, *The Critique of Judgment*, trans. J. H. Bernard (New York: Hafner Press, 1951), p. 5.

6 Roman Ingarden, *The Literary Work of Art*, trans. George Grabowicz (Evanston: Northwestern UP, 1973), p. 4.

7 Hans-Georg Gadamer, "The Relevance of the Beautiful" in *The Relevance of the Beautiful and Other Essays*, ed. Robert Bernasconi (Cambridge: Cambridge UP, 1986, pp. 3–53), p. 19.

8 As we will see, Gadamer frequently insists on the "ontological autonomy" of a "presentation," but in doing so redefines the scope of the work of art itself, insisting that the entire situation belongs to it.

9 This statement reveals one important consequence of considering the linguistic work of art from an aesthetic rather than a textual perspective. When we approach the verbally determined object as one manifestation of an aesthetic sphere that combines various arts, we must consider something other than linguistic meaning central to it, whereas a textualist would view scientific,

literary, and other texts as equally determined by the structure of language. Both Gadamer and Ingarden appear to be situated at an intermediate point in the historical transition from aesthetic to textual modes of examination. Ingarden quite evidently does place words at the forefront of his investigation, but he still prioritizes the "aesthetic" qualities of the literary work of art, while Gadamer examines "plays" and other linguistically based forms along with visual works.

10 One might note that only Gadamer explicitly situates himself within the tradition of the aesthetic in order to reject its separation from the philosophical search for truth. Ingarden's work in one sense continues on past this point, having already assumed the end of an independent aesthetics; however, he also still accepts without question the validity of certain distinctions between the aesthetic and the cognitive, endorsing, for example, the division between the literary and the scientific text (pp. 328–330). Gadamer's apparent belatedness in this respect may correspond to some extent to his own feeling, expressed in the "Afterword" to *Truth and Method*, that, "When I finished the present book at the end of 1959, I wondered whether it had not come 'too late' – that is, whether its attempt to reassess the value of traditional and historical thought was not by then almost superfluous" (p. 551).

11 J. M. Bernstein, *The Fate of Art: Aesthetic Alienation from Kant to Derrida and Adorno* (University Park: Penn. State UP, 1992), pp. 3–4. It is well worth noting that Gadamer, like Bernstein, believes that Kant began an invidious trajectory with his *Critique of Judgment* but in many respects does not consider Kant's own statements to be incorrect. Rather, Gadamer sees those who interpreted and extended Kant as deviating from his own intentions.

12 Anna-Teresa Tymieniecka, "Beyond Ingarden's Controversy with Husserl – The New Contextual Phase of Phenomenology" in *Ingardeniana: A Spectrum of Specialized Studies Establishing the Field of Research* (Analecta Husserliana IV [1976]: pp. 241–418), p. 310.

13 Hans-Georg Gadamer, "Poetry and Mimesis," in *The Relevance of the Beautiful and Other Essays* (pp. 116–122), p. 116.

14 Several times, Ingarden, claiming that he cannot develop his comments, footnotes the term "perceptual" in this context, specifying that it is not precisely accurate; as he writes, "The word 'perceptual' cannot be used here without a certain essential reservation, and it should be understood only as a conventional shorthand expression. I do not have the space to elaborate on these issues, however" (p. 379). Earlier he had provided a slightly more precise formulation of the problem, explaining that "[F]or the result of these functions, i.e. the represented objects, to be, apprehended in a concrete performance, a 'spectator' must be present and must effect special apprehending experiences. These experiences do not, of course, belong either to the play itself or to its concretizations (performances). It must be noted that these experiences are not true perceptions, even though, with respect to the nature of their intuitiveness, they are similar to perception" (p. 319). This latter remark sheds light on the disparity Ingarden observes between the experience of the spectator and the activity of the reader. One reason Ingarden believes the "experience" of the viewer should be distinguished from perception may be that he realizes that the roles of spectator and reader are not completely distinct. Partially endorsing the opinion of L. Blaustein, who maintains that "these apprehending experiences" might be "imaginational representations" (p. 320), Ingarden classifies them outside any complete passivity associated with pure perception. Blaustein's renaming likewise serves to bring the "apprehending experiences" of the spectator more in line with the cognitive activity involved in reading.

15 Hans-Georg Gadamer, "Image and Gesture," in *The Relevance of the Beautiful and Other Essays* (pp. 74–82), p. 78.

16 Gerald Bruns, "On the Tragedy of Hermeneutical Experience," in *Rhetoric and Hermeneutics in Our Time: A Reader*, ed. Walter Jost and Michael J. Hyde (New Haven: Yale UP, 1997, pp. 73–89). Bruns's article is particularly interesting when one considers the links between

Gadamer – and those who have adopted his perspective – and Kant. Although he only once explicitly names Kant, Bruns incessantly invokes ideas developed by the critical philosopher. For example, he writes of "that which resists conceptual framing and leaves one dumbstruck with its evasions" (pp. 73–74), echoing the Kantian assertion that the beautiful cannot be subsumed under a "concept." The more commonly promulgated notion that the sublime inhibits one's power of speech also seems to underlie Bruns' assertion. The Kantian version of the sublime – in its involvement with "limits" – is more distinctly present, however, in his statement that, in hermeneutics, "one *always* confronts the limits – in Gadamer's language, the finitude or historicality, the situatedness – of understanding itself" (p. 74). Finally, in designating "the other" as absolutely other, Bruns shifts the problems with the concept that Kant expressed in the aesthetic sphere to the realm of the ethical, maintaining that "what happens in understanding is that I always experience the refusal of the other to be contained in the conceptual apparatus that I have prepared for it" (p. 74).

[17] At this point, comparing Gadamer's concept of "play" with the idea of "theatricality" that Michael Fried presents in "Art and Objecthood" becomes unavoidable. Published in *Artforum* in July 1967 (vol. 5, no. 10: pp. 12–23), Fried's article describes the same phenomena as Gadamer at numerous points, although it adopts a negative attitude towards all that Gadamer endorses. Before comparing Fried's perspective with the one we have already elaborated, let us introduce another comment that Gadamer articulated on contemporary art, the subject of Fried's piece. In "The Speechless Image," Gadamer expresses doubts about the continual possibility of the "picture," writing: "It is indeed the case that the very concept of the picture that was characteristic of the traditional museum has now become too restrictive. The creative artist has eliminated the frame, and the articulation of surface constitutive of the picture points beyond itself into other contexts. It used to be said as a criticism of a picture that it was too decorative, but this is slowly losing its pejorative meaning" (p. 89). Since Gadamer had already attempted to reclaim the decorative in *Truth and Method*, this statement does not represent a substantial innovation in his thought, but it does encapsulate the idea, also expressed by Fried, that the artwork must become integrated into its "situation." Whereas the "literalist" art that Fried discusses is designed to create a unique experience for the beholder, resulting in a subjective moment of integration involving the object and the individual inspecting it, Gadamer always considers the experience from the standpoint of its repeatability and structure. In other respects, the two thinkers articulate quite similar concepts. Towards the beginning of his article, Fried emphasizes the anthropomorphic tendencies inherent in the art that he discusses (p. 12), a characteristic gesture that we have already imputed to Gadamer, while Fried also treats the literalists' claim that painting has become all too "pictorial." Associating literalist art with "presence," just as Gadamer emphasizes the "presence" of the play, Fried also focuses on the situation in which literalist art appears. Fried finally distinguishes between literalist and modernist art on the basis of the former's "theatricality," asserting that,

> [T]he literalist espousal of objecthood amounts to nothing other than a plea for a new genre of theatre; and theatre is now the negation of art. Literalist sensibility is theatrical because, to begin with, it is concerned with the actual circumstances in which the beholder encounters literalist work. Morris makes this explicit. Whereas in previous art "what is to be had from the work is located strictly within [it]," the experience of literalist art is of an object *in a situation* – one that, virtually by definition, *includes the beholder*.... (p. 15).

As we have seen, this inclusion of the beholder is central to Gadamer's account of "play" as well, although his concept of "experience" diverges from the one articulated here since it focuses more on the object than the subject, if those terms can still be employed after he has renounced them.

Fried writes, "The object, not the beholder, must remain the center or focus of the situation; but the situation itself *belongs to* the beholder – it is *his* situation" (p. 15), contravening Gadamer's view that the situation is entirely contained within the work of art itself. Fried also considers the topic of "distance," about which Gadamer, as we shall see, presents ambiguous theses; he maintains that "here again the experience of being distanced by the work in question seems crucial: the beholder knows himself to stand in an undetermined, open-ended – and unexacting – relation *as subject* to the impassive object on the wall or floor. In fact, being distanced by such objects is not, I suggest, entirely unlike being distanced, or crowded, by the silent presence of another *person*…." (p. 16).

18 In "The Relevance of the Beautiful," as well as *Truth and Method*, Gadamer endorses Ingarden's concept of the "schema" entailed by the literary work of art. In the later essay, Gadamer declares: "The activity is not arbitrary, but directed, and all possible realizations are drawn into a specific schema. Let us consider the case of literature. It was the merit of the great Polish phenomenologist Roman Ingarden to have been the first to explore this" (p. 27).

19 In "The Aesthetic Object and the Work of Art: Reflections on Ingarden's Theory of Aesthetic Judgment," *Ingardeniana II: New Studies in the Philosophy of Roman Ingarden*, ed. Hans H. Rudnick (Analecta Husserliana 30 (1990): pp. 193–210), Wlodzimierz Galewicz discusses the distinctions between these two entities with a thoroughness impossible here.

HOWARD STEVEN MELTZER

INGARDEN: VIEWING ART AS EXISTENTIALLY AUTONOMOUS

If an object were existentially autonomous, it would exist independently, owing its existence to nothing else, or if a process were existentially autonomous, it would take place spontaneously. Common sense would suggest that this is not possible, that all existence is interdependent, all processes interlinked. I owe my existence to my parents' existence. Yet in the moment, I do indeed exist independently. My continued existence does not depend upon my parents' existence. If I am not autonomous in my origins, still my continued existence appears to be autonomous.[1]

How do musical works exist in the world? Having come into existence, do they now exist autonomously? What is their relationship to their composers and to their audience? What can we understand of a musical work's history from the musical work itself?

Roman Ingarden's notion of existential autonomy appears in *Time and Modes of Being*. Although this work is not directed towards the problematic existence of what we term "artworks," Ingarden's more general explication of time and existence has a direct bearing on both his understanding of the ontological status of artworks and the reception of artworks in aesthetic experience. As an aside, I should note that existential autonomy remains problematic for Ingarden throughout *Time and Modes of Being*, problematic in the sense that I suggested initially. While the difficulties of existential autonomy are significant for a larger consideration of Ingarden's solution of ontological problems, they remain peripheral to the case at hand.

The distinction Ingarden makes between existential autonomy and existential heteronomy is temporally bound (Ingarden, 1964: 43). Through action, that is to say through an event, an individual creates an artwork. Once created, the artwork becomes existentially autonomous, self-sufficient in its mode of being: it is not dependent upon the artist for its continued existence. Our encounters with the artwork occur as aesthetic experiences, temporally bound, existing in a "now" distinct from the moment in which the artwork came into being. Ingarden's work embraces both literary and musical works under the rubric of artworks, although it will become clear that literary works and musical works have different ontological statuses.

A.-T. Tymieniecka (ed), Analecta Husserliana LXXIII, 315–321.

In either instance, it is this notion of the artwork as perceived in a temporally bound "now" that is central to my argument. Because the continued existence of the artwork does not depend upon the artist, artworks appear as existentially autonomous at the moment of reception. We can "know" a work without "knowing" anything about the artist's life. Thus, the premises underlying Ingarden's work suggest that musical works are necessarily uninformative as biographical evidence, and do not serve as musical autobiographies.

Musical works comprise a special case for artworks which are distinctive in their existence. Thus our encounters with musical works differ from our encounters with other kinds of artworks. In Ingarden's discussion of art, the central case for aesthetic experience was literature. While the identities of literary works and musical works are at risk in ontological description, their possible status as autonomous or heteronomous differs. This is due to another aspect of their ontological status. Literary works exist bearing multiple layers, "strata." Music has only one layer.

In a historical sense, both musical works and literary works originate in memory, through oral reproduction. Writing changes the character of a literary work, due to the connection between verbal sound and printed word. Notation, perhaps imperfectly, does the same for music (Ingarden, 1973: 14–15). Once written, our cognition of a literary work is effected by writing, in part because in reading we experience a literary work mediated through the written rather than the spoken word. Our encounters with music are not necessarily mediated through notation.

As Ingarden describes the process of cognition, the stratified character of the literary work emerges. Cognition begins with the visual perception of signs, letters. The signs are subsumed into *verbal signs,* words themselves. In a literary work, different strata exist, from letters on a page, to verbal signs, to grammatical understanding, to contextual understanding, up through understanding of plot, theme, etc.

Commenting on the role of strata, Ingarden says: "Neither the strata of the work nor any of its parts are organs [as in a human body] with specific functions to perform. The influence of one stratum or one element on other strata manifests itself only in the ontic dependence of the latter on the former, as well as in the presence of certain relative determinations which have their foundation in the common appearance of various sorts of elements in the whole of the work" [Ingarden, 1973: 77]. Implicitly, the work as a whole is taken as existentially autonomous, but the strata are not – indeed I doubt that such a question really arises for Ingarden. It is the sum total of the strata which appear in aesthetic experience. Literary works produce, through their strata, a special type of aesthetic experience. Again citing from *The Cognition of the Literary Work* [Ingarden, 1973: 214]:

> In the aesthetic experience there appear various "thetic" moments, as Husserl would say, that is moments of assuming the existence of something. The first group of these relates to the objectivities portrayed in the work of art, in particular, the literary work of art, and is related to (even if not identical with) those existence-assuming moments which appear in the cognitive experiences and in particular, in the perception of real objects. But such existence-assuming moments, moments assenting to the existence of something, appear only in some aesthetic experiences, in direct aesthetic contact with works of sculpture, representational pictures, or literary works. Hence, they are not altogether indispensable; they are absent, for example, in the aesthetic apprehension of pure, non-representational music.

While the rationale for eliminating the possibility of thetic moments in music is not explicit in the previous statement, I would suggest that it rests on music's possessing only one stratum. This in turn suggests that in its autonomy, the musical work differs from the literary work. The literary work remains heteronomous in the sense that its cognition depends upon these "thetic moments," these resonances with the world "out there," so that it is dependent upon the world in its cognition. The same is not true of a musical work: it does not resonate with the world.

Why are musical works possessed of only a single stratum? This question rests on Ingarden's definition of strata, and his exclusion of the components of a musical work as stratified entities of the same order as those comprising a literary work. In his work on music, Ingarden does suggest a hierarchy of musical elements. He begins with the notion of tones as given [Ingarden, 1986: 29–30]. It seems as if Ingarden is positing a level of given constructs, objects, "tones, chords..." yet is not assigning them to a distinct stratum, an ontological status equivalent to the letters and words which comprise the lowest strata of the literary work.

Ingarden takes melody as a higher order construct [Ingarden, 1986: 31]. While this might be taken as a different "stratum" from tones, the distinction Ingarden makes seems to be of a different order. Tones and melodies are given to us in sense experience, and yet for Ingarden melodies are not fully present in conscious experience. Melodies seem to be constructed from our sense experience, and are thus intentional objects. In other words, the distinction rests in our apprehension of the work, rather than in its mode of being. A melody might seem to be a process, understood as it unfolds in time, rather than an object. Thus, melodies are not present at any given moment. In contrast, the musical work, in whatever sense it exists, exists in its entirety.

Ingarden makes a distinction between the phenomenological reduction, done in an attempt to understand the issue of musical identity, and our experience. In hearing, we do not reduce a work to its elements, nor do we listen as if we were taking a dictation. Ingarden suggests that we infer melody from our sense experience, but that time and motion are considered as non-sounding elements of music. Affect and emotion are also inherent in the work. This is not a purely "aesthetic" valuation, but seems to suggest something intrinsic. The work carries with it non-sounding elements, but these are not directly experienced. They are not properties of the performance. Again, they co-exist in their entirety with the musical work, but are not directly part of our aesthetic experience.

Ingarden turns from performance to score, again failing to locate the identity of the musical work. The score is presented as a system of signs, almost as if Ingarden were contemplating Nelson Goodman's work in *Languages of Art.* Ingarden considers the nature of the score itself, asking if it is primarily a physical object, akin to a book as a physical object. At the same time, he hesitates about claiming that the score's existence as an object is necessary for the existence of the musical work. Not all musical works are notated. In the end, Ingarden decides that a score contains imperative symbols. "The score is a way of revealing the composer's wishes as to what the work is to be like" [Ingarden, 1986: 38]. Later in the work, Ingarden decides that notation is incomplete. Though the work is conceived in all its aspects, not all of those aspects can be present in notation. Except for its evanescence, improvisation would seem to be more satisfactory as a means of conveying a musical work completely. Thus, the score is not a prerequisite for the listener's encounter with the musical work; it is possibly necessary for the performer.

Reading and listening differ as temporal experiences. Ingarden speaks of reading a poem in the "now," and suggests that this "now" appears as a unitary experience. The explanation may rest, though this is not made explicit, on the private nature of reading. Ingarden says that "the 'now' of the poem and the 'now' of its receiver coincide in all their qualifications and in what fills them as if they were one and the same 'now'" [Ingarden, 1973: 138].

"A poem is not necessarily performed in public. A musical work is made present to us in a performance, an 'individual occurrence,' and in a real sense, a public occurrence" [Ingarden, 1986: 10]. Though fixed in time and space, a performance does not endure, but is an auditory experience of a transitory nature, essentially not repeatable. Performances are "out there," existent in the world, and give the musical work *in concreto*. This distinction between musical works and performances is crucial to the notion of existential autonomy. The musical work and its performances do not share a common mode of being. The action, the result of which is the existence of the musical work, remains temporally distinct from any performance of that work. Performance does not condition the creation or existence of a musical work [Ingarden, 1986: 17].

In contrast to performances, every musical work is an object persisting in time, and Ingarden stresses the distinction between the enduring and the temporal. The musical work endures, but its creation and subsequent performances as events are temporally bound. To summarize, a drama, as a literary work, carries with it strata, layers of meaning, as does a poem. The musical work, illusive in its existence, does not.

Because our knowledge of the origins of a musical work is indirect, accessed only through performances in a temporally bound present, our knowledge of its history is indirect as well. The problem of the identity of a musical work is a problem of re-identification. Performances are immediate, individual and particular, but the musical work is neither [Ingarden, 1986: 61]. However, it would appear that because the musical work is essentially "transparent," that is without layers to be explored, there is nowhere in the work to locate its history.

Before continuing, I would note that readers of Heidegger may be struck by a similarity. In the divorce of the moment of performance from the moment of creation, Ingarden says: "Every real process in the real world takes place at a specific, determined time that is not and cannot be repeated" [Ingarden, 1986: 65]. Inherent is a distinction between seeming and actual time, a distinction between temporal span in work and temporal span in performance [Ingarden, 1986: 70]. Process remains separate from object. At the same time, processes, as they occur, modify the past through retention – the trace of the past, and protention – the anticipation of the future. In performance, we find expectation. All musical moments, except for the end, contain a "future." Time is immanent in the work, and the work becomes self-contained.

At the moment of our encounter with the work, Ingarden says,

> Directed toward this aesthetic experience, I seem to forget about my real surroundings and commune with something complete in itself. If it is a pure musical composition, and not one linked with a literary work... then its content does not connect with real processes or events in the world. [Ingarden, 1986: 56]

This will lead directly to the assertion that while produced by real individuals who are not isolated from the world and who are influenced by that world, the results of those influences are both unknowable and irrelevant to the work itself. This is a critique of music as a reflection of the composer's life, of the circumstances of the work's composition. Can we "know" a work without "knowing" the circumstances in which it arose? If music is "existentially autonomous," then it should be possible to answer in the affirmative.

Ingarden makes it clear that we know a musical work *per se,* independent of its background. The investigation of the work and the investigation of its background are different. They can inform each other. We have access to indirect knowledge of the origins of the piece. That information may be crucial to our re-identification of a work, but it remains accessible only through the persistence of physical objects and memories. We cannot live in the past, re-inhabit the era in which Beethoven or Chopin lived. Again, for me this resonates with Heidegger's insistence on the nature of *Dasein* in the world.

Ingarden insists on immediacy of performance, and the ability of the work itself to more immediately invoke the past, rather than any attempt to reconstruct biography or history around it. Ingarden considers the practical aspects of performance, including technical differences in current and past instruments.

To conclude, there are two distinct groups of issues which flow out of Ingarden's view of musical works as self-contained, of performances as unique, temporally bound processes. The first concerns musicians and music historians. Think of the implications for the attempt to create authentic performance practice. Ingarden's view suggests that this is largely futile, as the past is not accessible in this sense. All performances will be bound to the time in which they occur.

The work, incomplete in notation, is constructed in part by the performer and the listener. The emotional content of the work is located in the work itself, not in the listener. The work is recognizable as an artwork because the listener recognizes his experience as an aesthetic experience.

What are the consequences? I will suggest that this insistence on a work as self-contained, and only received in the now, seems to separate the artwork from the artist: traces of biography, the composer's social status, gender, ethnicity, race are not present in the performance, and cannot be found in the work itself. There is no "place" for them to exist. In turn, musical works are divorced from moral structure: the notion of a great work of art created by an "immoral artist" has no power, and since there is no biography inherent in the work, this stricture becomes irrelevant. The appearance of autonomy suggests that there is no possibility for complete knowledge of an artwork, at least not in the sense of knowing the circumstances of the work from the work itself. Ingarden suggests that this might be true of the composer as well, as an artist does not necessarily have the "complete" or "authoritative view of his work."

Ingarden's ontology counters any attempt to construct an institutional theory of art as the idea that artworks, including musical works, exist because certain objects have been considered as artworks or musical works by society, or that artworks are the product of social convention. Artworks exist in the world because individuals have aesthetic experiences. At the same time, Ingarden rejects "native realism," a belief that all sensible qualities of an object are real.

The end result is the musical work posited as a moment of self-containment, received by the listener. It is ahistorical, atemporal. To know a musical work is to experience it, to experience it is to know. Thus, separated from its composer in its recurrent "now," the musical work becomes the model for existential autonomy.

NOTE

1 Ingarden makes a further distinction between the "existentially derivative" and "existentially original." Again, while the distinction is significant in *Time and Modes of Being*, it is outside the scope of this paper.

WORKS CITED

Ingarden, Roman, *Time and Modes of Being,* translated by Helen R. Michejda, Springfield, Illinois: Charles C. Thomas Publisher, 1964.

Ingarden, Roman, *The Cognition of the Literary Work of Art*, translated by Ruth Ann Crowley and Kenneth R. Olson, Evanston, Illinois: Northwestern University Press, 1973.

Ingarden, Roman, *The Work of Music and the Problem of Its Identity*, translated by Adam Czerniawski, Berkeley, California: University of California Press, 1986.

JIUAN HENG

RITUAL AND THE BODY IN LITERATI PAINTING

INTRODUCTION

One of the most fascinating aspects of Chinese literati painting, which crystallized in its mature form in the Yuan Dynasty (1279-1368) and continued through the Qing Dynasty (1643-1912), is the use of ritual to cultivate, to define, and thereby to express oneself. Ritual connects the individual with a tradition and a community through psychosomatic performance. Where we would expect ritual to bring about conformity and uniformity, Chinese literati painting assumes that what is most individual in a person can only emerge in and through ritualized expression. I relate this to the aesthetic revolution of Yuan literati, who systematized brushwork along two ethical axes – the martial and the genteel – and forced scholar-painters to align themselves with the "tradition" of Dong Yuan and Ju Ran of the tenth century by selecting from a limited range of brushwork derived from calligraphy. This aesthetic choice enabled scholar-painters to avail themselves of several discourses – the criticism and theories of literature, calligraphy and medicine. A survey of the ways of appraising the body in each of these disciplines reveals that there is no firm distinction between the body and the psyche, that both are fundamentally manifestations of *qi*, the vital energy that is both universal and individual. As such, insofar as the brush stroke embodies the *qi* of an artist, the sensitive critic can read him by reading the brushwork. This mode of reading a work liberates painting from its representational function, which marks the most visible change in the Yuan revolution. In the following essay, I will trace the aesthetic moves by which the literati – essentially masters of the arts of the brush – constituted painting as a form of ritualized self-expression by integrating three hitherto separate arts: poetry, painting and calligraphy, into a unified format.

1. *SHI YAN ZHI*: POETRY ARTICULATES INTENT

As early as possibly the first century A.D. in China, the *Great Preface to the Book of Odes* had canonised literature as the arena of self-definition, when it defined poetry as the articulation of intent. Because *zhi* refers to the heart/mind taking aim at an as-yet *unrealized goal*, much as the archer's aim is trained on a target, the relation between *zhi* and its expression is not that of

A.-T. Tymieniecka (ed), Analecta Husserliana LXXIII, 323–346.

representing deeds already undertaken, roles one has assumed, nor yet a persona already crystallised. Poetry as the articulation of *zhi* promises to reveal to the attuned reader the very source of oneself, a personality in the making. As such, it trades on a gap between the manifest self and the true self. Steven van Zoeren spells out the stakes:

> The revelation of the *zhi* was an assertion that one had depths as yet unplumbed, ambitions as yet unrealized, complexities and values and abilities as yet unperceived and unappreciated.[1]

Correspondingly, the appreciation of a person's *zhi* in the reading of poetry is a form of moral hermeneutic, rather than a form of "aesthetic" judgment in the Kantian sense. Poetry is of moral interest as its canonical definition situates it in the intermediate interval between the incipient tendency of the heart-and-mind, and its eventual manifestation in deed. And perhaps that is just why poetry should be singled out for the privilege of manifesting the true self. It is not so much because conversation and action may be regarded as inadequate, but because poetry has more depth. We may recall Mencius' understanding of his advantage: he had insight into speech[2] and could see where a man was coming from, as it were, by reading his sub-text:

> From biased words I can see wherein the speaker is blind; from immoderate words, wherein he is ensnared; from heretical words, wherein he has strayed from the right path; from evasive words, wherein he is at his wits' end.

Perhaps the importance of this passage is not that it contradicts the claim of poetry. Rather, when Mencius cites his adeptness at reading persons as his moral excellence, he attests to the influence of a hermeneutic which eventually finds its way into the Great Preface by equating insight into words with insight into persons.

It is worth quoting in full the famous passage in order to appreciate the profoundly *somatic* sense in which poetry is the enactment and embodiment of mind:

> Poetry is where the intent of mind goes. Lying in the mind, it is "intent"; when uttered in words, it is "poetry." When an emotion stirs inside, one expresses it in words; finding this inadequate, one intones and exclaims; not content with this, one chants and sings it in poetry; still not satisfied, one unconsciously waves hands, gesturing, and moves feet, dancing.[3]

The author is concerned with poetry in a catholic sense, as the discharge of an internal energy which cannot be exhausted by the voice and spills over into bodily movement. Kao Yu-Kung has argued that to interpret this passage as a comment on the inadequacy of language for expression would be implausible, as that could hardly serve as the canonical basis of poetry. He suggests that it is a manifesto proclaiming "the impact of intent upon the total realization of lyric art. What is at stake is the transformation [of the] original intent as it assumes new forms."[4] It may be argued that the earliest definition of poetry as the bodily manifestation of the mind's intent is elastic enough to encompass all modes of expression. Self-expression can be considered as the conversion of psychic energy into bodily movement, ranging from the subtle movement of the vocal cords to the movement of dancing feet. Since the canonical definition of poetry sees a seamless transition between forms of expression we would today regard as disparate, the function of articulation [*yan*] is carried out not only by words, but also by gestures and overtones. The expressive function of poetry is a trajectory of inner energy from the core – the heart/mind – to the periphery of the body. As poetry takes shape, the "inner stirring" becomes conscious, and the emotions are articulated through the body.

Feelings are literally embodied. Poetry transforms the embodiment of feeling from a physical to a literary body. The search for an ideal medium of self-expression in China – one that reveals the mind to itself as it expresses itself – was also a search for a mode of expression that retains the connection between the literary and human body.

Almost two millenia later, the French phenomenologist Merleau-Ponty would rediscover the body as the site of meaning and subjectivity, finding in the gesture and style of speech, primary meaning from which conceptual meaning could be derived.[5] Speech, he says, "does not translate ready-made thought, but accomplishes it."[6] To speak of speaking as thinking is to reject the quasi-Cartesian thought that language represents ideas. Instead, the speaking subject creates meaning by directing his aim at "a certain lack which is asking to be made good." To express oneself, therefore, is to transform oneself and the world at which one directs oneself. It invites comparison with the idea of realizing the indeterminate *zhi* in expressing oneself. Just as striking, however, is the location of the distinction, however elusive, between the body and its expression. For Merleau-Ponty, following the tradition he critiques, "speech or gesture transfigures the body";[7] in the Chinese literary tradition, even when speech is refined into style, it remains the body – both are denoted by the character *ti*. While he insists on the unity of the body, Merleau-Ponty treats motility, sexuality and speech as *different* modes of expression, implying that they are different phenomena. The Chinese literary tradition, while maintaining a conspicuous silence on the matter of sexuality, constantly seeks to relate movement, vitality and expression. To trace these connections is to stray beyond the scope of this essay, so I will simply state that underlying the ideal of a unified lyric aesthetic is the assumption that both *soma* and *psyche* are manifestations of *qi*: energy, whose movement characterises life, and whose stagnation characterises diminished vitality. As such, the body as patterns of energy flow is not so much the origin of subjectivity, as it is the circuit of subjectivity. Articulate speech is not the terminus of the act of meaning, but intermediate between the inarticulate stirring that is felt within and the gestures articulating and reshaping the expressed intention in the vital, kinetic body.

II. INNER AND OUTER: ROOTS OF METAPHORS OF SELF

I will now take a detour, exploring the metaphors of inner and outer, in order to relate them to the question of what is an adequate expression of the inner. Our ways of talking about expression are rooted in the idiom of the stage. Expression projects something "inner," communicates it to others. The division between the inner and the outer is a metaphor which we are inclined to represent to ourselves by analogy to the inside and outside of objects. We can see and touch what is outside, while the inside is hidden from view. Analogously, the body is the outside which encloses the mind within. Self-expression in this account of the inner, is the disclosure of contents which are revealed when we shine an imaginary light inward, as it were. Representational painting depicts objects of sensory perception; self-expressive painting depicts objects of self-observation, the emotions. In others, the emotions are observable through symptom, and in ourselves, they are immediately felt. In his survey of theories of artistic expression, Gombrich observes that

> When we speak of 'expression' in ordinary life, we mostly think of the visible or audible signs of emotion in man and also in animals, such as the symptoms of joy or of rage, the howl of pain, … the melancholy sigh or the radiant smile. There is no strict dividing line between these outlets of emotion in life and in the various arts.[8]

His classification of expressive theories into three categories – art as signal, symbol and symptom of emotion[9] – indicates the common tendency of Western theories of expression to equate expression with the communication of emotion. Self-expression, accordingly, is treated as the communication of one's own emotions. Insofar as these emotions are first felt within, then recreated via or in a work of art, self-expression is a variation on mimesis, a representation of the emotions.

The Chinese term for self-expressive, literati painting, *xie yi*: sketching intent/ideas, does not have explicitly emotional overtones. *Yi*: idea, is poised between an indeterminate intuition of something and its determinate expression in a medium. It undergoes transformation during the creative process, and the poet-critic Sikong Tu speaks of the *yixiang*: idea-image, that is *about to emerge* in a poem.[10] To sketch ideas connotes giving shape to the indeterminate stuff of poetry in a calligraphic form, to transform images into imagery. The relevant contrast is with *xie sheng*: sketching from life, with attention to verisimilitude.

In his study of Chinese musical aesthetics, Kenneth de Woskin points out that while Western metaphors of the mind have tended to be visual, the Chinese equivalents have been aural. He cites such figures of speech as "men of vision," "enlightened," the irony of "blind seers" and the etymological relationship of "vision" with "wit," "wisdom" and "wise" as evidence of a marked preference for the eyes as windows to the mind in the West. In ancient China, he shows, a parallel situation existed with respect to the ears. In text after classical text, the sage is credited with the ability to apprehend the nature of things by listening to sounds, by "hearing the soundless." De Woskin argues that the connection is founded on the appearance of the "ear" radical in both the characters for "sage" and "sound."[11] The character "intent," which is in the realm of the mental, the emotional and the spiritual, he notes, is constructed from the characters *yin,* tone over *xin,* mind. Xu Shen defines this in his etymological dictionary: "*Yi* is intent; it comes from 'tone' and 'mind.' Words [expressed outwardly] are examined to know [inner] intent."

We may extend de Woskin's observation. "Inner" and "outer" are visual and tactile attributes of an object. The sense of hearing does not discriminate between the inside and the outside of things. When we perceive three-dimensional objects, the sense of touch often confirms that of sight, but neither is coextensive with that of hearing. When we see ourselves in the mirror or see parts of our bodies, we experience ourselves as objects. When we hear ourselves speak, we experience ourselves as agent and object simultaneously.

The inner/outer distinction which is so natural to visual representation[12] suggests a vision of the inner as that which is closed in on itself, whose nature can only be indicated by outward signs. Leonardo da Vinci was of the opinion that the mind can only be represented by outward signs of emotion such as gestures and movements. Even the Romantics, whose view of art Wordsworth captured in his definition of poetry as "the spontaneous overflow of powerful feeling," were not immune from the thought that when the work of art expresses the mind, it exists as a copy of an original. M. H. Abrams has characterised the transition from the mimetic to the expressive orientation in Romantic literature as a change in the figure of art from mirror to lamp. Both are visual metaphors. And to the extent that the visual metaphor raises the question of the work's fidelity to the mind in terms of the perfection of the mirror-image of the heart, the Romantics still faced the problem of making images in order to match. Expression is cast as the mimesis of inner states:

> The first test any poem must pass is no longer, 'Is it true to nature?' … but a criterion looking in a different direction; namely, 'Is it sincere? Is it genuine? Does it match the intention, the feeling, and the actual state of mind of the poet while composing?' The work ceases then to be regarded as primarily a reflection of nature, actual or improved; the mirror held up to nature becomes transparent and yields to the reader insights into the mind and heart of the poet himself.[13]

On the aural metaphor of the mind, self-expression would be better understood as the projection of the voice, since sound is neither inside nor outside even though it issues from the body and is heard outside of it. The unity of sound reconciles the dichotomy of inner and outer as predicates applicable to oneself. The work of art as a resonance of the self implies an expansion of correspondences amongst self and things when the source reverberates. The eye discerns sameness when it encounters at least two things, while the ear hears a continuum of the same sound as it evolves. To cast the self in aural terms, regarding the work of art as a resonance of the self, has profound consequences for criticism: the work and its author are not separate entities.

If the voice supplies the figure of an integrated self which presents itself to itself as an undifferentiated reality, the point is not that self is identical with the voice. Rather, sound which is audible to us exists within a continuum of the universal medium of *qi*, "vital energy," that in which everything lives and moves and has its being. In its furthest ranges it presents itself as patterns of *qi* movement – the soundless music which the sage hears. As a configuration of *qi*, the self is not essentially different from other things which are also configured from *qi*. Yet, the *qi* which fills the body is as individual as the intake of breath which differs from musician to musician, giving rise to differences in performance. Cao Pi's (187–226) statement on *qi* was the first of many long and varied uses of *qi* in literary theory:

> The important thing in literature is *qi*.… It cannot be had through striving for it. To compare it with music, though the score may be the same, and the rhythm may have a given measure, inasmuch as the drawing of breath is unequal, the degree of skill [in performance] is predetermined: though the father or elder brother may have it, it cannot be passed on to the children or the younger brothers.[14]

The attributes of mind which are appraised in criticism are precisely the qualities and quantity of *qi*. The precise functions and uses of *qi* in literary theory have to be determined historically.[15] Underlying the different determinations is the protean quality of *qi*, spanning categories of self and other, of psyche and body, of culture and cosmology, which accounts for its broad appeal in Chinese criticism as the connecting thread between man and his work. *Qi* has two aspects which will be of particular interest to us: the understanding that it can be cultivated, and the notion that it can be transferred from a man to a medium which embodies it and thereby prolongs the man. If *qi* is of aesthetic interest to the Chinese precisely because it is protean, the most pertinent question to ask about self-expression as an art is not how the work matches the emotions of the artist, but how a medium can be sensitive to the movement of *qi*.

III. EMOTION AND MOOD

It should come as no surprise that the self is not identified with the emotions, and self-expression is not concerned chiefly with the expression of emotion in the Chinese tradition. Indeed, the emotions reinforce the dichotomy between self and other, inner and outer, when one should register them as transient states. The Neo-Confucian Cheng Hao addressed this in the following terms:

> Denying outer things and affirming inner ones is not as good as forgetting both outer and inner. When both are forgotten, one's mind is cleansed and uncluttered; uncluttered, it will be clear. Once one's mind is clear, how can any further responses to external things become an involvement?[16]

The idea that preoccupation with details, identification with things obscures one's sense of the larger pattern underlies the literati aesthetic which eschews the faithful representation of the likenesses of things. The ideal mind is the tranquil mind and its metaphor is water, which reflects the images of things without retaining them. The ideal quality of a work is *ping dan*, "even and placid," an appreciation whose closest equivalent is the proverb "still waters run deep." Ouyang Xiu first used this term to describe the poetry of Mei Yaochen, and it was especially associated with the paintings of Ni Zan. Ni said of himself,

> I do bamboo simply to express the untrammeled spirit in my breast. Then how can I judge whether it is like something or not; whether its leaves are luxuriant or sparse, its branches slanting or straight?[17]

One who has never seen a Ni Zan painting might well be misled by the notion of expressing "the untrammeled spirit" into expecting an exuberant expressionist painter who wields his brush the way a Romantic poet uses words – to create the impression of an emotional outpouring. Ni Zan, however, was so restrained that he was known as one who used ink as sparingly as if it were gold. Endlessly imitated by always inimitable, Ni Zan's forbearance is epitomized by the story of Shen Zhou's teacher, who would correct the student producing his *nth* copy after Ni Zan, "No, no, you've overdone it again!"[18] To debate whether his bamboo resembles real bamboo is somewhat beside the point. It is not representation versus expressionism which concerns the Yuan artist, but how his bamboo conveys his lightness of touch, how his painting registers his sensibility. By *removing* vivid detail from his painting, Ni Zan defined something that cannot be defined positively – the aftertaste of emotion, distilled into a mood. By leaving out people from his landscapes and fading the tones, he succeeds in painting desolate loneliness perhaps even better than he did by conveying it directly in poetry. The following poem on one of these paintings tells of the particular circumstances which prompted the painting *Wind among the Trees on the Stream Bank*:

River bank – the evening tides have started to ebb.
Wind-swept woods – frosty leaves are thinning out.
I lean on my cane; the bramble gate is quiet.
I long for my friend: mountain colors – dim and faint.[19]

The idea of leaving out certain qualities in order to suggest something otherwise indescribable was first developed in poetry. According to the Early Sung poet Mei Yaochen, the poet has "to express inexhaustible meaning which exists beyond the words themselves."[20] He exalts the quality of *ping dan*, "even and bland," as the pinnacle of expressive poetry in the following poem:

Poetry is basically stating one's feelings;
There's no need to shout them out loud!
When you realize that the poem should be even and bland,
You'll devote yourself to Yuan-ming morning and evening.[21]

In contrast to the intimacy and understatement of literati painting, which turned from the large-scale hanging scroll of the Sung to the handscroll and album leaf format in the Yuan, the abstract expressionists who "worked from within"[22] to express feelings and emotions, "to discover myself,"[23] went for the grand gesture, expanding the canvas beyond human proportions. The emotions vary, the intentions differ, yet the desire to overwhelm the spectator is shared. Rothko's avowed "passion for self-assertion,"[24] even when it emerged from religious feeling, required the large canvas, no less than de Kooning's need to give vent to "the melodrama of vulgarity."[25] Rothko explains in "I Paint Very Large Pictures":

> I realize that historically the function of painting large pictures is painting something very grandiose and pompous. The reason I paint them, however – I think it applies to other painters I know – is precisely because I want to be very intimate and human. To paint a small picture is to place yourself outside your experience, to look upon an experience as a stereopticon view with a reducing glass. However you paint the larger pictures, you are in it. It isn't something you command.[26]

Perhaps the same absorption of the viewer took place by other means in later Chinese painting, when the calligraphic brush stroke would demand attention to itself.

It would be interesting to draw up a list of feelings expressed in Chinese literati painting, and also in expressionist paintings. On the Chinese side, we encounter elegance, refinement, parting thought, sadness, loneliness, melancholy thoughts, remembrances, stillness, sobriety. On the modernist side, we find, above all, freedom, followed by tragedy, ecstasy, doom, drama, pain, anger, love – raw, palpable emotion as if of the moment. Jackson Pollock speaks for both sides when he explains that his own art "was not about the narration of feelings but was the expression of them,"[27] whereby "something new" comes about. The expression differs in a fundamental way; one is reflective, introspective, while the other is dramatic, projective. The former attempts to re-establish harmony when emotion has faded into poeticized feeling, while the latter enacts the tension. The critic Harold Rosenberg put his finger on its pulse and diagnosed the new work as a record of the process. The canvas had become a spectacle:

> an arena in which to act – rather than as a space in which to reproduce, redesign, analyze or "express" an object, actual or imagined. What was to go on a canvas was not a picture but an event.[28]

Emotion is specific, confrontational; mood is diffuse, somewhat evasive. Heidegger, one of the rare Western philosophers to have written about mood (his predecessor being Aristotle) writes,

> The way in which the mood discloses is not one in which we look at thrownness,[29] but one in which we turn towards or turn away.[30] States of mind disclose Dasein in its thrownness, and – proximally and for the most part – in the manner of an evasive turning away.[31]

Moods disclose things "primordially," that is to say, phenomenologically prior to the division into subject and object. Moods are "pallid," a surprising way of describing the state whereby *Dasein* "becomes satiated with itself."[32] Chaves has suggested that the quality of *ping dan*, "even and bland," which is prized in poetry, has its roots in the evocation of the ineffable *dao*. Tracing its use in a non-literary context, Chaves relates a passage by the Wei scholar Liu Shao's *Monograph on Personalities,* which reads:

In a man's character, it is balance and harmony that are most prized. A character which is balanced and harmonious must be even, bland and flavourless. Thus, such a man is able to develop in equal measure the five virtues [i.e. courage, wisdom, humanity, faithfulness and loyalty] and adapt himself flexibly to the situation. For this reason, in observing a man and judging his character, one must first look for the "even and bland," and then seek intelligence.[33]

The bland mind is balanced because no quality of mind predominates to the exclusion of any other, but all blend in harmony.

Not all literati paintings conformed to the *ping dan* ideal. Perhaps most did not. The remarkable thing is that the artist who pushed the unassertive, bland *ping dan* ideal to its limit in painting should be held up as the quintessentially self-expressive painter in the tradition. That is due in no small part to the recognition of the difficulty of describing something which is subtle, and has to be evoked through the absence of qualities. The scholar-statesman Ouyang Xiu remarked:

Loneliness and tranquility are difficult to paint. Even if an artist captures them, the viewer will not necessarily recognize them. Thus the pace of flying birds or running animals, things of superficial significance, are easy to perceive, while serenity and deep quietude, feelings of far-reaching subtlety, are hard to describe.[34]

We usually think of depth of feeling in terms of intensity, but there is another kind of depth which comes when the passionate feelings have subsided, and life returns to normal. The person who has integrated his passions into his examined life has grown, as we say, in depth. Emotion that is vented prematurely may well mitigate against depth of feeling. That kind of depth is bittersweet, whereas strong emotion tends to be either bitter or sweet, but not both.

The expression of the inner is subject to cultural, philosophical models of subjectivity. With the abstract expressionists, we find an emphasis on autonomy, on a rupture with the past[35] and on the virtue of self-assertion. In the scholar-artist tradition, subjectivity is found in extending the self in conversation with the past, with the community, with the learning acquired.

Individuality emerges through and not in spite of tradition, and individuality is consistent with self-effacement. When the self is regarded as a nexus of communication, the works acquire meanings in the context of a tradition of reading imagery.

IV. INTEGRAL ART

Historically, the transition from representational painting to self-expressive painting in China coincides with the emergence of the integral poem-painting in which image and text coexist within the same work. The refrains that calligraphy and painting have the same origin and that poetry and painting are equivalent, are constantly reiterated, but not much has been said about the nature of their equivalence, or common origin.[36]

I would like to suggest that the relation between text and image, which is always construed as an enhancement of the self-expressive possibilities of the image, should be understood in terms of their meta-structural similarities. Painting became self-expressive when it took upon itself the structure of a [ritual] practice, applying the principled choices of calligraphy to brushwork. A ritual practice is a set of routines which one performs repeatedly and regularly until they can be done without conscious choices. It can be as casual as reading the morning papers during breakfast, as demanding as the practice of *yoga asanas*, or as spiritual as meditation. The more structured the ritual, the more focused the practitioner-actor's attention on the forms. When he has internalised the forms, that is to say, when the mastery of these forms has shifted from focal to subsidiary awareness,[37] the narrow range of individual choice, paradoxically, highlights the individual's contribution. Nothing is as natural as walking, and yet nothing is as expressive as the *kabuki* actor's deliberate, stylized walk. Someday, a student of ritual may undertake a study of the changing states of awareness and the way they map onto changing physiological states in the ritual performer. In the meantime, I may make the banal observation that just as the intentional ritual acts transfigure the body, so the body also transfigures consciousness in ritual performance. The kind of mastery that counts is not intellectual consciousness, but bodily consciousness[38] which enables one to move rhythmically, fluently,[39] in order to inhabit the desired frames of mind implicit in particular rituals. In claiming that these meta-textual issues set the parameters of self-expression, I am trying to argue for a connection between what is expressed, the sense of self that is involved, and how one expresses oneself.

V. CALLIGRAPHIC AND SOMATIC EXPRESSION

The human body, viewed as a configuration of energy, does not coincide with the shape of the physical body, but rather with psychosomatic processes. Since the human body, in classical Chinese medicine, is understood in terms of the quality and flow of energy, it is not difficult to see the work of art as an extension of the body, or, in the *topos* of the Chinese artists, as the mind "lodged" in the painting. The body as the site and pathways of vital energy provided the framework of values for the appreciation of calligraphy and poetry. In calligraphy, both the morphology of the individual characters and the structure of a string of characters replicate the energy patterns of the human body. Zhang Huaiguan, a ninth-century theorist, adumbrates the connection between calligraphy (the grass script: *cao shu*) and physiology:

> The body and energetic configuration of a character is complete with a single stroke. There may happen a passage where the brush line is not continuous but the blood-artery is uninterrupted. Where there is continuity, then the energy communicates from one line to the next. Only Wang [Xianzhi] understood this profound principle, thus the character at the top of a line in his calligraphy often continues [the energy] from the character at the bottom of the preceding line. What is called the "one stroke writing" which originated with Zhang [Zhi] is this.[40]

Just as writing is not the representation of speech, so calligraphy is not the representation of writing. Zhang Huaiguan also revived a distinction between the meaning of the gesture of writing (*shu*) and the meaning of the word (*zi*).[41] To have made this distinction is to have prised apart, in principle, the aesthetic and utilitarian functions of calligraphy. The difference between viewing calligraphy aesthetically and reading it functionally is a difference between finding meaning in the gesture of writing itself and regarding the inscription as a means for grasping thoughts expressed in words. In the latter process, it is indifferent to the meaning of the thought whether it is written, recited or "composed in the head," because meaning resides in the grasp of the language in which the thought is expressed. The thought, *pace* Fenellosa, is no more intrinsically connected to the form of Chinese characters than it would be for any other language's writing system. Calligraphy as a vehicle for expressing thoughts does not impinge on the content of the thought; good calligraphy cannot redeem a bad poem or spurious philosophy. The act of expressing ourselves in words is like the face we present to the world. To paraphrase the Japanese psychologist Doi, we inevitably reveal even as we conceal ourselves, and conceal even as we reveal. But the body has fewer such guiles, is less able to play such games of hide and seek. It is a point which underlies Zhang's defense of calligraphy, that the gesture of writing could achieve *directly* what the sign system does indirectly:

Literary composition needs several characters to complete the meaning, whereas calligraphy can reveal the mind with only one character. This is certainly the ultimate attainment of economy and simplicity.[42]

The aesthetic conception of calligraphy is essentially one in which gesture becomes the grammar of the body and the mind simultaneously, insofar as the movement of the calligrapher's mind is connected simultaneously to the movement of his *qi* energy and the brush. The connection between the state of mind of the calligrapher and his calligraphy is not a linguistic structure, but kinetic: the brush and its traces are manifestations of the body and its energy. As John Hay puts it,

If there is any single, fundamental characterization of calligraphy, it is that of a line of energy, materializing through the brush into the ink-trace.[43]

In calligraphy, the movements of the body no longer play passing roles in accompanying the recitation of poetry and story-telling, but become gestures significant in themselves, and so transcend the impermanence of performance, while retaining its temporality.

Calligraphy is not midwife to thought, a medium through which thought is represented, but is itself the site of feeling. If only one character is necessary to reveal the calligrapher's mind, then which particular character he chooses does not matter much. The most significant monuments of calligraphy are personal letters and essays, not transcriptions of canonical texts. An aesthetic appreciation presupposes that the content of the written word can be abstracted from the significance of the representational content of writing so that calligraphy could be valued as a medium for displaying the calligrapher's ideas saturated with *farbung*.[44] Once *farbung* could be displayed, rather than merely evoked, it could not be dismissed as the private theatre of the individual mind, secondary to the objective thoughts held in language.

Calligraphy is a more economical expression of the heart-and-mind than literary composition to the extent that the dispositions of the body reflect the dispositions of the mind without representations. The premise is that mind and body are symbiotically related, that intelligence and consciousness is present throughout the body. Or, to borrow a phrase from a bioenergetics practitioner, consciousness is projected simultaneously on two screens – the body and the mind. The body is inherently expressive and spontaneous: the way we walk is uniquely individual and revealing of ourselves. Gesture, a language of the body, lies between spontaneous bodily expressions such as crying and the linguistic expressions we learn to replace them with. It retains the unmediated individuality of bodily movements while acquiring the symbolic function of language. To the accustomed eye, a calligrapher can be identified through the particular movements of his arm, hand and wrist gestures. A gesture supervenes on the natural bodily expression of feeling. The meaning of a gesture cannot be elucidated in a description of its content – it can make a point which, if translated into words, would undercut it.

VI. TRANSFORMING PAINTING INTO CALLIGRAPHIC EXPRESSION

What sets literati painting apart from any other type of painting is its conception of painting as an extension of calligraphy. Calligraphic painting can be described as sharing with calligraphy a common kinaesthetic basis which results in the same brush movements and brush strokes and the same aesthetic/symbolic values which assign a hierarchy of ethical values to certain brush-modes. A representative view, expressed by the Qing painter Wang Xuehao, equates literary painting with writing (*xie*), as opposed to describing (*miao*):

> Wang Hui said, "Someone asked me what scholar painting is. The answer is that it can be summed up in a single word – writing." This is most to the point. Characters must be written, not described; painting is no different. One who paints by describing is a vulgar craftsman.[45]

The aesthetics of calligraphy was grafted onto painting during the Yuan Dynasty by means of a transformation of the role of brushwork in painting. A consciousness of brushwork as a technical means of achieving artistic goals was always present in the appreciation of Chinese art, but was never the point of painting. A world of difference stands between respecting the brush stroke as a technical medium and elevating it to the status of iconography. And therein lies much of the difficulty of understanding Chinese scholar painting, for the brush stroke as iconography without icons is not easily identifiable, as scenes of the Annunciation are, nor is its significance as easy to communicate to someone unschooled in calligraphy, as an explanation to one lacking Biblical knowledge, of what the Annunciation means to a Christian. Yet, it is precisely because the calligraphic brush stroke can be meaningful without symbolizing particular events, and can embody intentions without representing particular states of affairs, that it was adopted by Chinese scholar artists as the medium of self-expression. The problem of self-expression, as Gombrich points out,[46] is misconstrued when it is posed as a choice between the unmediated outpouring of emotion, or the model of language which claims to represent the structure of thought but leaves no place for *farbung*. Unless a line, a dot, a colour is contained within a scale of values, it is impossible to assign any value to it. Without a grasp of the vocabulary in which an artist expresses himself, the nuances of feeling are lost on the audience. *A priori*, the artist himself can only express nuances of feeling within the range of a common medium whose values are shared by others.

VII. THE COMMON ORIGIN OF PAINTING AND CALLIGRAPHY: BRUSHWORK AS TECHNIQUE AND BRUSHWORK AS ICONOGRAPHY

As early as the 9th century, Zhang Yanyuan, the Tang critic, had called attention to the common origin of painting and calligraphy, namely the brushwork: a standard repertoire of dots, hooks and linear strokes, from which Chinese characters are formed. It was probably not the repertoire as such that suggested a fundamental similarity between painting and calligraphy, but rather that calligraphy and its criticism had reached a stage where dots and strokes were no longer mere dots and dabs, but images of dynamic energy in nature, the very vital forces which painters tried to depict. Zhang explicitly praises the 6th-century painter Zhang Sengyu for having "made his dots, dragged strokes, hacking strokes and sweeping strokes in accordance with Lady Wei's *Battle Strategy of the Brush*."[47] The centerpiece of that essay compares seven fundamental calligraphic strokes to the following images:

> Like a cloud formation stretching a thousand *li*, indistinct, but not without form.
> Like a stone falling from a high peak, bouncing and crashing, about to shatter.
> The tusk of an elephant or rhinoceros thrust into and broken by the ground.
> Fired from a three-thousand-pound crossbow.
> A withered vine, ten thousand years old.
> Crashing waves or rolling thunder.
> The sinews and joints of a mighty bow.[48]

Important as brushwork was, it was but one constituent of painting. What differentiated painting from calligraphy was perhaps as important as what they share: painting consisted of an interplay of "brush and ink," whereas calligraphy only had brush, but not ink. The technical terms "brush" and "ink" are defined in Jing Hao's *Bifa Ji*:

> Brush is obtained when you handle the brush freely, applying all the varieties of strokes in accordance with your purpose, although you must follow certain basic rules of brushwork. *Here you should regard brushwork neither as substance nor as form but rather as a movement, like flying or driving*. Ink is obtained when you distinguish between higher and lower parts of objects with a gradation of ink tones and represent clearly shallowness and depth, thus making them appear natural as if they had not been done with a brush.[49]

Elsewhere, he contrasts Xiang Rong, who mastered ink but not brush, with Wu Daozi who possessed brush but not ink.[50]

While "ink" relates to tonal contrasts which model an illusion of space, "brush" is less easy to translate. Obviously, a trace of brushwork must be visible as such to qualify as "brush," but brush, in addition, defines "the real substance of things," their "spiritual character" which an excess of "ink" obscures. None of these attributes may be rendered as "texture," even if "texture strokes" fall within the category of brush. Most importantly, it is a kinaesthetic category, which has no obvious equivalent outside Chinese painting. Brushwork is indicative of the responsiveness of hand to mind, and both are evaluated in terms of the quality of the "flow of energy" which determines how lifelike the painting looks. Guo Ruoxu listed the following faults of brushwork in his "Painting Treatise":

> … the first is "board-like," the second, "engraved," and the third, "knotted." In "board-like" [brushwork], the wrist is weak and the brush sluggish, completely lacking in give and take. The forms of objects are flat and mean, and there is no ability to turn and bend. If "engraved," the movement of the brush is uncertain, and mind and hand are at odds. In delineating an outline, one will produce sharp angles at random. If "knotted," one wishes to go ahead but does not or fails to break off when one should. It seems as if things are congested or obstructed, unable to flow freely.[51]

Someone whose brushwork suffers from these defects is not simply lacking in technical mastery of the brush; the fault ultimately lies in his not having the "idea before the brush," a failure of imagination.

While Guo I, the Sung painter, insisted that good calligraphers made good painters, he had in mind their dexterity of wrist, which made for good brushwork. He constantly relates the use of the brush and ink to specific *representational* effects:

> Use light ink in six or seven layers to achieve depth. Then the inktone will be moist and fresh and not dull and dry. Use concentrated ink and black ink especially to establish outlines, for without thick or black inks, pine cones and corners of rocks will not be clearly distinguished from the background. When the boundaries are thus clarified, go over them with repeated layers of indigo ink wash, then the inktones will become clear, and will appear to emerge from mist and haze.
>
> Light ink in overlapping layers applied very slowly is called *wo dan*. A sharply pointed brush, lightly drawn across is called *cun ca*. Shaking inkwash off the brush repeatedly and sprinkling the surface is called *xuan*. To dampen with inkwash is called *shua*. Bearing down straight with the brushtip is called *zu*. Dotting is used in figure-painting, also for leaves. Drawing the brush from one direction to the other is called *hua*. Lines are used in architectural drawing, also for pine needles.
>
> Snow shadings are indicated with light or dark ink. Inktones must not be monotonous. For shades of mist, leave silk untouched in its original color, and merely stain it with very light

inkwash, without leaving brush or ink traces. [i.e. do not let the movements of the brush become discernible].[52]

Joan Stanley-Baker has appraised the shift from Sung to Yuan painting in terms of the estimation of brushwork:

> To Sung art lovers brushwork had been a *sine qua non*, but not the ultimate goal; a painting had to succeed as a painting, not as an aggregate of brushwork.... With the Yuan literati, brush-consciousness sometimes might reach unhealthy proportions.[53]

A fundamental shift in painting and its relation to calligraphy came when Zhao Mengfu used only brushwork in his painting, thereby challenging the notion that good painting had to include both ink and brush. Moreover, the brushwork, selected and classified on the basis of calligraphic script, was applied consistently throughout the painting. Two late paintings, *Elegant Rocks and Sparse Trees* and *Twin Pines, Level Distance*[55] defined the prototype of literati painting for the rest of Chinese painting history. Every stroke in the painting retains its integrity, uncompromised by painterly techniques of the sort Guo Xi, the court painter, talks about: layering, shaking ink from the brush, going over a clear line to establish inktones, in short, effacing the traces of brush movement. In Zhao Mengfu's paintings, calligraphic strokes are systematically grafted onto pictorial forms, almost as an exercise in extending the pictorial possibilities of a certain brush mode, rather than because there is any inherent fit between the type of brush stroke and the objects they depict. In a colophon to *Elegant Rocks and Sparse Trees*, Zhao codified his program for integrating the vocabulary of painting and calligraphy in these terms:[56]

> Rocks are like flying white, trees are like seal script;
> When painting bamboo, one should master the spreading-eight method.[57]
> Those who understand this principle thoroughly
> Will recognize that calligraphy and painting have always been one.

An analogy may clarify the difference between the approaches of Zhao and Guo Xi to brushwork and the so-called similarity between calligraphy and painting. Whereas the pre-Yuan painter adapted brush strokes to nature, Zhao and literati painters following him adapted nature to the brush strokes of choice.

The calligrapher's contribution lies in the momentum, inflection and variety of his strokes and the individuality of his composition, for the same character written by different calligraphers, or by a calligrapher on different occasions, will vary in the placement of its radicals, its individual strokes.

Calligraphic painting applies the brush discipline and conventions of a particular script to forms which, unlike those of written characters, are not prescribed: therein lies the element of improvisation, of invention.

The choice of script type, or alternatively brush mode, was significant, for each type of script had a different function, a different image. Seal script was archaic, clerical script was monumental, regular script was for prosaic/official documents, cursive script was for personal communication. While these associations of script type with function may seem as familiar to us as the association of the Times Roman font with serious text and the Sans Serif font with the attention-grabbing look of advertising copy, the association of the brush mode with the moral character of the calligrapher is not. It is the symbolic content of the choice of brush mode in which meaning can be prised apart from pictorial content that is unique to literati painting.

The new consciousness of brushwork as iconography in the Yuan prompted a fundamental change in thinking about how to use brushwork in painting. The Yuan master Huang Gongwang advised artists to refrain from displaying an array of brush techniques in order to bring into focus the keynote of a painting: "First decide on the brush mode and do not mix them."[578]

This remarkable advice attests to the fact that by the Yuan Dynasty, painters had begun to codify the smorgasbord of brush techniques we found in Guo Xi's painting treatise and in his own work, into brush modes with an internal coherence. Even more amazingly, only two brush modes were recognized: the Dong Yuan-Ju Ran round brush mode and the Li Cheng-Guo Xi angular brush mode. Most remarkably of all, Huang could say unequivocally, "Those learning to paint landscapes must use Dong (Yuan) as a model."[59]

The brushwork of the four Masters of the Yuan – Wu Zhen, Ni Zan, Huang Gongwang and Wang Meng, all derived from Dong Yuan and Ju Ran. This is a significant example of glory by association. Here is a feat of classification which would have done Aristotle proud, all the more so because there is no particular theorist to which the classification could be attributed. The ideology underlying it is that of personal expression. Consistency of brushwork equals consistency of character. Brushwork has symbolic value apart from the pictorial content, or the representational content of the calligraphy.

The "round" brushwork of seal scripts and cursive scripts is so called because the extremities of lines are written with a reverse motion of the brush tip opposing the stroke, thus concealing the point tip. Holding the brush vertically while writing makes for a centered tip, whose general effect is that of rounded corners and edges. This creates the impression of energy being held in reserve as ink is deposited in knobs during the momentary halt, together with increased pressure on the brush at the extremities, and an overall impression of tensile strength within the stroke. The same principle applies when the basketball player squats slightly before springing off the ground, in order to gain momentum. The effect of evenness contributes to the impression of containment and unassertiveness, unforced "naturalness" – the very aesthetic qualities which were prized in poetry, music, and conduct befitting a Confucian gentleman. The centered brush stroke connotes poise, internal strength held in reserve, readiness to respond in any direction. In "hidden tip" brushwork, only about ten percent of the brush – its tip – is used, and this is considered the store of brush strength. John Hay sees the connection between physiology and calligraphy in terms of macrocosm and microcosm. The energetic orbs of the human body also store energy. Just as the "strongest pulse in the body, that of the *shen* orb, which unfolds through the bones, should not be overmanifested," so "the bone tip in calligraphy should not be overmanifested at the surface."[60] The "angular" brushwork of clerical script is produced by holding the brush at an angle, and has great dramatic potential. It is forceful, strong, monumental, emblematic of the martial. And just as importantly, calligraphic brush strokes are produced with the entire body, whereas painterly effects need not be. The question of which brush mode has greater potential for self-expression is not one which can be settled by appealing to the senses. Each brush mode has its inherent expressive potential, which is why Guo Xi, who was most sensitive to the individual expression of seasons, insisted that painters master all brush techniques. The crux of the matter is that when brushwork ceases to be nothing but technique – when it is received as a manifestation of character – it assumes a hierarchy of ethical values. Then, not all brushwork is equally expressive, for only certain values are worthy of expression, for one who regards himself as a person of cultivation.

Since it is *immensely difficult* for one who is unfamiliar with the literati aesthetic to differentiate between angular and square brushwork, much less read into them the particular aesthetic values which the Chinese literati associate with each category, it is difficult to sustain the claim that calligraphic brushwork is *inherently* expressive. That is why it was important to constitute painting as a practice which has the same structure as the textual arts in order to enable the painting to be read as text. It was within the compass of the common values of gentlemen-scholars, that the calligraphic lines of painting from the Yuan onwards were expressive, and expressive of a specific scale of values.

The beginning of the process is the decision, as Huang Gongwang recorded it, to choose one's medium of expression – brush mode – and to maintain it throughout the painting. Consistency of brush, it must be supposed, indicates consistency of character. It takes a person of a certain temperament, in possession of a certain degree of awareness, to maintain a constant pressure on the brush so as to produce strokes of even thickness, and constant saturation of the brush to produce even tonalities. It requires intensity of a certain pitch to maintain the same strokes without making them identical over the course of days (or four years, as in the case of Huang Gongwang's *Dwelling in the Fuchun Mountains*).[61] Anyone can paint a flat line by bearing down on the brush with pressure and dragging it across the paper, but it takes mastery of technique and spirit to *articulate* the brush for the effect of evenness. The aim, needless to say, is to make not a flat painting, but one whose interest is contained within as narrow a compass of technical possibilities as possible. A subtle calligraphic gesture, bonded to the contemplative spirit, is found in the paintings of Agnes Martin. Her *Night Sea*[62] uses the discipline of the grid pattern as a point of departure for the hand-drawn, slightly uneven graphite lines, regular, yet imprecise. Her lines register a sensitive touch. These qualities in her painting "enabled her to achieve a profound expression of herself" and she notes that the rejection of "objectivity" in abstract painting puts it on "the same pinnacle as music."[63] Martin took to heart Duchamp's dictum, that a title is like an invisible color. Her evocation of the oceanic feeling of infinity does not lie only in the subtle variations of the grid; her title *Night Sea* introduces a vastness onto the canvas, activates the darkness of the canvas, transforms the abstraction into an intimate graphic diary.

CONCLUSION

It may seem that in calling calligraphic painting a ritual practice, and saying that brushwork carries a set of symbolic values, I am reducing the meaning of a work to the artist's assertion (in not so many words) that he has such and such virtues. And that would be correct, except that if he had to state what he did in words, it would not be worth saying. It could be banal, yet it would not be too different from conversations in which participants recite baseball scores, discuss the weather, or dissect the peculiarities of the latest spectacle on television. Once painting is constituted as a practice, it can become a forum in which the audience participates in creating the meaning of the text, extending the painting in the colophons. If, like Zhao Mengfu, one is caught in the dilemma of serving a foreign government, the gesture of affirming one's Confucian values over and over is not banal, but a demonstration of one's principles. The problem of representation, in the context of a self-expressive practice, may be put this way: When every leaf and every blade of grass is investigated in the terms of the scholar's studio, then there is resonance of the self everywhere. The point is not whether the orchid is recognisable as an orchid, but that it could not be just an orchid: like poetry, painting must convey meanings beyond what meets the eye.

The marvel of later literati painting is that even without the coloration of titles, such as when paintings bear prosaic titles naming studios or describing contents of the painting, the coloration is imparted by the austere, monochrome brushwork and the painting becomes textual. The internal unity of mood, thought, self-recognition and expression which Hegel required of the art of inwardness is presented in Shen Zhou's *Night Vigil*[64] in which he relates and embodies the search for enlightenment. He wrote of how one night, he "acquired sounds and colours through a state of [outward quiescence and inward repose]" which had the effect of

> cleansing man's mind, spirit, and feelings and helping him to express and realize his intent and meaning.... Tonight's sounds and colours are no different from those of other nights, yet upon reaching my ears and eyes, they suddenly and wondrously become one with me.... When the sounds die down, and the colours disappear, my intent, brimming and overflowing, alone remains. What is this so-called intent? Is it, indeed, within me, or outside me? Is it found in external things, or is it expressed only when interacting with things? ... Through mental fasting and sitting alone by the light of a bright candle deep into the night, I must pursue the principles of affairs and things, as well as the wondrous workings of the mind and body; by using this method for self-cultivation and responding to things, I shall surely acquire understanding.[65]

In this painting, the older Song ideal of finding lodging in painting[66] is raised to new heights. That Shen Zhou could represent the evanescent experience of enlightenment in painting is a paradox. The painting is an expression of his intent. But his intent has now become a question. By asking what intent is, he implicitly asks if the sights and sounds which he paints are inner or outer. And he settles it by the act of painting, which internalises the outer at the same time as it renders the inner visible. Perhaps all painting does this. It is the wonder of Shen Zhou's painting that it can unsettle the boundaries of experience so powerfully while resolving them within the experience of the painting. In raising within itself the question of the status of painting, Shen Zhou's *Night Vigil* becomes, perhaps, the first philosophical Chinese painting.

National University of Singapore

NOTES

[1] Van Zoeren, Steven. *Poetry and Personality: Reading, Exegesis and Hermeneutics in Traditional China*. Stanford, California: Stanford University Press, 1991, 13.
[2] *Mencius*. Translated by D. C. Lau. 2 vols. Hong Kong: The Chinese University Press, 1979, 1984. 2A:3, 57, 59.
[3] Tr. Kao, Yu-kung. "Chinese Lyric Aesthetics." In *Words and Images: Chinese Poetry, Calligraphy and Painting*, ed. Fong and Murck, NY and Princeton, N.J.: Metropolitan Museum of Art and Princeton University Press, 1991, 61.
[4] Ibid., 61.
[5] Merleau-Ponty, M. 1962. *Phenomenology of Perception*. Translated by Colin Smith. London: Routledge & Kegan Paul, 174–199.
[6] Ibid., 178.
[7] Ibid., 197.
[8] Gombrich, Ernst H. "Four Theories of Artistic Expression." In *Gombrich on Art and Psychology*, ed. Richard Woodfield. Manchester and New York: Manchester University Press, 1996, 141.
[9] In Gombrich's terminology, the symbol indicates an inner state, the signal arouses an emotion, and the symbol represents or depicts emotional states.
[10] Sikong Tu. *Twenty-four Categories of Poetry*. "Category 14: Close-Woven and Dense," translated by Owen, Stephen, ed. *Readings in Chinese Literary Thought*. Cambridge, MA: Council on East Asian Studies, Harvard University, 1992, 334:

> This does possess genuine traces,
> But it is as though they cannot be known.
> As the [idea]-image is about to emerge,
> As the process of creation is already wondrous….

[11] De Woskin, Kenneth. *A Song for One or Two; Music and the Concept of Art in Early China*. Ann Arbor: Center for Chinese Studies, University of Michigan, 1982, 33–35.
[12] Representation appeals to vision even when the medium is literary, rather than plastic.
[13] Abrams, M. H. *The Mirror and the Lamp*. London, Oxford, U.K.: Oxford University Press, 1953, 23.
[14] Pollard, David. "Ch'i in Chinese Literary Theory." In *Chinese Approaches to Literature from Confucius to Liang Ch'i-ch'iao*, ed. Adele Rickett, 43–66. Princeton: Princeton University Press, 1978, 48.
[15] See ibid. Also Fong, Wen. "Ch'i-yün Sheng-tung: Vitality, Harmonious Manner and Aliveness." *Oriental Art*, n.s. 12, no. 3 (Autumn 1966): 159–64. And Soper, Alexander. "The First Two Laws of Hsieh Ho," *Far Eastern Quarterly*, vol. 8 (1949), 412–423.
[16] Cheng Hao. *Ming-dao Wen Ji*, 3: 1a–b.
[17] Bush, Susan, and Hsio-yen Shi. *Early Chinese Texts on Painting*. Cambridge, MA: Harvard University Press, 1985, 280.
[18] Fong, Wen C. *Images of the Mind: Selections from the Edward L. Elliot Family and John B. Elliot Collections of Chinese Calligraphy and Painting at the Art Museum*. Princeton University, Princeton, N.J.: Princeton University Press, 1984, 126.
[19] Ni Zan. "On the Fifteenth Day of the Ninth Month of the Year Kuei-mao of the Chih-cheng Period (Oct. 22, 1363), I Painted This to Send to the Summoned Scholar, Sheng-po, and Inscribed this Poem on It." Tr. Chaves, Jonathan. *The Columbia Book of Later Chinese Poetry*, NY: Columbia University Press, 1986, 70.

[20] Chaves, Jonathan. *Mei Yao-ch'en and the Development of Early Sung Poetry*. NY: Columbia University Press, 1971, 110.
[21] Chaves. Ibid., 116.
[22] Jackson Pollock used this expression (echoing Kandinsky before him, who expounded on "the art of inner necessity") to distinguish modern(ist) painters from their predecessors in an interview with William Wright in 1950. See Stiles, Kristine and Selz, Peter (eds.), *Theories and Documents of Contemporary Art*. California: University of California Press A:A: (1996) 33.
[23] Hans Hofmann, quoted in "Artists' Sessions at Studio 35" (1950), ed. Robert Goodnough, reprinted in Motherwell Reinhardt, et al., eds. *Modern Artists in America*, NY: Wittenborn Schultz (1951) 10.
[24] Rosenthal, Mark. *Abstraction in the Twentieth Century: Total Risk, Freedom, Discipline*. NY: Guggenheim Museum. 1996, 117.
[25] Ibid., 117.
[26] Rothko, Mark. "I Paint Very Large Pictures" (1951), reprinted in Stiles and Selz (eds.), *Theories and Documents of Contemporary Art*, 26.
[27] Rosenthal, Mark. *Abstraction in the Twentieth Century*, 117.
[28] Rosenberg, "The American Action Painters," *Art News*, 51, no. 8 (December 1952), 22.
[29] Being becomes temporal through being thrown into situations.
[30] Heidegger, Martin. *Being and Time*. Albany, NY: State University of New York Press, 1996, 174.
[31] Ibid., 175.
[32] Ibid., 173.
[33] Liu Shao, *Renwu ji* (in Sibu Congkan) 1/1b-2a. Cited in Chaves, *Mei Yao-Ch'en and the Development of Early Sung Poetry*, 118.
[34] Cited in Joan Stanley Baker, "The Development of Brush-Modes in Sung and Yuan," *Artibus Asiae*, 39, no. 1 (1977): 13–59, 20.
[35] Jackson Pollock expressed the quintessentially modernist view of art in an interview: "Modern artists have found new ways and means of making their statements… the modern painter cannot express this age… in the old forms of the Renaissance or of any other past culture." Stiles and Selz (eds.), *Theories and Documents of Contemporary Art*, 22.
[36] Michael Sullivan's *The Three Perfections*, NY: George Braziller, 1974, remains unsurpassed.
[37] Michael Polanyi coined these terms. See his *Personal Knowledge*, NY, NY: Harper Torchbooks, 1964.
[38] The distinction between intellectual and bodily consciousness may be compared to Gilbert Ryle's distinction between knowing that and knowing how.
[39] In sitting meditation, the body does not cease to move, but all movement is concentrated in the act of breathing.
[40] Cited in Hay, John, "The Body as a Microcosmic Source of Macrocosmic Values in Calligraphy," in Bush and Murck, *Theories of the Arts in China*, 87–88. (See also PWCSHP 1: 29.)
[41] Kao yu-kung, "Chinese Lyric Aesthetics," 75.
[42] Cited in Kao yu-kung, "Chinese Lyric Aesthetics," 75.
[43] Hay, ibid., 88.
[44] In his classic essay "Thoughts," Frege introduced *farbung* – coloration – as the third component of sentences, over and above the assertion and the thought. To judge is to acknowledge the truth of a thought, and to assert is to manifest this judgment in a sentence. *Farhbung* act on the feelings and mood of the hearer, and arouse his imagination.

They intimate something which cannot be conceptually grasped. Frege, *Collected Writings on Mathematics, Logic and Philosophy*, ed. McGuinness, Oxford: Basil Blackwell, 1985, 356–7.

[45] Wang Xuehao. *Shannan Hualun*. In Yu Jianhua, *Zhongguo Hualun Leibian*, Hong Kong: Zhongguo Shuju (2nd printing), 1973, 248.

[46] Gombrich, E. H. *Meditations on a Hobby Horse*. London: Phaidon Press Ltd., 1963, 67–69.

[47] Bush and Shih. *Early Chinese Texts on Painting*, 61.

[48] Barnhart, Richard. "Wei Fu-jen's *Pi Chen T'u* and Early Texts on Calligraphy." *Archives of the Chinese Art Society of America*, 18 (1964): 13–25, 16.

[49] Bush and Shih. *Early Chinese Texts on Painting*, 110. Emphasis added.

[50] Ibid., 159.

[51] Ibid., 97.

[52] Stanley-Baker, Joan. "The Development of Brush-Modes in Sung and Yüan." *Artibus Asiae*, 39, no. 1 (1977): 13–59, 23–24.

[53] Ibid., 18.

[54] Reproduced in Fong, *Images of the Mind*, 105 (ca. 1310, handscroll, ink on paper, 27.3 × 92.8 cm, Palace Museum, Beijing).

[55] Reproduced in Fong, ibid., 104 (ca. 1310, handscroll, ink on paper, 26.7 × 107.3 cm., The Metropolitan Museum of Art, New York).

[56] Zhao Mengfu. "Songxue Lunhua Zhu in Zhongguo Hualun Leibian," 1063. Translated in Fong, *Beyond Representation*, 440.

[57] Official script.

[58] Stanley Baker, *op. cit.*, 24.

[59] Stanley Baker, *op. cit.*, 21.

[60] Hay, John. "The Body as a Microcosmic Source of Macrocosmic Values," 89.

[61] Reproduced in Fong, *Images of the Mind.*

[62] Reproduced in Rosenthal, Mark, *Abstraction in the Twentieth Century: Total Risk Freedom and Discipline*. New York: Guggenheim Museum, 1996, plate 172.

[63] Rosenthal, Mark. *Abstraction in the Twentieth Century*, 163.

[64] Reproduced in Fong and Watt, *Possessing the Past*: *Treasures from the National Palace Museum, Taipei*. The Metropolitan Museum of Art, NY, and National Palace Museum, Taipei, 1996, 377. (Dated 1492. Hanging scroll, ink and color on paper, 84.8 × 21.8 cm. National Palace Museum, Taipei.)

[65] Fong, Wen. *Images of the Mind*, 147.

[66] Guo Xi articulated for the connoisseur the ontological hierarchy: "It is generally accepted opinion that in landscapes there are those through which you may travel, those in which you may sightsee, those through which you may wander, and those in which you may lodge. Any paintings attaining these effects are to be considered excellent, but those suitable for travelling and sightseeing are not as successful in achievement as those suitable for wandering and living." Tr. Bush and Shih, *Early Chinese Texts on Painting*, 151–2.

WILLIAM V. DAVIS

THE PRESENCE OF ABSENCE: MIRRORS AND MIRROR IMAGERY IN THE POETRY OF R. S. THOMAS

Life may certainly not enter into philosophical discourse other than as *presence* to a reflection. – Emmanuel Levinas

He scratched on invisible glass. – R. S. Thomas

I

Throughout his long career – now in its sixth decade – R. S. Thomas has grown accustomed, and has accustomed his readers, to seeing things in terms of mirrors and mirror images. Such images are, as he himself has acknowledged, one of his obsessions.[1] Even if such "reflections" are as old as poetry itself, Thomas has worked on his own very modern and quite sophisticated variations on them, utilizing them as a way of defining his most insistent theme and as a means of elaborating his rather unique technique, and, thus, he has brought forward into our day new ways of "looking" at this long poetic tradition as well as having provided us with a fascinating way of looking at, of evaluating, his own considerable poetic output.

Of the many ways of reflecting on the mirrors and the omnipresent mirror images in Thomas's poetry, some are old, obvious, and well-known, and have frequently been seen, and often been used, before. There is, for instance, what might be called his use of human "mirrors" – characters who are simply seen as emblems of the self. One such character, clearly the most obvious and probably the most important in Thomas's work, is the character of Iago Prytherch, the Welsh peasant farmer Thomas has contemplated closely over the years, beginning with a series of powerful early poems about him. Indeed, in "A Peasant," Thomas describes Prytherch as that "ordinary man of the bald Welsh hills" with his "frightening … vacancy" of mind, and calls him "your prototype."[2]

There are also many literal mirrors in Thomas's canon. Some of them are natural mirrors, such as pools of water, replete with their inevitably mythic or romantic trappings, which afford glimpses of *doppelgängers* or of Narcissus-like characters lurking within them. In the autobiographical poem "A Life," for instance, we find Thomas, "Visionary only / in his perception of a horizon / beyond the horizon," personally and poetically appearing as a Narcissus-

A.-T. Tymieniecka (ed), Analecta Husserliana LXXIII, 347–360.

like figure, "Saving his face / in verse," but nevertheless "tortured" (in a rather elaborately "tortured" metaphor) "by the whispers behind / the mirror" (CP, 516).[3]

And there are poems based on paintings, poems which reflect back and forth between poet and painter as if one or the other (or both) were a mirror. "Gallery" is an obvious example. In it, Thomas tells us that "It is not they [the paintings that] // are being looked at / but *we* by faces / which over the centuries // keep their repose" (CP, 455, italics added). Likewise, in "Similarities," we find "the face / that is life's trophy," starting from another "gallery," before which, "corrected by a resemblance," one becomes "silent" (CP, 494). One of the most interesting of the portrait poems is "Self-Portrait." In it, a mirror reflects an image quite surprisingly different from the one the spectator expected to find there – not at all "the portrait / ... posed for."[4]

But perhaps the most important and the most fascinating mirrors in Thomas's poetry are those mirrors and that mirror imagery in which, through the "reflections" that come from and through them, Thomas grapples with the relationship between God and the man "made in his own image" – as if the metaphor itself were, or had become, a mirror. It is these mirrors and this mirror imagery, these "reflections," that I want to "look" at most specifically today.

Absence – the presence of an absence, a *Deus absconditus* – has haunted Thomas's mind and imagination throughout his career, but in recent years the insistence on the *presence* of this absence has taken on a double urgency – and a doubled ambiguity. The empty mirror, like "The Empty Church" of one of his most famous poems, is seen or described as a trap tempting God to come into it, to make himself visible, and thus available, to man. But, we are told, God, having "burned himself / before in the human flame ... will not come any more" to such a "lure" (CP, 349). Even as man waits "upon / him as a mirror / ... waits upon absence"[5] – even if one awaits or anticipates, looks forward to the possibility of some presence that might or may appear in, or out of, the mirror's "absence" – the absence, just as insistently as man anticipates, expects, or waits for it to appear or to come forth, itself waits, holds itself back from appearing, from presenting itself as presence. Indeed, it is as if any such "presence" were as "illimitable / as its absence."[6] And thus the presence of this absence vanquishes any, or all, prospects of God's appearance, of His "presence." Therefore, the only hope for the "believer" is the prospect that he might be "ambushed in a mirror"[7] by the "concealed likeness" in his "looking-glass" – a "likeness" that is "always ahead [of him] in its [own] ambush." However, "with the refinement of the mirror," there occurs only "the refinement of [the] dilemma."[8]

From such a "concealed" position, where any suspected "presence" within the mirror is only "visible" as absence, the "presence" outside the mirror, the presence looking into it, lurking we might say within it, must be examined and evaluated. But this is to court Narcissism, egotism, or illusion – perhaps all three. In such a situation one must imagine that *all* "presences" are their own mirrors or, as Thomas says in a poem entitled "Present," "the mirror[s]/ of a mirror, effortlessly repeating / [one's own] reflections" (CP, 325).

II

> This otherness, this
> "Not-being-us" is all there is to look at
> In the mirror, though no one can say
> How it came to be this way.
>
> – John Ashbery

These kinds of "reflections" then have become, in the second half of Thomas's career, his insistent thesis, his dominant theme. It is a theme that has significant, multifold ramifications – poetically, philosophically, and theologically. And when we begin to think of these ramifications, to think them through, and to think *through* them, we rather quickly think of a philosopher like Emmanuel Levinas. Levinas, a student of both Husserl and Heidegger, carried out the phenomenological agenda in his own rather uniquely complicated way, but in a way that might fruitfully serves to illuminate R. S. Thomas's central theological thesis in terms of his obsession with such mirrored reflections.[9]

Levinas sought to describe and to delineate an encounter with an "Other" that is both immediate and "present," at the same time that it is absent and transcendent. This, of course, caused him to court criticism in terms of thinkers primarily obsessed with questions of ontology, and it also took him toward phenomenological limits. At the risk of drastically oversimplifying his complex, complicated thought, let me try to briefly summarize Levinas's description and analysis of "otherness" and of "the Other" in terms of the ways that I think his thought might fruitfully be applied to Thomas's poetry.

Going back to Husserl's *Cartesian Meditations* and to his *Phenomenological Psychology*[10] as primary sources, Levinas extracted for his own use Husserl's liberation of philosophy from the grip of "naturalist" epistemology through his reconsideration of the "phenomenon." However, Levinas criticized Husserl's solipsistic description of the Other as a reflection of the self, and he attempted to show that the "existence of transcendental

Egos other than [one's] own in effect leaves no place for the Other *as* Other" (italics added).[11]

Furthermore, Levinas argued that there is "a relationship irreducible to adversity" in which "the Other [*Autre*], instead of alienating the uniqueness of the Same ... only calls the Same from the depths of himself toward what is deeper than himself"; that is, "the Other" calls "'the Same' at and to the depths of himself."[12] Starting from this Other, then, Levinas (by way of Husserl) describes "transcendental subjectivity" as the "tearing [of] the I from its isolation in itself."[13] Therefore, the attempt to "reach the other," to attempt communication, can only be "realized" in terms of a "relationship with the Other that is cast in ... language" even though the "other is maintained and confirmed in his heterogeneity as soon as one calls upon him."[14] But, Levinas asks, "What then can this relationship be with an absence radically withdrawn from disclosure and from dissimulation?" And he answers, "The other proceeds from the absolutely absent. His relationship with the absolutely absent from which he comes *does not indicate, does not reveal* this absent; and yet the absent has a meaning...."[15] In short, the Other remains Other at the same time that it seems, simultaneously, to be part of the self, a self that Levinas calls "the Same."[16]

For Levinas then, God, as "the Other," can be approached, but He can never be reached. God's "Otherness" is not a presence nor an absence, but a "trace" – a trace that is "transcendent to the point of absence."[17] In short, "God is not the supreme Other, but rather the absent condition – or the incondition as Levinas frequently writes – of the encounter with the Other."[18] That is, "Levinas does not offer a personal God"; instead, "he depicts a trace or near-absence." And thus his "is no comfortable religion providing divine succour in need, or answers to questions beyond our competence." Indeed, for Levinas, "the question is more important than the answer, the search more urgent than the solution."[19]

For Levinas then, philosophy, initially born of religion, becomes with it one of "two distinct moments" of a single spiritual process which he describes as "*the approach* of transcendence."[20] And perhaps then, finally, it is the case that discussions of such relationships, and attempts to understand them, can only be done (if they can be done or understood at all) in philosophy or theology – or in poetry.

III

The mirror melts and moulds itself and moves
And catches from nowhere brightly-burning breath.

> And one trembles to be so understood and, at last,
> To understand....
>
> – Wallace Stevens

And thus, with these trappings in mind, let me turn back to R. S. Thomas, who, in his own way, combines the disciplines of philosophy and theology as he considers such questions – or as he questions such answers. The process is, often enough for Thomas, as it is for phenomenological philosophy and for theology, a manipulation of metaphors and of the meanings of metaphors.

Mirrors and mirror imagery appear early on in Thomas's work. Usually, in the early work, this imagery is used in the rather obvious and automatic ways that I have already suggested and briefly described above. And, indeed, perhaps the first crucial instance of Thomas's use of a mirror image in terms of what I want to turn to now occurs in his first book, *The Stones of the Field* (1946), when he refers to an "unscrubbed" floor that "Is no mirror for the preening sun / at the cracked lattice" of the window.[21] This is an interesting image in several senses, but it is perhaps most significant in that it is ultimately a double – and negative – use of the mirror, an example of a mirror that fails in its primary purpose. The prospect of the illuminating light of the "preening" anthropomorphic sun, even though it shines through a "cracked lattice," is eliminated by the "unscrubbed," dirty floor it shines in on. And thus this prospective mirror is made useless, blank or empty, a void, voided. In short, it becomes a vehicle whose purpose is defeated by being deflected.[22]

But perhaps the first important poem in Thomas's canon to make overt use of the theological implications of mirror imagery is "Judgment Day" (CP, 105). "Judgment Day" is based on the reference in Genesis to man having been made in God's image. But Thomas conflates the description of man's creation as it is described in the Biblical story with man's final "judgment day" in terms of a dual, mirror-like, doubled "reflection" on the nature of God and man and on the reciprocal relationship between them. The poem begins in a self-reflective moment with a man seeing himself with clean eyes, with clarified vision:

> Yes, that's how I was,
> I know that face.

This man goes on to describe himself accurately and honestly in terms of "the knot of life" that he was tied to at the time of his creation. And then, suddenly, this man turns to God and prays to be taken into death on this day of his judgment as someone else, an other and new man, a man different from, and better than (he now sees) the one he has been. And in envisioning his new "birth" at the very moment of death – this rebirth into another (after)life – he envisions and describes it in terms of a rather amazing and complicated ontological mirror image. He describes man's creation as God's breathing on a mirror, and man's death as God breathing again on that same mirror, erasing it, emptying it, voiding it in the same way that He first filled it. The basis for Thomas's metaphor is, of course, biblical. In Genesis, God is described as having created the world through speech, breath – "And God *said*...."[23] But Thomas here uniquely adds the complicating metaphor of the mirror, a metaphor which suggests rather startling theological ramifications, since the metaphor seems to suggest that God looked into a mirror, saw *himself*, and then decided to *make man in His own image*. He did so by breathing on the mirror. In terms of this trope, God, in creating man by *breathing* him in the mirror, by breathing into him the breath of life, simultaneously obliterates or replaces His own image in this process of creating an "other" – an *other* that is initially an image in a mirror; an *other* that is similar to but separate from himself; an *other* that He will name "Man."

But then, in the "now" of the poem, at the time of his final "Judgment Day," as he is about to be taken back by the God who created him, man, seeing himself as he is, or as he has become, willingly wishes to be "blurred out," killed off, replaced, by God's *re*-doing of the original act of creation: by breathing again on the mirror and, by that second breath – which is both a second, creative, but simultaneously destructive act – destroying His own earlier creation. What Thomas suggests here is literally quite intriguing from a theological point of view – since when God breathes on the mirror this second time, He blurs out the first image, the image of man, but, when the mirror clears of God's second blown breath, the "original" image of God Himself will (apparently) be restored to the mirror, will replace or take the place of the *displaced* image of man. Either that, or the mirror will become empty, will hold only the presence of absence. In one sense, then, the man making this request on this "judgment day," is simply asking to be "breathed away" so that God, source of both breath and life, can be "reborn" or restored. In this way man can again become one with God. (This process, of course, reverses the act of man's creation by God's breath having been blown on the mirror, and also reverses the sacrifice that God made for man by becoming man. But, at the same time, it shows that man has learned the lesson, that man has finally "found" the God hidden in the mirror – where before, when he looked into the mirror, he had only seen himself.) Even so, at the "judgment day" of "Judgment Day," man seems not fully or entirely content with his own final request. And Thomas, in his typically complex way, and even though he acknowledges that these are rather "bleak reflections," puts the man's final request in the form of a prayer: "Lord, breathe once more / On that … mirror, / Let me be lost / … for ever…. / Let me go back / … to undo / The knot of life / That was tied there."

IV

I would be the mirror
of a mirror, effortlessly repeating
my reflections.
– R. S. Thomas

There are many poems between "Judgment Day" and Thomas's most recent work that suggest the sequences and the consequences of such obsessions with mirrors and mirror imagery in terms of the turns of the development of his theological themes and theses, but, for our purposes here, a brief sampling of them will have to suffice.

Perhaps Thomas puts these "bleak reflections" most clearly in a poem entitled "Scenes," in *Laboratories of the Spirit* (1975), one of his first books to grapple meaningfully with his dominant theological theme. He says:

So in the huge night,
awakening, I have re-interpreted
the stars' signals and seen the reflection
in an eternal mirror of the mystery
terrifying enough to be named Love.[24]

In another poem in *Laboratories of the Spirit*, one appropriately enough entitled "Probing," Thomas intriguingly combines a mirror-like image with an epistemological twist to suggest a phenomenological thesis, as he argues that "we never awaken / from the compulsiveness of the mind's / stare into the lenses' furious interiors" (CP, 279). In the same book we encounter a "pure mirror / of water" to which "a face" comes which looks into it, and sees only absence, as if it were an empty mirror (CP, 298). In "Present," Thomas, who "would be the mirror / of a mirror, effortlessly repeating / my reflections," (CP, 325) speaks of "engaging" with philosophy and thus of being "at the switchboard / of the exchanges of the people / of all time." Then, bringing together many of these "reflections," catching up the imagery of "Judgment Day," and looking forward to other poems and "reflections" still to come, in a poem enigmatically entitled "Perhaps" (CP, 353), Thomas raises a series of questions that he will spend much of the rest of his career trying to answer – or asking further questions about.

His intellect was the clean mirror
he looked in and saw the machinery of God
assemble itself? It was one that reflected
the emptiness that was where God
should have been. The mind's tools had
no power convincingly to put him
together. Looking in that mirror was a journey
through hill mist where, the higher
one ascends, the poorer the visibility
becomes. It could have led to despair
but for the consciousness of a presence
behind him, whose breath clouding
that looking-glass proved that it was alive.
To learn to distrust the distrust
of feeling – this then was the next step
for the seeker? To suffer himself to be persuaded
of intentions in being other than the crossing
of a receding boundary which did not exist?
To yield to an unfelt pressure that, irresistible
In itself, had the character of everything
But coercion? To believe, looking up
Into invisible eyes shielded against love's
Glare, in the ubiquity of a vast concern?

And thus, sitting "down by the still pool / in the wind, waiting for the unknown / visitant's quickening of its surface,"[25] after having "crawled out … / far as I dare on to a bough / of country that is suspended / between sky and sea," Thomas wonders if he can "console [him]self / with reflections?" as he acknowledges that "There are // times even the mirror / is misted as by one breathing / over my shoulder" (CP, 503). Still, Thomas finds that even "With the refinement of the mirror"[26] – and in spite of the various games he plays with mirrors, the metaphors he makes, "as though to take by surprise // the self that is my familiar" (CP, 505) – "there occur[s] only the refinement of [the] dilemma" that "life" must still be "brought round to confront" an "image in an oblique / glass."[27] In *Counterpoint* (1990), Thomas asks:

What is the virginity
of mirrors? Are they surfaces
of fathoms which mind
clouds when examining itself
too closely?[28]

And then, in lines that point forward toward his fullest investigation of this theme in terms of his use of mirrors and mirror images, Thomas, "hungry for meaning / at life's pane" (an interesting pun), begins to believe that the face he sees disappearing, a face that blurs or that moves in and out of focus, is both God Himself, glimpsed briefly, and his own face, in life and in death, "a skull / brushed by a smile," these two "faces" superimposed upon one another. This image of a "face, vague / but compelling" is, however, bleared "over as much with my shortness / of faith as of breath,"[29] as Thomas turns the image, adding an intriguing twist, an "infinite counterpoint / between mirror and mirror,"[30] on the "judgment day" referred to above.

No Truce with the Furies, Thomas's most recent book, which brings so many of his themes and theses to climax, likewise brings to climax his use of mirrors and mirror metaphors in terms of his obsession with the Other in both philosophical and theological ways. There are several poems in *No Truce* that we ought to *look* at.

The first of these is "Fathoms." "Fathoms" is a brief autobiography. It begins with a natural mirror image, "Young I visited / this pool," and then "In the middle years / visited it again."[31] Then Thomas jumps to the present, and describes himself as "on the margin / of eternity," and contemplating his own "dissolution." This is a fascinating word in the context, and in terms of the mirror metaphor that the poem is built around. "Dissolution," related to both "dissoluble" and to "dissolve," is a word that clearly fits a context of watery pools in which things appear and disappear, into which, Narcissus-like, one peers and sees oneself or in which other presences and reflections can be seen, singly or simultaneously. Thomas, "on the margin / of eternity," facing *disill*usion and *dissol*ution, and fearing misapprehension, seeks to see and to find in the mirror of contemplation here beside this pool, the "truth" of his life and of the human condition. He hopes, he says, to be able to "dredge up the truth." He describes his situation as "nothing but [a] self / looking up at the self / looking down."

In addition to the suggestion of a bifurcated self, there also seems to be in these lines a second self, an Other, or another "self" that, as Thomas looks "up," looks "down" at him. As I have tried to show elsewhere, these doubled "selves" here in "Fathoms" may well have behind them (since Kierkegaard is invoked at the end of Thomas's poem) Kierkegaard's dual notions of "double reflection," "inwardness," and "indirect communication."[32] But, more importantly for our purposes, Thomas's "starings" here seem to suggest an "other" who is not simply "self" but something or someone beyond the self – something or someone who is as intent on staring back at him as he, Thomas, is on staring "up" at him.

In "S. K,"[33] another poem in *No Truce with the Furies*, Kierkegaard again provides Thomas with a way of reflecting on such questions. With "truth" as his goal, as it was at the end of "Fathoms," Thomas here again considers the question of the self in terms of a mirrored self-image and also in terms of an "other" who may be either a mirror image of the self as "Same," or a literal "other," or the "Other" which is God – or all three simultaneously.

"Imagining / from his emphasis on the self // that God is not other," Thomas, with Kierkegaard, discovers that "thought" is "brought to bay by a truth / as inscrutable as its reflection." If such dazzlingly explosive thoughts are "truths" as "inscrutable" as any "reflections" that can be made about them, Thomas continues to contemplate and explore such thoughts and such truths, and he comes to his conclusions in interestingly complex theological statements – posed, as is typical of him, in a series of questions which often describe the relationships between man and God in the context of overt mirror images. First, following both Kierkegaard and his own sequence of speculations in "Judgment Day," Thomas raises the "heresy" of seeing "the self / as God." Then he immediately puts everything into the context of prayer, in which, he says, there occurs an "exchange / of places between I and [a] thou."[34] Then he asks whether such an "emphasis / on the subjective" results only in "soliloquy." The final question, the final lines of the poem "answer" these questions – as they ask another – couched in yet another mirror image:

> Is prayer
> not a glass that beginning
> in obscurity …
> the longer we stare
> into the clearer becomes
> the reflection of a countenance
> in it other than our own?

Thomas here "seems to suggest that prayer *is* a 'glass,'" a mirror, "and that this glass, stared into long enough or looked at hard enough, may finally begin to clear, and to suggest the possibility of a presence in it – a 'countenance' … 'other than our own.'"[35] This "presence" is simultaneously present and absent. It is an "Other" that is both self and other; an "Other" that may well be God; an "Other" who, although not present, *is present as absence* in the mirror of man's, or of R. S. Thomas's, mind.

Baylor University

NOTES

[1] See R. S. Thomas, "Autobiographical Essay," in *Miraculous Simplicity: Essays on R. S. Thomas*, ed. William V. Davis (Fayetteville, AR: University of Arkansas Press, 1993), p. 17.

[2] R. S. Thomas, *Collected Poems 1945–1990* (London: Dent, 1993), p. 4. Whenever possible, all references to Thomas's poems will be to this substantial – although *not* complete – "Collected Poems," and will be included in parentheses in the text, marked CP, and followed by the relevant page number. It should be noted, however, that Thomas has the curious habit of excluding major poems from his various collections. It is almost as if he wants to "hide" some of his most important poems. Indeed, such a procedure might be seen as an instance of the "presence of absence" in his work. This intriguing habit, together with the fact that Thomas sometimes uses the same title for more than one poem, and that he frequently revises poems from one published version to another, increases the complications of reading and commenting on his work.

[3] Cf. Thomas's "Hark": "You were wrong, Narcissus. / The replica of the self / is to be avoided. Echo / was right, warning you against // the malevolence of mirrors." [See *Mass for Hard Times* (Newcastle upon Tyne: Bloodaxe Books, 1992), p. 38.]

[4] R. S. Thomas, *Laboratories of the Spirit* (London: Macmillan, 1975), p. 27. Interestingly enough, this poem was one of the poems from *Laboratories* that Thomas failed to include in his *Collected Poems*. In addition, as a slight variation on this theme, there are numerous mirror-like windows in Thomas's poems, windows which give off glimpses of a "reflected" place or of a person and of the semi-hidden self which stares through them, the window being, or becoming, a mirror – a means of both seeing the self as it is and of seeing beyond the self.

[5] R. S. Thomas, *Counterpoint* (Newcastle upon Tyne: Bloodaxe Books, 1990), p. 45.

[6] *Ibid*., p. 48.

[7] R. S. Thomas, *The Echoes Return Slow* (London: Macmillan, 1988), p. 72.

[8] *Ibid*., p. 108.

[9] I have no knowledge that Thomas had read Levinas, although given Thomas's eclectic habit of mind and his wide-ranging reading in many fields, especially in philosophy and theology, and because Levinas would be, for Thomas (as indeed Thomas would be for Levinas!) an illuminating, mirror-like thinker, it would only be surprising if he had *not* read him – and *vice versa*. We do have Thomas's testimony that he "engage[s] with philosophy / in the morning[s]" (CP, 325) and there are numerous references to philosophers and philosophic thought scattered throughout his work. The poem "I," for instance, begins, "Kierkegaard hinted, Heidegger / agreed…." (See Thomas, *Mass for Hard Times*, p. 58.)

[10] Levinas says that in his *Phenomenological Psychology*, Husserl shows that "the subjective modes of appearing of the world and of nature, the *Erscheinungeweisen* or the aspects of the real are still a part of being" in such a way that the "sphere of the world swims in the subjective." (See Emmanuel Levinas, *Of God Who Comes to Mind*, trans. Bettina Bergo (Stanford, CA: Stanford University Press, 1998), pp. 17–18.) Speaking of the essay quoted from here, the cover blurb for this collection of Levinas's work suggests that this essay "illuminates Levinas's relation to Husserl and thus to phenomenology, which is always his starting point, even if he never abides by the limits it imposes."

[11] Colin Davis, *Levinas: An Introduction* (Notre Dame, IN: University of Notre Dame Press, 1996), p. 28. As Davis says, "The rest of Levinas's philosophical career will, then, be dominated by one question: what does it mean to think of the Other as Other?" (p. 33). Levinas said, "The relationship with the other is not an idyllic and harmonious relationship of communion or a

sympathy [*sympathie* is Levinas's translation of Husserl's *Einfühlung*] through which we put ourselves in the other's place; we recognize the other as resembling us, but exterior to us; the relationship with the other is a relationship with a Mystery." (See Emmanuel Levinas, *Time and the Other*, trans. Richard A. Cohen (Pittsburgh, PA: Duquesne University Press, 1987), p. 75.) Davis says that "This passage indicates the crux of Levinas's dispute with phenomenology...." (See C. Davis, *Levinas: An Introduction*, p. 31). In my summaries of Levinas's thinking, here and elsewhere, I am indebted to Davis's lucid summarizerations of Levinas's thought.

12 Levinas, *Of God...*, p. 24.

13 *Ibid.*, p. 26. Cf. "Between me and the other there gapes a difference which no unity of transcendental apperception could recover" (p. 71).

14 Emmanuel Levinas, *Totality and Infinity: An Essay on Exteriority*, trans. Alphonso Lingis (Pittsburgh, PA: Duquesne University Press, 1969), p. 69. Cf. "The intelligibility of transcendence is not ontological. The transcendence of God can neither be said nor thought in terms of being." (See Levinas, *Of God...*, p. 77.)

15 Emmanuel Levinas, "The Trace of the Other," in *Deconstruction in Context: Literature and Philosophy*, ed. Mark C. Taylor (Chicago: The University of Chicago Press, 1986), pp. 354–55.

16 Levinas says, "The self is the site where the Same identifies itself as such." (See C. Davis, *Levinas, An Introduction*, p. 42).

17 Levinas says, "God is not simply the 'first other' [*premier autrui*] or the 'other par excellence' [*autrui par excellence*] or the 'absolutely other' [*absolument autrui*], but other than the other, other otherwise, other by an alterity prior to the alterity of the other [*autre qu'autrui, autre autrement, autre d'altérité préalable à l'altérité d'autrui*], prior to the ethical obligation to the neighbour, transcendent to the point of absence...." See Levinas, *En découvrant l'existence avec Husserl et Heidegger* (Paris: Vrin, 1974), pp. 199–202; *De Dieu qui vient à l'idée* (Paris, Vrin, 1992), p. 115. Cf. Levinas, *Of God*, p. 69. The word "trace" is a particularly important word in this philosophical tradition. It is also a word that links these philosophical considerations with specifically literary ones. Such a linkage has perhaps been most overtly made by Jacques Derrida, who brings the two disciplines together through his deconstructive approach to both. And Derrida is indebted to Levinas as well as having been an influence on him. The first and arguably the most incisive analysis of *Totality and Infinity* is Derrida's "Violence and Metaphysics: An Essay on the Thought of Emmanuel Levinas" (see *Writing and Difference*, trans. Alan Bass (Chicago: The University of Chicago Press, 1978), pp. 79–153). As Colin Davis says, "Derrida's engagement with Levinas clearly does go far deeper than a simple critique. His intense and ongoing studies of Husserl and Heidegger put him in an almost unique position to assess Levinas's relationship to his most evident sources. Most importantly, Derrida's essay is also to some extent an act of philosophical self-recognition, as Levinas's ambiguous relationship to the language and values of the philosophical tradition is a reflection of Derrida's own position. Derrida perceives that Levinas's writing anticipates the difficulties faced by deconstruction. According to Feron, 'The essential point of Derrida's argument consists in recognizing that philosophical discourse can only say the Other in the language of the Same' (Étienne Feron, *De l'idée de transcendance à la question du language: L'Itinéraire philosophique d'Emmanuel Levinas* (Grenoble: Jérôme Millon, 1992), p. 260); and this can be taken to summarize Derrida's dilemma as much as that of Levinas," (See C. Davis, *Levinas: An Introduction* p. 66.) Derrida's essay appeared before Levinas's *Otherwise than Being or Beyond Essence* and some critics argue that, in that book, Levinas attempted to take Derrida's criticism of his earlier work into account. At the very least, it seems certain that Derrida influenced Levinas's later thought and brought it more specifically into the realm of literary criticism *per se*. In another crucial text, *Of*

Grammatology, trans. Gayatri Chakravorty Spivak (Baltimore, MD: The Johns Hopkins University Press, 1976), Derrida traces the word "trace" through the philosophy of Heidegger and Levinas before coming to his own conclusions with respect to it in terms of this tradition. Derrida describes this "trace" as a metaphysical concept based on the notion of presence. For him, "the word designates something of which the metaphysical concepts of trace and presence are the erasure." Therefore, it "names an originary tracing and effacement" and becomes "the trace of effacement"; "it names something of which ... self and Other are the erasure...." (See Rodolphe Gasché, *The Tain of the Mirror: Derrida and the Philosophy of Reflection* (Cambridge: Harvard University Press, 1986), pp. 186–87.) For more on the relationship between Derrida and Levinas, see Levinas's "Jacques Derrida: Wholly Otherwise" in *Proper Names*, trans. Michael B. Smith (Stanford, CA: Stanford University Press, 1996), pp. 55–62: Robert Bernasconi, "The Trace of Levinas in Derrida," in *Derrida and Différence*, David Wood and Robert Bernasconi, eds. (Evanston, IL: Northwestern University Press, 1988), pp. 13–29; Robert Bernasconi and Simon Critchley, eds. *Re-Reading Levinas* (Bloomington, IN: Indiana University Press, 1991), pp. 3–10, 11–48, 162–189, *passim*; Simon Critchley, *The Ethics of Deconstruction: Derrida and Levinas* (Blackwell: Oxford, 1992).

18 C. Davis, *Levinas: An Introduction*, p. 99.

19 *Ibid.*, p. 100. As Levinas says, "One may wonder whether the true God can ever discard His incognito." (See Emmanuel Levinas, *Entre Nous: On Thinking-of-the-Other*, trans. Michael B. Smith and Barbara Harshav (New York: Columbia University Press, 1998), p. 56.) Cf. "... one may wonder whether the *incognito* should not be the very mode of revelation...." (See Levinas, *Proper Names*, p. 78.)

20 See Emmanuel Levinas, *A l'heure des nations* (Paris: Minuit, 1988), p. 204.

21 R. S. Thomas, *The Stones of the Field* (Carmarthen: The Druid Press, Ltd., 1946), p. 40.

22 There are several other instances of images of "cracked mirrors" in Thomas's canon, but perhaps the most interesting in terms of my purposes here is one that combines several references to be noted later. This reference occurs in a poem called "Pardon." The reference is to "homo sapiens, that cracked mirror, / mending himself again and again like a pool." (See Thomas, *Experimenting with an Amen* (London: Papermac, 1988), p. 29.) Another early poem makes a similar, equally interesting and rather complex use of this imagery; in it Thomas describes "an old farmer" who "plays" the "neglected music" of a woman's "thin body" as she, watching him, sees in the mirror of "his eye / The light of the cracked lake // That once she had propped to comb / Her hair in." (See Thomas, *Tares* (Chester Springs: Dufour Editions, 1961), p. 22.)

23 See Genesis 1:3–31, especially 26–27: "Then God *said*, 'Let us make man in our image, after our likeness....' So God created him in his own image ..." (italics added). It would seem obvious that Thomas initially found his metaphor of the mirror in this biblical passage, and based his subsequent contemplations on it.

24 Thomas, *Laboratories*, p. 44. This is another of Thomas's important poems that he failed to include in the *Collected Poems*.

25 Thomas, *Experimenting*, p. 43.

26 Thomas, *Echoes*, p. 108.

27 *Ibid.*, pp. 108–109.

28 Thomas, *Counterpoint*, p. 12.

29 *Ibid.*, p. 46.

30 See Thomas, *Mass for Hard Times*, p. 82. *Mass* is a book that, in several ways, itself serves as a mirror or a "counterpoint" to *Counterpoint*.

31 R. S. Thomas, *No Truce with the Furies* (Newcastle upon Tyne: Bloodaxe Books, 1995), p. 10.

[32] See William V. Davis, "'At the Foot of the Precipice of Water ... Sea Shapes Coming to Celebration': R. S. Thomas and Kierkegaard," *Welsh Writing in English: A Yearbook of Critical Essays*, 4 (1998), pp. 104–105. It is perhaps worth stating the not surprising fact that Levinas has also often mentioned Kierkegaard, a philosopher who clearly interested him. See, for instance, Levinas's essay, "Kierkegaard: Existence and Ethics," where he remarks that "Kierkegaard's philosophy has marked contemporary thought so deeply that the reservations and even the rejections it may elicit are yet forms of that influence" (*Proper Names*, p. 71). And, he adds, no one "has developed with greater rigor" than Kierkegaard the "phenomenology" of "the correlate of truth crucified" (*Proper Names*, p. 70). In "A Propos of 'Kierkegaard vivant'" Levinas says, "Thus Kierkegaard brings something absolutely new to European philosophy: the possibility of attaining truth through the ever-recurrent inner rending of doubt…." (*Proper Names*, p. 77).

[33] Thomas, *No Truce*, pp. 15–17. The fact that Thomas, who is usually referred to by his initials, chooses to refer to Kierkegaard as "S. K." suggests Thomas's indebtedness to Kierkegaard. For more on the relationship between Thomas and Kierkegaard, see my "At the Foot of the Precipice…," p. 106.

[34] Thomas may well be thinking here of Martin Buber, a thinker to whom he frequently refers. Likewise, Levinas regularly mentions Buber, especially his seminal book *Ich und Du* (*I and Thou*). However, as Colin Davis points out, "Levinas rejects Buber's I-Thou relationship because it implies too much familiarity with the Other, which should be addressed with the more formal *vous*; but this *vous* is in turn too familiar, too direct an address for God. God is glimpsed only in the third person, neither a presence nor an absence, but a trace (*En découvrant l'existence avec Husserl et Heidegger*, pp. 199–202), infinitely close and absolutely distant" (see C. Davis, *Levinas: An Introduction*, p. 99). See also Levinas's essays on Buber: "Martin Buber and the Theory of Knowledge," and "Dialogue with Martin Buber" (both in Levinas, *Proper Names*, pp. 17–35 and 36–39).

[35] William V. Davis, "At the Foot of the Precipice…," p. 113.

INDEX OF NAMES

Analecta Husserliana

The Yearbook of Phenomenological Research

Editor-in-Chief

Anna-Teresa Tymieniecka

The World Institute for Advanced Phenomenological Research and Learning, Belmont, Massachusetts, U.S.A.

1. Tymieniecka, A-T. (ed.), *Volume 1 of Analecta Husserliana.* 1971 ISBN 90-277-0171-7
2. Tymieniecka, A-T. (ed.), *The Later Husserl and the Idea of Phenomenology.* Idealism – Realism, Historicity and Nature. 1972 ISBN 90-277-0223-3
3. Tymieniecka, A-T. (ed.), *The Phenomenological Realism of the Possible Worlds.* The "A Priori', Activity and Passivity of Consciousness, Phenomenology and Nature. 1974 ISBN 90-277-0426-0
4. Tymieniecka, A-T. (ed.), *Ingardeniana.* A Spectrum of Specialised Studies Establishing the Field of Research. 1976 ISBN 90-277-0628-X
5. Tymieniecka, A-T. (ed.), *The Crisis of Culture.* Steps to Reopen the Phenomenological Investigation of Man. 1976 ISBN 90-277-0632-8
6. Tymieniecka, A-T. (ed.), *The Self and the Other.* The Irreducible Element in Man, Part I. 1977 ISBN 90-277-0759-6
7. Tymieniecka, A-T. (ed.), *The Human Being in Action.* The Irreducible Element in Man, Part II. 1978 ISBN 90-277-0884-3
8. Nitta, Y. and Hirotaka Tatematsu (eds.), *Japanese Phenomenology.* Phenomenology as the Trans-cultural Philosophical Approach. 1979 ISBN 90-277-0924-6
9. Tymieniecka, A-T. (ed.), *The Teleologies in Husserlian Phenomenology.* The Irreducible Element in Man, Part III. 1979 ISBN 90-277-0981-5
10. Wojtyła, K., *The Acting Person.* Translated from Polish by A. Potocki. 1979 ISBN Hb 90-277-0969-6; Pb 90-277-0985-8
11. Ales Bello, A. (ed.), *The Great Chain of Being* and *Italian Phenomenology.* 1981 ISBN 90-277-1071-6
12. Tymieniecka, A-T. (ed.), *The Philosophical Reflection of Man in Literature.* Selected Papers from Several Conferences held by the International Society for Phenomenology and Literature in Cambridge, Massachusetts. Includes the essay by A-T. Tymieniecka, *Poetica Nova.* 1982 ISBN 90-277-1312-X
13. Kaelin, E. F., *The Unhappy Consciousness.* The Poetic Plight of Samuel Beckett. An Inquiry at the Intersection of Phenomenology and literature. 1981 ISBN 90-277-1313-8
14. Tymieniecka, A-T. (ed.), *The Phenomenology of Man and of the Human Condition.* Individualisation of Nature and the Human Being. (Part I:) Plotting the Territory for Interdisciplinary Communication. 1983 *Part II* see below under Volume 21. ISBN 90-277-1447-9

Analecta Husserliana

15. Tymieniecka, A-T. and Calvin O. Schrag (eds.), *Foundations of Morality, Human Rights, and the Human Sciences.* Phenomenology in a Foundational Dialogue with Human Sciences. 1983 ISBN 90-277-1453-3
16. Tymieniecka, A-T. (ed.), *Soul and Body in Husserlian Phenomenology.* Man and Nature. 1983 ISBN 90-277-1518-1
17. Tymieniecka, A-T. (ed.), *Phenomenology of Life in a Dialogue Between Chinese and Occidental Philosophy.* 1984 ISBN 90-277-1620-X
18. Tymieniecka, A-T. (ed.), *The Existential Coordinates of the Human Condition: Poetic – Epic – Tragic.* The Literary Genre. 1984 ISBN 90-277-1702-8
19. Tymieniecka, A-T. (ed.), *Poetics of the Elements in the Human Condition.* (Part 1:) The Sea. From Elemental Stirrings to Symbolic Inspiration, Language, and Life-Significance in Literary Interpretation and Theory. 1985
For Part 2 and 3 *see below* under Volumes 23 and 28. ISBN 90-277-1906-3
20. Tymieniecka, A-T. (ed.), *The Moral Sense in the Communal Significance of Life.* Investigations in Phenomenological Praxeology: Psychiatric Therapeutics, Medical Ethics and Social Praxis within the Life- and Communal World. 1986
ISBN 90-277-2085-1
21. Tymieniecka, A-T. (ed.), *The Phenomenology of Man and of the Human Condition.* Part II: The Meeting Point Between Occidental and Oriental Philosophies. 1986 ISBN 90-277-2185-8
22. Tymieniecka, A-T. (ed.), *Morality within the Life- and Social World.* Interdisciplinary Phenomenology of the Authentic Life in the "Moral Sense'. 1987
Sequel to Volumes 15 and 20. ISBN 90-277-2411-3
23. Tymieniecka, A-T. (ed.), *Poetics of the Elements in the Human Condition.* Part 2: The Airy Elements in Poetic Imagination. Breath, Breeze, Wind, Tempest, Thunder, Snow, Flame, Fire, Volcano . . . 1988 ISBN 90-277-2569-1
24. Tymieniecka, A-T., *Logos and Life*. Book I: Creative Experience and the Critique of Reason. 1988 ISBN Hb 90-277-2539-X; Pb 90-277-2540-3
25. Tymieniecka, A-T., *Logos and Life*. Book II: The Three Movements of the Soul. 1988 ISBN Hb 90-277-2556-X; Pb 90-277-2557-8
26. Kaelin, E. F. and Calvin O. Schrag (eds.), *American Phenomenology.* Origins and Developments. 1989 ISBN 90-277-2690-6
27. Tymieniecka, A-T. (ed.), *Man within his Life-World.* Contributions to Phenomenology by Scholars from East-Central Europe. 1989 ISBN 90-277-2767-8
28. Tymieniecka, A-T. (ed.), *The Elemental Passions of the Soul.* Poetics of the Elements in the Human Condition, Part 3. 1990 ISBN 0-7923-0180-3
29. Tymieniecka, A-T. (ed.), *Man's Self-Interpretation-in-Existence.* Phenomenology and Philosophy of Life. – Introducing the Spanish Perspective. 1990
ISBN 0-7923-0324-5
30. Rudnick, H. H. (ed.), *Ingardeniana II.* New Studies in the Philosophy of Roman Ingarden. With a New International Ingarden Bibliography. 1990
ISBN 0-7923-0627-9

Analecta Husserliana

31. Tymieniecka, A-T. (ed.), *The Moral Sense and Its Foundational Significance: Self, Person, Historicity, Community.* Phenomenological Praxeology and Psychiatry. 1990 ISBN 0-7923-0678-3
32. Kronegger, M. (ed.), *Phenomenology and Aesthetics.* Approaches to Comparative Literature and Other Arts. Homages to A-T. Tymieniecka. 1991 ISBN 0-7923-0738-0
33. Tymieniecka, A-T. (ed.), *Ingardeniana III.* Roman Ingarden's Aesthetics in a New Key and the Independent Approaches of Others: The Performing Arts, the Fine Arts, and Literature. 1991
Sequel to Volumes 4 and 30 ISBN 0-7923-1014-4
34. Tymieniecka, A-T. (ed.), *The Turning Points of the New Phenomenological Era.* Husserl Research – Drawing upon the Full Extent of His Development. 1991 ISBN 0-7923-1134-5
35. Tymieniecka, A-T. (ed.), *Husserlian Phenomenology in a New Key.* Intersubjectivity, Ethos, the Societal Sphere, Human Encounter, Pathos. 1991 ISBN 0-7923-1146-9
36. Tymieniecka, A-T. (ed.), *Husserl's Legacy in Phenomenological Philosophies.* New Approaches to Reason, Language, Hermeneutics, the Human Condition. 1991 ISBN 0-7923-1178-7
37. Tymieniecka, A-T. (ed.), *New Queries in Aesthetics and Metaphysics.* Time, Historicity, Art, Culture, Metaphysics, the Transnatural. 1991 ISBN 0-7923-1195-7
38. Tymieniecka, A-T. (ed.), *The Elemental Dialectic of Light and Darkness.* The Passions of the Soul in the Onto-Poiesis of Life. 1992 ISBN 0-7923-1601-0
39. Tymieniecka, A-T. (ed.), *Reason, Life, Culture, Part I.* Phenomenology in the Baltics. 1993 ISBN 0-7923-1902-8
40. Tymieniecka, A-T. (ed.), *Manifestations of Reason: Life, Historicity, Culture.* Reason, Life, Culture, Part II. Phenomenology in the Adriatic Countries. 1993 ISBN 0-7923-2215-0
41. Tymieniecka, A-T. (ed.), *Allegory Revisited.* Ideals of Mankind. 1994 ISBN 0-7923-2312-2
42. Kronegger, M. and Tymieniecka, A-T. (eds.), *Allegory Old and New.* In Literature, the Fine Arts, Music and Theatre, and Its Continuity in Culture. 1994 ISBN 0-7923-2348-3
43. Tymieniecka, A-T. (ed.): *From the Sacred to the Divine.* A New Phenomenological Approach. 1994 ISBN 0-7923-2690-3
44. Tymieniecka, A-T. (ed.): *The Elemental Passion for Place in the Ontopoiesis of Life.* Passions of the Soul in the *Imaginatio Creatrix.* 1995 ISBN 0-7923-2749-7
45. Zhai, Z.: *The Radical Choice and Moral Theory.* Through Communicative Argumentation to Phenomenological Subjectivity. 1994 ISBN 0-7923-2891-4
46. Tymieniecka, A-T. (ed.): *The Logic of the Living Present.* Experience, Ordering, Onto-Poiesis of Culture. 1995 ISBN 0-7923-2930-9

Analecta Husserliana

47. Tymieniecka, A-T. (ed.): *Heaven, Earth, and In-Between in the Harmony of Life.* Phenomenology in the Continuing Oriental/Occidental Dialogue. 1995 ISBN 0-7923-3373-X
48. Tymieniecka, A-T. (ed.): *Life. In the Glory of its Radiating Manifestations.* 25th Anniversary Publication. Book I. 1996 ISBN 0-7923-3825-1
49. Kronegger, M. and Tymieniecka, A-T. (eds.): *Life. The Human Quest for an Ideal.* 25th Anniversary Publication. Book II. 1996 ISBN 0-7923-3826-X
50. Tymieniecka, A-T. (ed.): *Life. Phenomenology of Life as the Starting Point of Philosophy.* 25th Anniversary Publication. Book III. 1997 ISBN 0-7923-4126-0
51. Tymieniecka, A-T. (ed.): *Passion for Place. Part II.* Between the Vital Spacing and the Creative Horizons of Fulfilment. 1997 ISBN 0-7923-4146-5
52. Tymieniecka, A-T. (ed.): *Phenomenology of Life and the Human Creative Condition.* Laying Down the Cornerstones of the Field. Book I. 1997 ISBN 0-7923-4445-6
53. Tymieniecka, A-T. (ed.): *The Reincarnating Mind, or the Ontopoietic Outburst in Creative Virtualities.* Harmonisations and Attunement in Cognition, the Fine Arts, Literature. Phenomenology of Life and the Human Creative Condition. Book II. 1997 ISBN 0-7923-4461-8
54. Tymieniecka, A-T. (ed.): *Ontopoietic Expansion in Human Self-Interpretation-in-Existence.* The I and the Other in their Creative Spacing of the Societal Circuits of Life. Phenomenology of Life and the Creative Condition. Book III. 1997 ISBN 0-7923-4462-6
55. Tymieniecka, A-T. (ed.): *Creative Virtualities in Human Self-Interpretation-in-Culture.* Phenomenology of Life and the Human Creative Condition. Book IV. 1997 ISBN 0-7923-4545-2
56. Tymieniecka, A-T. (ed.): *Enjoyment.* From Laughter to Delight in Philosophy, Literature, the Fine Arts and Aesthetics. 1998 ISBN 0-7923-4677-7
57. Kronegger M. and Tymieniecka, A-T. (eds.): *Life. Differentiation and Harmony... Vegetal, Animal, Human.* 1998 ISBN 0-7923-4887-7
58. Tymieniecka, A-T. and Matsuba, S. (eds.): *Immersing in the Concrete.* Maurice Merleau-Ponty in the Japanese Perspective. 1998 ISBN 0-7923-5093-6
59. Tymieniecka, A-T. (ed.): *Life - Scientific Philosophy/Phenomenology of Life and the Sciences of Life.* Ontopoiesis of Life and the Human Creative Condition. 1998 ISBN 0-7923-5141-X
60. Tymieniecka, A-T. (eds.): *Life - The Outburst of Life in the Human Sphere.* Scientific Philosophy / Phenomenology of Life and the Sciences of Life. Book II. 1998 ISBN 0-7923-5142-8
61. Tymieniecka, A-T. (ed.): *The Aesthetic Discourse of the Arts.* Breaking the Barriers. 2000 ISBN 0-7923-6006-0
62. Tymieniecka, A-T. (ed.): *Creative Mimesis of Emotion.* From Sorrow to Elation; Elegiac Virtuosity in Literature. 2000 ISBN 0-7923-6007-9

Analecta Husserliana

63. Kronegger, M. (ed).: *The Orchestration of The Arts – A Creative Symbiosis of Existential Powers.* The Vibrating Interplay of Sound, Color, Image, Gesture, Movement, Rhythm, Fragrance, Word, Touch. 2000 ISBN 0-7923-6008-7
64. Tymieniecka, A-T. and Z. Zalewski (eds.): *Life - The Human Being Between Life and Death.* A Dialogue Between Medicine and Philosophy, Recurrent Issues and New Approaches. 2000 ISBN 0-7923-5962-3
65. Kronegger, M. and Tymieniecka, A-T. (eds.): *The Aesthetics of Enchantment in the Fine Arts.* 2000 ISBN 0-7923-6183-0
66. Tymieniecka, A-T. (ed.): *The Origins of Life, Volume I: The Primogenital Matrix of Life and Its Context.* 2000 ISBN 0-7923-6246-2; Set ISBN 0-7923-6446-5
67. Tymieniecka, A-T. (ed.): *The Origins of Life, Volume II: The Origins of the Existential Sharing-in-Life.* 2000 ISBN 0-7923-6276-4; Set ISBN 0-7923-6446-5
68. Tymieniecka, A-T. (ed.): *PAIDEIA.* Philosophy / Phenomenology of Life Inspiring Education of our Times. 2000 ISBN 0-7923-6319-1
69. Tymieniecka, A-T. (ed.): *The Poetry of Life in Literature.* 2000 ISBN 0-7923-6408-2
70. Tymieniecka, A-T. (ed.): *Impetus and Equipoise in the Life-Strategies of Reason.* Logos and Life, volume 4. 2000 ISBN 0-7923-6731-6; HB 0-7923-6730-8
71. Tymieniecka, A-T. (ed.): *Passions of the Earth in Human Existence, Creativity, and Literature.* 2001 ISBN 0-7923-6675-1
72. Tymieniecka, A-T. and E. Agazzi (eds.): *Life – Interpretation and the Sense of Illness within the Human Condition.* Medicine and Philosophy in a Dialogue. 2001 ISBN Hb 0-7923-6983-1; Pb 0-7923-6984-X
73. Tymieniecka, A-T. (ed.): *Life – The Play of Life on the Stage of the World in Fine Arts, Stage-Play, and Literature.* 2001 ISBN 0-7923-7032-5

Kluwer Academic Publishers – Dordrecht / Boston / London